T0177396

Geophysical Monograph Series

Including
IUGG Volumes
Maurice Ewing Volumes
Mineral Physics Volumes

geo press.

Geophysical Monograph 192

Antarctic Subglacial Aquatic Environments

Martin J. Siegert
Mahlon C. Kennicutt II
Robert A. Bindschadler
Editors

American Geophysical Union
Washington, DC

Published under the aegis of the AGU Books Board

Kenneth R. Minschwaner, Chair; Gray E. Bebout, Kenneth H. Brink, Jiasong Fang, Ralf R. Haese, Robert B. Jackson, Yonggang Liu, W. Berry Lyons, Laurent Montési, Nancy N. Rabalais, Todd C. Rasmussen, A. Surjalal Sharma, David E. Siskind, Rigobert Tibi, and Peter E. van Keken, members.

Library of Congress Cataloging-in-Publication Data

Antarctic subglacial aquatic environments / Martin J. Siegert, Mahlon C. Kennicutt II, Robert A. Bindschadler, editors.
   p. cm. — (Geophysical monograph ; 192)
  Includes bibliographical references and index.
  ISBN 978-0-87590-482-5 (alk. paper)
1. Subglacial lakes—Antarctica—History. 2. Subglacial lakes—Antarctica—Discovery and exploration. 3. Subglacial lakes—Polar regions—History. 4. Subglacial lakes—Polar regions—Discovery and exploration. 5. Aquatic ecology—Antarctica.
I. Siegert, Martin J. II. Kennicutt, Mahlon C. III. Bindschadler, R. A. (Robert A.)
  GC461.A596 2011
  551.48'2—dc22

                                     2011007605

  ISBN: 978-0-87590-482-5
  ISSN: 0065-8448

Cover Image: The location (red triangles) of 387 subglacial lakes superimposed on the BEDMAP database depiction of Antarctic sub-ice topography. (top) The ice sheet surface is illustrated, which is used along with basal topography to predict (bottom) hydrological pathways (blue lines). Image credit: Andrew Wright and Martin Siegert.

Copyright 2011 by the American Geophysical Union
2000 Florida Avenue, N.W.
Washington, DC 20009

Figures, tables and short excerpts may be reprinted in scientific books and journals if the source is properly cited.

Authorization to photocopy items for internal or personal use, or the internal or personal use of specific clients, is granted by the American Geophysical Union for libraries and other users registered with the Copyright Clearance Center (CCC) Transactional Reporting Service, provided that the base fee of $1.50 per copy plus $0.35 per page is paid directly to CCC, 222 Rosewood Dr., Danvers, MA 01923. 0065-8448/11/$01.50+0.35.

This consent does not extend to other kinds of copying, such as copying for creating new collective works or for resale. The reproduction of multiple copies and the use of full articles or the use of extracts, including figures and tables, for commercial purposes requires permission from the American Geophysical Union. geopress is an imprint of the American Geophysical Union.

Printed in the United States of America.

CONTENTS

PREFACE

Between 15 and 17 March 2010, 83 scientists from 11 nations gathered in Baltimore, Maryland, United States, for an AGU Chapman Conference to discuss the current status of knowledge about, and future exploration plans for, Antarctic subglacial aquatic environments. This was the fifth in a series of international conferences. In response to recent funding of two new major exploration programs and the continuation of work at Vostok Subglacial Lake, this meeting focused attention on emerging scientific frontier and the attendant environmental stewardship issues. The chapters of this book expand on keynote presentations and are augmented by selected invited authors to produce the first comprehensive summary of research on, and planning for, the exploration of subglacial aquatic environments. The chapters include summaries of the most recent identification, location, and physiography of 387 subglacial lakes; a detailed analysis of the results, from years of study, from Vostok Subglacial Lake; the rationale for subglacial lakes as analogues for extraterrestrial environments; protocols for the protection and stewardship of these unique environments; critiques of the technological issues facing future exploration programs; and, finally, summaries of the three projects that will enter and sample subglacial aquatic environments in the next 3 to 5 years. This book serves as a benchmark in subglacial aquatic environmental research, marking the beginnings of the main phase of exploration for this new frontier in Antarctic science.

Martin J. Siegert
University of Edinburgh

Mahlon C. Kennicutt II
Texas A&M University

Robert A. Bindschadler
NASA Goddard Space Flight Center

Antarctic Subglacial Aquatic Environments
Geophysical Monograph Series 192
Copyright 2011 by the American Geophysical Union
10.1029/2010GM001005

Subglacial Aquatic Environments: A Focus of 21st Century Antarctic Science

Mahlon C. Kennicutt II

Department of Oceanography, Texas A&M University, College Station, Texas, USA

Martin J. Siegert

School of GeoSciences, University of Edinburgh, Edinburgh, UK

In 1996, growing evidence suggested a massive lake of liquid water had pooled beneath the East Antarctic Ice Sheet. This feature became known as "Lake Vostok." Early on, two hypotheses were posed: the lake contained microbial life that had evolved over millions of years in isolation beneath the ice and lake sediments contained records of past climate change obtainable nowhere else in Antarctica. Many subglacial lakes, in a number of locales, have been identified, suggesting that studies at multiple locations will be needed to fully understand the importance of subglacial aquatic environments. As of 2010, more than 300 lakes have been identified; this will increase as surveys improve spatial coverage. Given the likely pristine nature of these environs and the low levels of microbial life expected, exploration must be done in a manner that causes minimal impact or contamination. It has been shown that many of these lakes are part of an active, sub-ice hydrological system that experiences rapid water flow events over time frames of months, weeks, and even days. Microbial life in subglacial environments has been inferred, and is expected, but it has yet to be directly confirmed by in situ sampling. Current understanding of subglacial environments is incomplete and will only be improved when these subglacial environments are entered and sampled, which is projected to occur in the next few years. This book synthesizes current understanding of subglacial environments and the plans for their exploration as a benchmark for future discoveries.

1. INTRODUCTION

The recent study and exploration of subglacial aquatic environments has transformed our understanding of processes operating at the Antarctic ice sheet bed and the role they have played in the evolution of the continental ice mass. In 1996, an article in *Nature* reported that a giant lake existed beneath ~4 km of ice in East Antarctica [*Kapitsa et al.*, 1996]. The lake was large enough to be visible in satellite altimetry data of the ice sheet surface [*Ridley et al.*, 1993]. The *Nature* article marked the beginning of modern subglacial aquatic environment research. Even though the original data had been collected in the 1960s and 1970s, these features had gone largely unnoticed by the broader scientific community for more than two decades [e.g., *Oswald and Robin*, 1973; *Robin et al.*, 1977]. While the existence of liquid bodies of water beneath ice sheets was in itself of interest to glaciologists, attention swiftly turned to whether these environments harbored unusual life forms. Some also conjectured that if sediments were preserved in the lake they would contain otherwise unavailable, valuable records of ice and climate change in the interior of Antarctica. *Kapitsa et al.* [1996] had

Antarctic Subglacial Aquatic Environments
Geophysical Monograph Series 192
Copyright 2011 by the American Geophysical Union
10.1029/2010GM001001

identified Vostok Subglacial Lake buried beneath ~4 km of ice. Not only was the lake's area an order of magnitude greater than any other subglacial lake known at the time, *Kapitsa et al.* [1996] also discoverd that the lake contained a water column at least 510 m deep, with an estimated volume of 1800 km^3.

An international community of scientists became convinced that subglacial lakes represented a new frontier in Antarctic research. Within a decade, this community developed the scientific rationale for the exploration and study of these environments through a series of international meetings. The first was convened in 1994 in expectation of the publication of the *Nature* article (Scott Polar Research Institute, University of Cambridge, 1994). The dimensions and setting of Vostok Subglacial Lake were discussed, and a preliminary inventory of Subglacial Lakes was presented. By the time of the publication of the *Nature* article, 77 subglacial lake features had been identified based on analysis of existing radio echo sounding records [*Siegert et al.*, 1996]. Three more workshops were convened in the late 1990s in quick succession: (1) "Lake Vostok Study: Scientific Objectives and Technological Requirements" (St. Petersburg, March 1998); (2) "Lake Vostok: A Curiosity or a Focus for Scientific Research?" (Washington D. C., United States, November 1998 [*Bell and Karl*, 1998]); and (3) "Subglacial Lake Exploration" (Scientific Committee on Antarctic Research (SCAR), Cambridge, September 1999). The general conclusion of these gatherings was that to adequately explore subglacial environments a major, sustained investment in time, resources, and scientific effort would be needed for at least a decade. In recognition of this emerging frontier, SCAR convened a forum for scientists and technologists to gather, exchange ideas, and plan for the future: the Subglacial Antarctic Lake Environments Group of Specialists (SALEGoS) (2000–2004). In due course, SALEGoS transformed into a major SCAR Scientific Research Program entitled Subglacial Antarctic Lake Environments (SALE) (2004–2010).

2. EARLY SCIENTIFIC DEVELOPMENTS

In the early stages, understanding of subglacial environments was refined by remote sensing studies and theoretical modeling [*Ridley et al.*, 1993; *Siegert and Ridley*, 1998; *Wüest and Carmack*, 2000; *Mayer and Siegert*, 2000]. The interface between the ice sheet and the underlying bed was shown to contain liquid water at many locations. Ongoing speculation about life in these lakes caught the imagination of not only scientists but the public in general. At the time, the only available samples were surrogates of lake water, the so-called "accreted ice" that forms as lake water adfreezes to the underside of the ice sheet [*Karl et al.*, 1999; *Jouzel et al.*,

1999; *Priscu et al.*, 1999; *Bell et al.*, 2002]. This "accreted ice" was unexpectedly encountered and recovered during deep coring of the Vostok ice borehole. The ice was recognized as unique because of its unusually large crystal sizes (feet in length), lack of meteoric gasses, and the purity of the water collected on melting [*Jouzel et al.*, 1999], compared with the meteoric glacier ice above that contained a well-characterized record of climate change [*Petit et al.*, 1999].

As additional geophysical surveys were conducted and integrated with previously collected data [*Tabacco et al.*, 2002; *Studinger et al.*, 2003; *Wright and Siegert*, this volume], it was established that subglacial lakes were common beneath thick (>2 km) ice sheets. In early inventories [*Siegert et al.*, 1996], the number and distribution of features was limited by the coverage of surveys. However, it was expected that identification of additional features would continue to mount as unexplored areas of Antarctica were surveyed. On the basis of fundamental considerations, subglacial lakes were expected to occur across the Antarctic continent wherever thick accumulations of ice occurred, a hydrological collection basin was accessible, and a source of water was available. Vostok Subglacial Lake dominated early discussions as it was the only lake whose shape and size were known well; it remains the largest known subglacial lake with an area of about 14,000 km^2 and water depths reaching >1000 m [*Siegert et al.*, this volume].

As the inventory of lakes expanded, it was apparent that subglacial aquatic features are not randomly distributed across Antarctica. Instead they are located in preferred settings suggesting that a spectrum of lakes exist that might well have differing histories, ages, origins, and possibly living residents [*Dowdeswell and Siegert*, 1999]. Clusters of lakes were documented in regions that exhibit distinct ice sheet dynamics in settings defined by the underlying basement morphology [*Dowdeswell and Siegert*, 2002]. In the vicinity of Dome C and Concordia Station, "lake districts" were identified where subglacial features clustered near ice divides and also at the heads of ice streams [*Siegert and Ridley*, 1998; *Siegert and Bamber*, 2000]. It was speculated that some lakes were hydrologically connected in a manner analogous to subaerial lake, stream, and wetland systems [*Dowdeswell and Siegert*, 2002]. The existence of sub-ice hydrological systems transformed ideas about the evolution and functioning of subglacial environments and redefined interests in these settings to include a wide variety of subglacial aquatic environments.

3. MOMENTUM BUILDS

Planning for, and discussions of, subglacial aquatic environment exploration and study gained additional momentum with the formation and approval of the International Polar Year 2007–2008 program "Subglacial Antarctic Lake Environments Unified International Team for Exploration and Discovery (SALE UNITED)." Together SCAR SALE and SALE UNITED served as forums to exchange information among those interested in the study of these environments. In combination, the programs included scientists and technologists from Belgium, Canada, China, France, Germany, Italy, Russia, the United Kingdom, and the United States. Meetings were convened in Austria (2005), France (2006), the United States (2007), Russia (2008), and Belgium (2009) to develop and refine plans for exploration and to share the latest geophysical, microbiological, and modeling information. An international workshop entitled "Subglacial Antarctic Lake Environment in the IPY 2007-2008: Advanced Science and Technology Planning Workshop" was convened in Grenoble, France, in 2006 bringing together 84 participants from 11 countries [*Kennicutt and Petit*, 2006]. During this period, understanding of subglacial environments took an unexpected turn.

Analyses of changes in ice sheet surface elevations in central East Antarctica, using satellite remote sensing, demonstrated that a lake in the Adventure Subglacial Trench discharged approximately 1.8 km^3 of water over a period of 14 months [*Wingham et al.*, 2006]. The water flowed along the axis of a trench and into at least two other lakes about 200 km downstream. The flux of water, ~ 50 m^3 s^{-1}, was equivalent to the flow of the River Thames in London. This discovery was particularly interesting as, up until then, the central East Antarctica Ice Sheet was considered to be undynamic compared with West Antarctica. If significant flow of water occurred at the center of East Antarctica, flows of subglacial water were thought to be commonplace in Antarctica. Subglacial aquatic features appeared to be linked by a network of hydrological channels that were defined by basal topography and surface ice sheet slope. *Siegert et al.* [2007] suggested that groups of lakes were likely to be joined in discrete clusters acting as a system. *Wright et al.* [2008] established that flow channels were sensitive to the ice surface slope, concluding that small changes in surface slope could result in major alterations of basal water flow. Periods of ice sheet changes, such as after the Last Glacial Maximum, or even as a consequence of global warming, might affect the frequency, magnitude, and direction of these flow events. Up topographic slope (uphill) flow could be expected as discharges were predicted to follow the hydrologic potential established by variations in overlying ice thickness interacting with underlying basement elevations.

Further satellite remote sensing analyses illustrated that subglacial discharge and water flow were indeed commonplace in Antarctica [*Smith et al.*, 2009]. It was confirmed that many newly identified lakes and discharge areas were preferentially located at the heads of ice streams [*Siegert and Bamber*, 2000; *Bell et al.*, 2007]. *Smith et al.* [2009] further suggested that lakes actively discharge water into ice stream beds in response to varying basal flows. Satellite investigations of the Byrd Glacier established that subglacial lake discharges coincided with variations in flow velocities observed at an outlet glacier that drained East Antarctica [*Stearns et al.*, 2008]. This inferred subglacial dynamics both influenced, and were influenced by, overlying ice sheet dynamics.

As subglacial lakes represent unique habitats, environmental stewardship during their eventual exploration was seen as a critical issue, and, early on (i.e., Cambridge, 1999), guiding principles were developed and adopted by the community. These concerns included the cleanliness of access techniques, contamination by the experiments that might be performed, the introduction of alien chemicals and biota, how to collect unadulterated samples for laboratory analysis (especially microbiological samples), and how best to protect subglacial aquatic environments as sites of scientific and public interest. The U.S. National Academies convened a committee to review aspects of subglacial lake exploration from an environmental protection and conservation perspective [*Committee on Principles of Environmental Stewardship for the Exploration and Study of Subglacial Environments, National Research Council*, 2007; *Doran and Vincent*, this volume]. The National Academy findings were introduced at the Antarctic Treaty Consultative Meeting in 2008 in Kiev, Ukraine, and SCAR subsequently provided guidance on these issues as a code of conduct for subglacial lake exploration [*Doran and Vincent*, this volume]. These deliberations serve as the basis for promulgating standards and procedures for the responsible conduct of subglacial aquatic environment study and exploration.

4. VOSTOK SUBGLACIAL LAKE

Vostok Subglacial Lake has been, and continues to be, a major focus of subglacial lake research [*Lukin and Bulat*, this volume]. A consortium of Russian research institutions led by the Arctic and Antarctic Research Institute of Roshydromet conducted extensive geophysical surveys of the Vostok Subglacial Lake area and its vicinity within the framework of the Polar Marine Geological Research Expedition and the Russian Antarctic Expedition (RAE) [*Masolov et al.*, 2006; *Popov et al.*, 2006, 2007; *Popov and Masolov*, 2007]. A series of 1:1,000,000 maps of Vostok Subglacial Lake's extent, ice and water body thicknesses, and bedrock relief were produced as well as maps of the spatial pattern of internal layers in the overlying ice sheet. From this work, the lake's dimensions were better defined, the inclination of the ice-water interface was confirmed, and it was recognized that Vostok Subglacial Lake lies in a deep trough. *Studinger et al.* [2003] collected more than 20,000 km of aerogeophysical data producing detailed assessments of the lake and its glaciological setting. The existence of two basins was confirmed by gravity modeling of lake bathymetry, and the southern basin of the lake was determined to be more than 1 km deep [*Studinger et al.*, 2004; *Masolov et al.*, 2001, 2006; *Siegert et al.*, this volume].

Geophysical, geodetic, and glaciological traverses, undertaken by RAE, measured ice flow lines starting at Ridge B and passing through the drilling site at Vostok Station. An Italian/French/Russian partnership also conducted traverses from Talos Dome via Dome C, Vostok Station, Dome B, and Dome A. Thermomechanical ice flow line models were further constrained by this new information [*Richter et al.*, 2008; *Salamatin et al.*, 2008] to yield accurate estimates of the distribution of accreted ice thickness and freezing rates, refined ice depth ages and temperature profiles, and estimated basal melt rates in the northern part of Vostok Subglacial Lake.

Continued deepening of the borehole at Vostok Station extended the ice core isotopic profiles revealing significant spatial and/or temporal variability in physical conditions during accreted ice formation [*Ekaykin et al.*, 2010]. Analysis of accreted ice revealed a distribution of helium isotopes in the lake water that could be explained by hydrothermal activity contributing to the lake water hydrochemistry [*Jean-Baptist et al.*, 2001; *Bulat et al.*, 2004; *de Angelis et al.*, 2004; P. Jean-Baptist, personal communication, 2009]. Although the lake is known to possess small tides [*Dietrich et al.*, 2001], geodetic GPS observations in the southern part of Vostok Subglacial Lake demonstrated that, on a time scale of 5 years, the lake and ice sheet in the vicinity of Vostok Station were in steady state in contrast to other subglacial lakes that were then known to be dynamic [*Richter et al.*, 2008].

5. LIFE IN SUBGLACIAL AQUATIC ENVIRONMENTS

As understanding of the physical conditions in subglacial environments (temperature, pressure, salinity, etc.) was being refined, the existence of life in the lakes remained a focus of great speculation [*Skidmore*, this volume]. A consensus grew that extremely low nutrient levels were to be expected, suggesting these habitats could be challenging for possible microbial inhabitants. Superoxic conditions caused by clathrate decomposition and formation, especially at the water-sediment interface on the lake floor, were also speculated, and it was suggested that these conditions would be toxic to organisms other than anaerobes [*Siegert et al.*, 2003].

At this time, the only clues about possible life in Vostok Subglacial Lake came from extrapolations based on the analyses of the accreted ice [*Karl et al.*, 1999; *Priscu et al.*, 1999]. Contamination of accreted ice samples during drilling, recovery, transportation, and analysis called these results into question as these samples were not originally retrieved for microbiological analyses. The effects of partitioning of lake water constituents during ice formation, under subglacial lake conditions, were poorly understood making inferences of lake water chemistry difficult. The outcome was conflicting evidence for life in the lake and ambiguity about the biogeochemistry of lake water. These uncertainties led to differing opinions about whether hydrothermal effluents contributed to Vostok Subglacial Lake waters. A general consensus evolved that these environments would most likely contain life and that organisms more complex than microbes were highly unlikely. The recognition of hydrological connections among these environments meant that water in many subglacial lakes was likely isolated for far fewer years than first speculated, decreasing the possibility of long-term (>1 Ma) isolation. Depending on the method, the turnover times for water in Vostok Subglacial Lake have been calculated to be between 50,000 and 100,000 years but certainly not millions of years [*Siegert et al.*, 2001; *Bell et al.*, 2002].

More recently, additional accretion ice, and samples of snow collected from layers deposited before the beginning of coring at Vostok Station, contributed further to the debate about possible life within the lake [*Bulat et al.*, 2004, 2007b; *Alekhina et al.*, 2007]. These results suggest that extremely low biomass of both atmospheric and lake water origins is present [*Bulat et al.*, 2009]. Similar studies by United States and United Kingdom researchers confirmed the low cell numbers and low microbial diversity in glacial and accreted ice, though a range of cell numbers and greater diversity have been detected by some investigators [*Christner et al.*, 2006]. The few bacterial phylotypes recovered from accreted ice were isolated from ice layers that contain mineral inclusions raising further questions about their origin [*Bulat et al.*, 2009].

Current knowledge of the lake conditions, inferred from the chemistry of accretion ice studies and from modeling, suggests that the Vostok Subglacial Lake may be inhabited by chemoautotrophic psychrophiles that can tolerate high pressures and possibly high oxygen concentrations, though no conclusive evidence of such microorganisms has yet been found because of a lack of direct sampling of lake water [*Bulat et al.*, 2007a]. The presence of a thermophilic, chemoautotrophic bacterium, *Hydrogenophilus thermoluteolus* (previously identified in other areas influenced by hydrothermal activity remote from Antarctica), has been reported

[*Bulat et al.*, 2004; *Lavire et al.*, 2006]. It has been speculated that water in Vostok Subglacial Lake will be an extremely dilute biological solution suggesting that life, if present, may be primarily restricted to lake sediments and the basal water interface [*Bulat et al.*, 2009]. While studies of the Vostok Subglacial Lake accretion ice have improved comprehension of physical, chemical, and biological processes in the lake, considerable debate continues as to the level and type of life expected in these environments. The debate will not be resolved until direct measurement and sampling of these environments has taken place.

6. OTHER SUBGLACIAL LAKES

A major set of subglacial lakes was recently identified at the onset of the Recovery Ice Stream a major East Antarctic ice flow unit [*Bell et al.*, 2007]. Three or possibly four large subglacial lakes (smaller than Vostok Subglacial Lake but, nonetheless, far larger than most) are thought to be coincident with the onset of rapid ice flow. The lakes exhibit distinctive ice surface morphologies including extensive, relatively flat featureless regions bounded by upstream troughs and downstream ridges generated by changes in bottom topography. The Recovery subglacial lakes are hypothesized to contain water derived from basal melting routed to the lake from a large upstream catchment area. To study the Recovery lakes region a U.S.-Norway traverse conducted surface geophysical surveys and installed GPS stations. Ice sheet motion was quantified by collecting gravity magnetics, laser, and radar data over the two southernmost Recovery lakes [*Block et al.*, 2009]. Once fully interpreted, these data will clarify the dynamics of the origins of subglacial water in the lakes and the upstream catchment as well as evaluate the geologic setting of these features. All four of the Recovery lakes were crossed by the U.S.-Norwegian traverse in January 2009, and low-frequency radar was used to map the morphology of the subglacial lakes and image the ice sheet bed of the lakes identified by *Smith et al.* [2009]. In the coming years, as these data sets are processed, the role that subglacial lakes play in controlling the onset of fast ice flow will be better defined.

7. NEW FRONTIER

Significant progress in the study of subglacial aquatic environments is now at hand with the initiation of an important phase with three exploration programs likely to advance understanding of these environments over the next 3 to 5 years. A United Kingdom–led international program has completed a full survey of Ellsworth Subglacial Lake located in West Antarctica, and plans to undertake direct clean measurement sand sampling of the lake in 2012/2013 are in place [*Ross*

et al., this volume]. The United States has launched a major program to survey, enter, instrument, and sample an "actively discharging" subglacial aquatic system beneath Whillans Ice Stream in West Antarctica at around the same time [*Fricker et al.*, this volume]. Russian researchers are developing further strategies for penetration of Vostok Subglacial Lake, and lake entry is expected in the next few field seasons [*Lukin and Bulat*, this volume].

In the past decade, our understanding of the importance of subglacial aquatic systems as habitats for life, and of their influence on ice sheet dynamics, has been greatly advanced. Subglacial features that contain liquid water are now known to be common beneath the ice sheets of Antarctica. A spectrum of subglacial environments exists as connected subglacial hydrologic systems and water movement beneath ice sheets can and does occur over a range of spatial and temporal scales. The location of subglacial aquatic accumulations and the onset of ice streams have been shown to be linked in some areas, suggesting that ice sheet dynamics can be affected by hydrological systems at the base of the ice sheet.

The exploration and study of subglacial aquatic environments remains at its early stages and if the major advances realized to date are an indication of what is to come, even more fundamental discoveries will be realized in the years ahead. In little more than a decade, findings regarding subglacial aquatic systems have transformed fundamental concepts about Antarctica and its ice sheets. Ice sheet bases are now seen as being highly dynamic at their beds, involving a complex interplay of hydrology, geology, glaciology, tectonics, and ecology now and in the past. Ongoing and planned projects to directly sample these environments will ultimately determine if subglacial waters house unique and specially adapted microbiological assemblages and records of past climate change. The most remarkable advances to be realized from the study of this next frontier in Antarctic science will probably be wholly unexpected as these recently recognized environments are explored.

This volume serves as a benchmark for knowledge about subglacial aquatic environments and as an update on the latest research developments, setting the stage for major new exploration efforts.

REFERENCES

Alekhina, I. A., D. Marie, J.-R. Petit, V. V. Lukin, V. M. Zubkov, and S. A. Bulat (2007), Molecular analysis of bacterial diversity in kerosene-based drilling fluid from the deep ice borehole at Vostok, East Antarctica, *FEMS Microbiol. Ecol.*, *59*, 289–299.

Bell, R., and D. M. Karl (1998), Lake Vostok Workshop: "A Curiosity or a Focus for Interdisciplinary Study," final report, Natl. Sci. Found., Washington, D. C.

Bell, R. E., M. Studinger, A. Tikku, G. K. C. Clarke, M. M. Gutner, and C. Meertens (2002), Origin and fate of Lake Vostok water frozen to the base of the East Antarctic ice sheet, *Nature*, *416*, 307–310.

Bell, R. E., M. Studinger, C. A. Shuman, M. A. Fahnestock, and I. Joughin (2007), Large subglacial lakes in East Antarctica at the onset of fast-flowing ice, *Nature*, *445*, 904–907.

Block, A. E., R. E. Bell, and M. Studinger (2009), Antarctic crustal thickness from satellite gravity: Implications for the Transantarctic and Gamburtsev subglacial mountains, *Earth Planet. Sci. Lett.*, *288*, 194–203.

Bulat, S. A., et al. (2004), DNA signature of thermophilic bacteria from the aged accretion ice of Lake Vostok, Antarctica: Implications for searching life in extreme icy environments, *Int. J. Astrobiol.*, *3*(1), 1–12, doi:10.1017/S1473550404001879.

Bulat, S. A., I. A. Alekhina, V. Y. Lipenkov, V. V. Lukin, and J. R. Petit (2007a), Microbial life in extreme subglacial Antarctic lake environments: Lake Vostok, in *Bresler Memorial Lectures II*, pp. 264–269, Petersburg Nucl. Phys. Inst., Russ. Acad. of Sci., St. Petersburg, Russia.

Bulat, S. A., I. A. Alekhina, J.-R. Petit, V. Y. Lipenkov, and V. V. Lukin (2007b), An assessment of the biogeochemical potential of subglacial Lake Vostok, East Antarctica, in relation to microbial life support (in Russian), *Probl. Arktiki Antarktiki*, *76*, 106–112.

Bulat, S. A., I. A. Alekhina, V. Y. Lipenkov, V. V. Lukin, D. Marie, and J. R. Petit (2009), Cell concentration of microorganisms in glacial and lake ice of the Vostok ice core, East Antarctica (in Russian), *Mikrobiologiya*, *78*(6), 850–852.

Christner, B., G. Royston-Bishop, C. M. Foreman, B. R. Arnold, M. Tranter, K. A. W. B. Welch, W. B. Lyons, A. I. Tsapin, M. Studinger, and J. C. Priscu (2006), Limnological conditions in subglacial Lake Vostok, Antarctica, *Limnol. Oceanogr.*, *51*(6), 2485–2501.

Committee on Principles of Environmental Stewardship for the Exploration and Study of Subglacial Environments, National Research Council (2007), *Exploration of Antarctic Subglacial Aquatic Environments: Environmental and Scientific Stewardship*, 152 pp., Natl. Acad. Press, Washington, D. C.

de Angelis, M., J. R. Petit, J. Savarino, R. Souchez, and M. H. Thiemens (2004), Contributions of an ancient evaporitic-type reservoir to subglacial Lake Vostok chemistry, *Earth Planet. Sci. Lett.*, *222*, 751–765.

Dietrich, R., K. Shibuya, A. Pötzsch, and T. Ozawa (2001), Evidence for tides in the subglacial Lake Vostok, Antarctica, *Geophys. Res. Lett.*, *28*, 2971–2974, doi:10.1029/2001GL013230.

Doran, P. T., and W. F. Vincent (2011), Environmental protection and stewardship of subglacial aquatic environments, in *Antarctic Subglacial Aquatic Environments*, *Geophys. Monogr. Ser.*, doi: 10.1029/2010GM000947, this volume.

Dowdeswell, J. A., and M. J. Siegert (1999), The dimensions and topographic setting of Antarctic subglacial lakes and implications for large-scale water storage beneath continental ice sheets, *Geol. Soc. Am. Bull.*, *111*, 254–263.

Dowdeswell, J. A., and M. J. Siegert (2002), The physiography of modern Antarctic subglacial lakes, *Global Planet. Change*, *35*, 221–236.

Ekaykin, A. A., V. Y. Lipenkov, J. R. Petit, S. Johnsen, J. Jouzel, and V. Masson-Delmotte (2010), Insights into hydrological regime of Lake Vostok from differential behavior of deuterium and oxygen-18 in accreted ice, *J. Geophys. Res.*, *115*, C05003, doi:10.1029/2009JC005329.

Fricker, H. A., et al. (2011), Siple Coast subglacial aquatic environments: The Whillans Ice Stream Subglacial Access Research Drilling project, in *Antarctic Subglacial Aquatic Environments*, *Geophys. Monogr. Ser.*, doi: 10.1029/2010GM000932, this volume.

Jean-Baptiste, P., J. R. Petit, V. Y. Lipenkov, D. Raynaud, and N. I. Barkov (2001), Constraints on hydrothermal processes and water exchange in Lake Vostok from helium isotopes, *Nature*, *411*, 460–462.

Jouzel, J., J. R. Petit, R. Souchez, N. I. Barkov, V. Y. Lipenkov, D. Raynaud, M. Stievenard, N. I. Vassiliev, V. Verbeke, and F. Vimeux (1999), More than 200 meters of lake ice above subglacial Lake Vostok, Antarctica, *Science*, *286*(5447), 2138–2141.

Kapitsa, A. P., J. K. Ridley, G. Q. Robin, M. J. Siegert, and I. A. Zotikov (1996), A large deep freshwater lake beneath the ice of central East Antarctica, *Nature*, *381*, 684–686.

Karl, D. M., D. F. Bird, K. Bjorkman, T. Houlihan, R. Shackelford, and T. Tupas (1999), Microorganisms in the accreted ice of Lake Vostok, Antarctica, *Science*, *286*(5447), 2144–2147.

Kennicutt, M. C., and J. R. Petit (2006), Subglacial Antarctic lake environments (SALE) in the International Polar Year: Advanced science and technology planning - Grenoble France, April, 2006, workshop report, 43 pp., Natl. Sci. Found., Washington, D. C.

Lavire, C., P. Normand, I. Alekhina, S. Bulat, D. Prieur, J. L. Birrien, P. Fournier, C. Hänni, and J. R. Petit (2006), Presence of *Hydrogenophylus thermoluteolus* DNA in accretion ice in the subglacial Lake Vostok, Antarctica, assessed using rrs, cbb and hox, *Environ. Microbiol.*, *8*(12), 2106–2114.

Lukin, V., and S. Bulat (2011), Vostok Subglacial Lake: Details of Russian plans/activities for drilling and sampling, in *Antarctic Subglacial Aquatic Environments*, *Geophys. Monogr. Ser.*, doi: 10.1029/2010GM000951, this volume.

Masolov, V. N., V. V. Lukin, A. N. Sheremetyev, and S. V. Popov (2001), Geophysical investigations of the subglacial Lake Vostok in Eastern Antarctica (in Russian), *Dokl. Akad. Nauk*, *379*(5), 680–685.

Masolov, V. N., S. V. Popov, V. V. Lukin, A. N. Sheremet'ev, and A. M. Popkov (2006), Russian geophysical studies of Lake Vostok, central East Antarctica, in *Antarctica: Contributions to Global Earth Sciences*, edited by D. K. Fütterer et al., pp. 135–140, Springer, New York.

Mayer, C., and M. J. Siegert (2000), Numerical modelling of ice-sheet dynamics across the Vostok subglacial lake, central East Antarctica, *J. Glaciol.*, *46*, 197–205.

Oswald, G. K. A., and G. d. Q. Robin (1973), Lakes beneath the Antarctic ice sheet, *Nature*, *245*, 251–254.

Petit, J. R., et al. (1999), Climate and atmospheric history of the past 420,000 years from the Vostok ice core, Antarctica, *Nature*, *399*, 429–436.

Popov, S. V., and V. N. Masolov (2007), Forty-seven new subglacial lakes in the 0–110° sector of East Antarctica, *J. Glaciol.*, *53*, 289–297.

Popov, S. V., A. N. Lastochkin, V. N. Masolov, and A. M. Popkov (2006), Morphology of the subglacial bed relief of Lake Vostok basin area (central East Antarctica) based on RES and seismic data, in *Antarctica: Contributions to Global Earth Sciences*, edited by D. K. Fütterer et al., pp. 141–146, Springer, New York.

Popov, S. V., V. Y. Lipenkov, A. V. Enalieva, and A. V. Preobrazhenskaya (2007), Internal isochrone layers in the ice sheet above Lake Vostok, East Antarctica (in Russian), *Probl. Arktiki Antarktiki*, *76*, 89–95.

Priscu, J. C., et al. (1999), Geomicrobiology of subglacial ice above Lake Vostok, Antarctica, *Science*, *286*(5447), 2141–2144.

Richter, A., S. V. Popov, R. Dietrich, V. V. Lukin, M. Fritsche, V. Y. Lipenkov, A. Y. Matveev, J. Wendt, A. V. Yuskevich, and V. N. Masolov (2008), Observational evidence on the stability of the hydro-glaciological regime of subglacial Lake Vostok, *Geophys. Res. Lett.*, *35*, L11502, doi:10.1029/2008GL033397.

Ridley, J. K., W. Cudlip, and S. W. Laxon (1993), Identification of subglacial lakes using ERS-1 radar altimeter, *J. Glaciol.*, *39*, 625–634.

Robin, G. d. Q., D. J. Drewry, and D. T. Meldrum (1977), International studies of ice sheet and bedrock, *Philos. Trans. R. Soc. London*, *279*, 185–196.

Ross, N., et al. (2011), Ellsworth Subglacial Lake, West Antarctica: Its history, recent field campaigns, and plans for its exploration, in *Antarctic Subglacial Aquatic Environments*, *Geophys. Monogr. Ser.*, doi: 10.1029/2010GM000936, this volume.

Salamatin, A. N., E. A. Tsyganova, S. V. Popov, and V. Y. Lipenkov (2008), Ice flow line modeling in ice core data interpretation: Vostok Station (East Antarctica), in *Physics of Ice Core Records*, vol. 2, edited by T. Hondoh, Hokkaido Univ. Press, Sapporo, Japan.

Siegert, M. J., and J. L. Bamber (2000), Subglacial water at the heads of Antarctic ice-stream tributaries, *J. Glaciol.*, *46*, 702–703.

Siegert, M. J., and J. K. Ridley (1998), An analysis of the ice-sheet surface and subsurface topography above the Vostok Station subglacial lake, central East Antarctica, *J. Geophys. Res.*, *103*, 10,195–10,208.

Siegert, M. J., J. A. Dowdeswell, M. R. Gorman, and N. F. McIntyre (1996), An inventory of Antarctic sub-glacial lakes, *Antarct. Sci.*, *8*, 281–286.

Siegert, M. J., J. C. Ellis-Evans, M. Tranter, C. Mayer, J.-R. Petit, A. Salamatin, and J. C. Priscu (2001), Physical, chemical and biological processes in Lake Vostok and other Antarctic subglacial lakes, *Nature*, *414*, 603–609.

Siegert, M. J., M. Tranter, C. J. Ellis-Evans, J. C. Priscu, and W. B. Lyons (2003), The hydrochemistry of Lake Vostok and the potential for life in Antarctic subglacial lakes, *Hydrol. Processes*, *17*, 795–814.

Siegert, M. J., A. Le Brocq, and A. Payne (2007), Hydrological connections between Antarctic subglacial lakes and the flow of water beneath the East Antarctic ice sheet, in *Glacial Sedimentary Processes and Products*, edited by M. J. Hambrey et al., *Spec. Publ. Int. Assoc. Sedimentol.*, *39*, 3–10.

Siegert, M. J., S. Popov, and M. Studinger (2011), Vostok Subglacial Lake: A review of geophysical data regarding its discovery and topographic setting, in *Antarctic Subglacial Aquatic Environments*, *Geophys. Monogr. Ser.*, doi: 10.1029/2010GM000934, this volume.

Skidmore, M. (2011), Microbial communities in Antarctic subglacial aquatic environments, in *Antarctic Subglacial Aquatic Environments*, *Geophys. Monogr. Ser.*, doi: 10.1029/2010GM000995, this volume.

Smith, B. E., H. A. Fricker, I. R. Joughin, and S. Tulaczyk (2009), An inventory of active subglacial lakes in Antarctica detected by ICESat (2003–2008), *J. Glaciol.*, *55*, 573–595.

Stearns, L. A., B. E. Smith, and G. S. Hamilton (2008), Increased flow speed on a large East Antarctic outlet glacier caused by subglacial floods, *Nat. Geosci.*, *1*, 827–831.

Studinger, M., et al. (2003), Ice cover, landscape setting and geological framework of Lake Vostok, East Antarctica, *Earth Planet. Sci. Lett.*, *205*, 195–210.

Studinger, M., R. E. Bell, and A. A. Tikku (2004), Estimating the depth and shape of subglacial Lake Vostok's water cavity from aerogravity data, *Geophys. Res. Lett.*, *31*, L12401, doi:10.1029/2004GL019801.

Tabacco, I. E., C. Bianchi, A. Zirizzotti, E. Zuccheretti, A. Forieri, and A. Della Vedova (2002), Airborne radar survey above Vostok region, East Antarctica: Ice thickness and Lake Vostok geometry, *J. Glaciol.*, *48*, 62–69.

Wingham, D. J., M. J. Siegert, A. Shepherd, and A. S. Muir (2006), Rapid discharge connects Antarctic subglacial lakes, *Nature*, *440*, 1033–1036.

Wright, A., and M. J. Siegert (2011), The identification and physiographical setting of Antarctic subglacial lakes: An update based on recent discoveries, in *Antarctic Subglacial Aquatic Environments*, *Geophys. Monogr. Ser.*, doi: 10.1029/2010GM000933, this volume.

Wright, A. P., M. J. Siegert, A. Le Brocq, and D. Gore (2008), High sensitivity of subglacial hydrological pathways in Antarctica to small ice sheet changes, *Geophys. Res. Lett.*, *35*, L17504, doi:10.1029/2008GL034937.

Wüest, A., and E. Carmack (2000), A priori estimates of mixing and circulation in the hard-to-reach water body of Lake Vostok, *Ocean Modell.*, *2*(1), 29–43.

M. C. Kennicutt, Department of Oceanography, Texas A&M University, College Station, TX 77843-3146, USA.

M. J. Siegert, School of GeoSciences, University of Edinburgh, Edinburgh EH9 3JW, UK. (M.J.Siegert@ed.ac.uk)

The Identification and Physiographical Setting of Antarctic Subglacial Lakes: An Update Based on Recent Discoveries

Andrew Wright and Martin J. Siegert

Grant Institute, School of GeoSciences, University of Edinburgh, Edinburgh, UK

We investigate the glaciological and topographic setting of known Antarctic subglacial lakes following a previous assessment by Dowdeswell and Siegert (2002) based on the first inventory of 77 lakes. Procedures used to detect subglacial lakes are discussed, including radio echo sounding (RES) (which was first used to demonstrate the presence of subglacial lakes), surface topography, topographical changes, gravity measurements, and seismic investigations. Recent discoveries of subglacial lakes using these techniques are detailed, from which a revised new inventory of subglacial lakes is established, bringing the total number of known subglacial lakes to 387. Using this new inventory, we examine various controls on subglacial lakes, such as overlying ice thickness and position within the ice sheet and formulate frequency distributions for the entire subglacial lake population based on these (variable) controls. We show how the utility of RES in identifying subglacial lakes is spatially affected; lakes away from the ice divide are not easily detected by this technique, probably due to scattering at the ice sheet base. We show that subglacial lakes are widespread in Antarctica, and it is likely that many are connected within well-defined subglacial hydrological systems.

1. INTRODUCTION

A variety of methods have been used in the discovery and characterization of subglacial lakes and the identification of subglacial water movement in Antarctica (Figure 1). The first inventory of subglacial lakes, recording 77 lake locations, used the technique of radio echo sounding [*Siegert et al.*, 1996]. This was later updated to 145 lakes by *Siegert et al.* [2005]. Several other techniques are available for the detection of subglacial lakes, including surface topography, topographical changes, gravity survey, and seismic investigations. The aim of this paper is to detail all the techniques available for the detection of subglacial lakes, and to pull together recent information regarding lake locations. In doing so, a revised inventory is established, from which an assessment of the dimensions and topographic setting of subglacial lakes is updated [from *Dowdeswell and Siegert*, 2002].

2. DISCOVERY, IDENTIFICATION, AND CHARACTERIZATION OF SUBGLACIAL LAKES

2.1. Radio Echo Sounding (RES)

2.1.1. Development of the technique. The technique of RES takes advantage of a window in the radio part of the electromagnetic spectrum within which emitted waves will travel freely through both ice and air. As with all E-M waves, reflections occur at boundaries between materials with different dielectric properties and therefore different speeds of

Antarctic Subglacial Aquatic Environments
Geophysical Monograph Series 192
Copyright 2011 by the American Geophysical Union
10.1029/2010GM000933

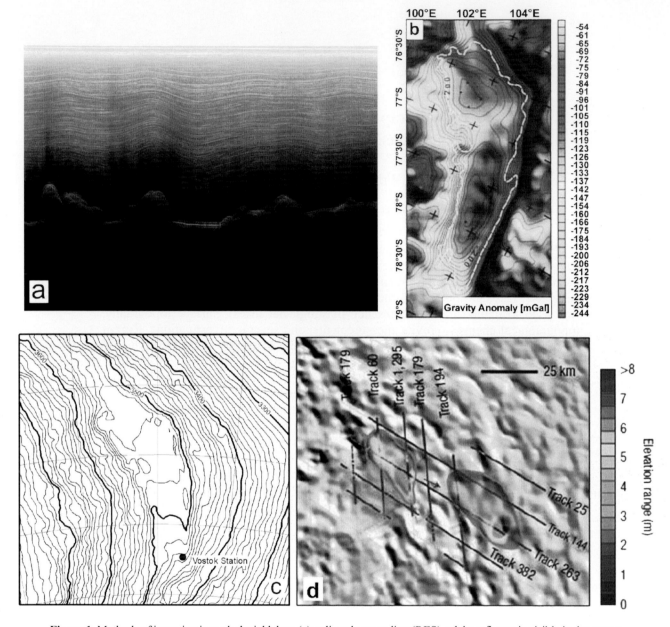

Figure 1. Methods of investigating subglacial lakes: (a) radio echo sounding (RES), a lake reflector is visible in the center; (b) free-air gravitational anomaly (with ice sheet effects removed) detected from the air above Vostok Subglacial Lake [*Studinger et al.*, 2004b]; (c) prominent flat spot in the ice sheet extending north of Vostok Station, 10-m contours from ERS-1 radar altimetry [*Siegert and Ridley*, 1998b]; and (d) vertical surface elevation changes measured by ICESat used to identify two lakes within the catchment of the Byrd Glacier, East Antarctica. Reprinted by permission from Macmillan Publishers Ltd: *Nature Geoscience* [*Stearns et al.*, 2008], copyright 2008.

wave propagation. On entering ice, the speed of the radio wave drops by nearly half, from 300 to 168 m μs^{-1} [*Glen and Paren*, 1975]. An active transmit/receive radar antenna, mounted either near the surface [e.g., *Popov et al.*, 2003] or on an airborne platform [e.g., *Blankenship et al.*, 2001], can

therefore be used to detect reflections originating from both within and at the base of a glacier or ice sheet.

The initial investigations of the base of the Antarctic ice sheet were carried out as part of a joint Scott Polar Research Institute (SPRI), National Science Foundation (NSF), and

Technical University of Denmark (TUD) airborne RES campaign. Between 1967 and 1979, this project completed over 400,000 km of line transects, spaced by an average of 50 km, and covering approximately half of the total area of Antarctica [*Robin et al.*, 1977; *Drewry*, 1983].

The SPRI-NSF-TUD survey was able to penetrate even the thickest ice in Antarctica using a 60-MHz frequency radar. By measuring the time elapsed between transmit and receive, it was shown that an ice thickness greater than 4 km existed over large parts of East Antarctica [*Drewry*, 1983]. Nearly 40 years later, this is still the greatest aerial coverage of any single aerogeophysical survey in Antarctica. More recent airborne RES surveys have been characterized by smaller spatial coverage but much higher spatial resolution [e.g., *Rémy and Tabacco*, 2000; *Popov et al.*, 2002; *Rippin et al.*, 2003; *Studinger et al.*, 2003a, 2004a; *Holt et al.*, 2006a; *Vaughan et al.*, 2006]. In several instances, these surveys have targeted areas not covered, or only sparsely covered by the SPRI-NSF-TUD data, with the result being a patchwork of bed information. Despite these efforts, however, a number of large gaps still exist where our knowledge of the ice sheet bed remains poor [e.g., *Le Brocq et al.*, 2008].

Several reviews detail both developments in the techniques of radioglaciology and the resulting enhancements in our understanding of the ice sheets [e.g., *Plewes and Hubbard*, 2001; *Dowdeswell and Evans*, 2004; *Bingham and Siegert*, 2007]. This subject will therefore not be discussed further here.

2.1.2. The discovery of subglacial lakes.

The discovery of the first subglacial lake occurred near the Russian station at Sovetskaya during the 1967/1968 season of SPRI-NSF-TUD radio echo sounding [*Robin et al.*, 1970]. An area of unusually low signal fading and short duration of the returned pulse, indicating a specular reflection, was found to coincide with low attenuation of the transmitted signal and a near-horizontal, flat bed geometry. This was, at first tentatively, best explained as the result of a sub-ice water body [*Robin et al.*, 1970]. During the 1971/1972 season, an extension of the RES survey over the Dome C area identified a further 16 similar locations, indicating that the occurrence of pockets of liquid water, or subglacial lakes, beneath the central regions of the ice sheet might be relatively commonplace [*Oswald and Robin*, 1973]. Owing to the lack of penetration through water of radio waves at megahertz frequencies, the depths of these newly discovered features could not be determined, only that a sufficient depth (i.e., a few meters or more) must exist to permit the continuous, strong, and flat echo returns observed [*Oswald and Robin*, 1973].

2.1.3. Characterization of bed reflections.

Radio echo sounding has been used to gather much of what is currently known about the subglacial environment of Antarctica. When the velocity of the radar pulse is known, or can be estimated, the thickness of the ice can be calculated by measuring the time difference between echoes received from the air/surface and ice/bed interfaces. A time series of such echoes recorded as the observer moves over the ice surface can be used to create a pseudo-cross-section of the ice sheet and of the underlying bed. These data combined with measurements of the surface elevation can be used to reconstruct the topography of the underside of the ice sheet [*Drewry*, 1983; *Lythe et al.*, 2001].

It was soon realized that much more information about the subglacial environment could be obtained from an analysis of the echo returns. In particular, the strength [*Neal*, 1976] and shape [*Berry*, 1973, 1975] of the returned pulse is related to the degree of scattering at the interface and therefore to the microtopography of the subglacial surface. Early studies were limited to the use of "incoherent" radar-sounding apparatus. Modern RES equipment can record both amplitude and phase of reflected pulses ("coherent" radar) allowing a moving platform to operate in the Synthetic Aperture (SAR) mode [*Gogineni et al.*, 1998]. Coherent integration both allows the detection of radar reflections where they would otherwise be obscured by scattering from crevasses, etc. and improves the ability to quantify reflection and scattering from a subglacial interface [*Peters et al.*, 2005].

Radar reflections within ice are caused by changes in dielectric permittivity (ε_r) due to changing density or crystal fabric orientation and by changes in electrical conductivity, due largely to varying acidity associated with the fallout of volcanic aerosols [*Fujita et al.*, 1999]. Basal reflections are generally caused by the large difference in dielectric impedance between ice and the basal material, the magnitude of the reflection being proportional to the change in impedance. Dielectric constants for the various types of bedrock, observed in Antarctica, range from a minimum of ~4 to a maximum of ~9. This is very much closer to the value for glacier ice ($\varepsilon_r = 3.2$) than is the dielectric constant of pure water ($\varepsilon_r = 80$). For this reason, a basal reflection from an ice-water subglacial interface is much brighter than the equivalent reflection from a dry interface or from frozen sediments [*Bogorodskiy et al.*, 1985]. *Shabtaie et al.* [1987] showed that the minimum sub-ice water thickness required for a water-dominated reflection is between a few tens of centimeters to a few meters, depending on salinity.

The shape of a smooth ice/bed or ice/water interface is an additional factor which can affect the strength of the returned echo in the same way that a concave or convex mirror acts to focus light [*Tabacco et al.*, 2000]. In tests conducted over a floating ice tongue, isolated geometrical effects have been shown to influence the total received power by ±6–8 dB [*Bianchi et al.*, 2004].

2.1.4. Identifying subglacial lakes by RES. The strength of the radio echo from the base of the ice sheet has been used by several authors to infer information about basal conditions in various glaciated regions [e.g., *Bentley et al.*, 1998; *Gades et al.*, 2000; *Catania et al.*, 2003; *Peters et al.*, 2005, 2007]. To do this, the proportion of energy reflected at the bed (the basal reflection coefficient) must be distinguished from the many other factors which can affect the strength of the signal received at the antenna. Probably, the most significant of these is the dielectric power loss during transmission through the ice. This depends sensitively on ice temperature and can vary spatially by 15–20 dB km^{-1} [*Peters et al.*, 2007]. While subglacial water will always produce a bright radar reflection, an "absolute brightness" criteria for lake identification can be misleading. Rather, it is the brightness of a particular feature "relative" to its surroundings, which can be more useful in identifying subglacial lakes [*Carter et al.*, 2007].

Amplitude fading is the fluctuation in radio echo amplitude as the observer moves at a fixed distance from an interface, it is caused by interference from different scattering centers fore and aft of the observers position and can be used to obtain useful information about bed roughness [*Oswald*, 1975]. Very low fading (or alternatively, a very large "fading distance") implies a continuous, flat, mirror-like or "specular" reflection. A purely specular reflection can only occur where the interface is smooth on the scale of the radar footprint. A substantial body of water at the bed of an ice sheet will exhibit a smooth ice-water interface that will also satisfy the criteria of hydrostatic equilibrium. This states that due to the different densities of ice and water, and assuming that the water supports the full overburden pressure of the ice, the ice-water interface will have a slope 11 times greater and in the opposite direction to the slope of the ice surface. Calculations of the hydrological potential field can therefore be useful in evaluating subglacial lake candidates from their radar profile [*Oswald*, 1975; *Carter et al.*, 2007].

The electrical properties of liquid water act to inhibit the transmission of electromagnetic waves. For this reason, RES cannot normally be used to determine the depths of subglacial lakes. An exception to this has been found in shallow regions of some lakes surveyed with the SPRI-NSF-TUD radar, where bottom reflections have been recorded from depths of up to 21 m below the lake's surface [*Gorman and Siegert*, 1999]. These observations confirm that these lakes are, in fact, substantial bodies of water, but indicate only their minimum depths.

2.2. Identification of Subglacial Lakes From Ice Surface Topography

Even before subglacial lakes had been firmly identified in RES records, their surface expressions had been noted by pilots traversing the center of the continent. Unusually flat areas of the ice sheet were often referred to as "lakes" and were frequently used as landmarks for navigation before any connection was made to the subglacial environment [*Robinson*, 1960].

When an ice sheet flows over a localized body of water, the weight of the ice is taken by the incompressible fluid. Provided that there is no outlet channel for the water to escape, this will lead to the establishment of local hydrostatic equilibrium. This has a significant affect on the flow regime of the ice and, for a large enough lake, can result in the morphological expression of an extremely flat and featureless ice surface, similar to that of a floating ice shelf.

Satellite observations with the Seasat radar altimeter identified a prominent flat area, in Terre Adelie, East Antarctica, the position of which was shown to correspond to a subglacial reflector identified in the SPRI-NSF-TUD radar record [*Cudlip and McIntyre*, 1987]. This technique achieved greater success when several RES lake reflectors in the area to the north of Vostok Station [*Robin et al.*, 1977] were shown to lie beneath a single, continuous flat surface area observed with the ERS-1 satellite [*Ridley et al.*, 1993; *Kapitsa et al.*, 1996]. A finding later confirmed and elaborated on using more sophisticated radar altimetry techniques [*Roemer et al.*, 2007] and laser altimetry [*Studinger et al.*, 2003a].

Subsequent analysis of ERS-1 data identified flat surface features associated with a further 28 subglacial lakes known from RES records in the Dome C and Terre Adelie regions [*Siegert and Ridley*, 1998a]. Small subglacial lakes (dimensions <4 km) are generally not found to have a corresponding flat surface feature. Furthermore, flat areas of ice, meeting the criteria for identification of a subglacial lake, have also been shown to occur where no lake exists [*Siegert and Ridley*, 1998a]. Water-saturated sediments can cause a similar reduction in basal stress and, therefore, induce a similar surface expression. For this reason, a surface flat area alone is not normally sufficient evidence for a lake discovery [*Siegert and Ridley*, 1998a].

In addition, for floating ice, the retarding force of the basal shear stress is reduced to zero. As a result, ice flowing from a solid bed onto a subglacial lake experiences acceleration. The resulting extensional flow has been shown to cause a local thinning of the ice and a lowering of the surface on the upstream side of the lake [*Shoemaker*, 1990; *Gudmundsson*, 2003; *Pattyn et al.*, 2004]. Conversely, a thickening of the ice can occur over the downstream lake shore, where the return of basal drag causes compressive flow.

Until recently, it was thought that no other subglacial lakes of a similar scale to Vostok Subglacial Lake (hereinafter referred to as Lake Vostok) existed beneath Antarctica [*Siegert*, 2000]. Imagery from the Moderate Resolution Imaging Spectroradiometer (MODIS) satellite, however, has now been used to determine large surface areas for two lakes (90°E Lake and Lake Sovetskaya) that were previously known only from relatively short sections of RES survey [*Bell et al.*, 2006]. Further discoveries of four large lakes in the upstream region of the Recovery Ice Stream have been made using MODIS imagery to locate the lake surfaces, which have then been shown to possess surface ridge and trough features consistent with the direction of ice flow [*Bell et al.*, 2007].

2.3. Discovery of Active Subglacial Lakes by Measurements of Surface Height Change

It is now becoming widely recognized that subglacial drainage systems are generally dynamic in nature. Changes associated with the movement of water either between known lakes or between lakes and a distributed hydrological system are apparently common occurrences [*Siegert et al.*, 2007]. Movement of subglacial water has been known, or suspected, as the cause of vertical displacements of the ice surface of valley glaciers for a number of years [e.g., *Iken et al.*, 1983; *Fatland and Lingle*, 2002]. Not until recently, however, have the means, in the form of repeat satellite measurements, been available to observe such local surface height changes in the remote regions of the Antarctic plateau.

Gray et al. [2005] were the first to identify vertical movement of the Antarctic ice sheet that could be attributed to the movement of subglacial water. They used the technique of Interferometric Synthetic Aperture Radar with repeat-passes of the RADARSAT satellite to detect areas of vertical displacement in upstream areas of Ice Streams C ("Kamb Ice Stream") and D ("Bindschadler Ice Stream") in the Siple Coast region of West Antarctica. The surface height changes measured in this study have several features in common with a large number of events identified by other authors since. The measured vertical displacements averaged ~0.5 m, they occurred within an orbital period of 24 days, and were smoothly varying in amplitude over roughly spherical regions 10–20 km across. In the case of the Ice Stream D event, an upstream surface lowering was observed over the same time period as a downstream surface rise representing an approximately equivalent change in volume [*Gray et al.*, 2005].

Further evidence for subglacial water movement beneath the Siple Coast ice streams was presented by *Fricker et al.* [2007] and *Fricker and Scambos* [2009]. This was obtained from repeat-pass laser altimetry with the ICESat mission to

identify areas of raising and lowering together with differencing of MODIS images in which discrete areas of surface change are visible. While these features demonstrated the scale and activity of the subglacial water system beneath the fast flow features of West Antarctica, *Wingham et al.* [2006] used the radar altimeter on the ERS-2 satellite to identify similar phenomena occurring beneath the thick interior ice of the Adventure Subglacial Trench in East Antarctica. In this case, a single lake was discovered to be deflating (~3 m drop over a surface area of ~600 km^2) upstream along a predicted flow path from several smaller lakes found to be inflating. Unlike the RADARSAT data, this work showed lakes draining and filling over a period of many months and inferred water transport over a distance of around 260 km. Further work has shown that a significant fraction of the water discharged from the upstream lake during this event was retained by the downstream lakes and that a distributed model of subglacial water transport is needed to explain the observed travel times [*Carter et al.*, 2009].

The launch of the ICESat laser altimeter has greatly improved the spatial coverage of Antarctica and has resulted in a large number of new lakes discovered by the effects of their filling and draining on the ice sheet surface. *Smith et al.* [2009] analyzed all ICESat repeat tracks between 2003 and 2008 for indications of anomalous surface height changes that could not be explained by normal glaciological processes [see *Gudmundsson*, 2003; *Sergienko et al.*, 2007]. They detected new lakes throughout the continent (130 in total) with concentrations centered under several of the major outlet glaciers of both East and West Antarctica. Especially significant among them are two large lakes in East Antarctica, the discharge of which was found to have coincided with a 10% increase in the flow speed of the Byrd Glacier, located directly downstream of the lakes [*Stearns et al.*, 2008].

As a footnote to this section, *Richter et al.* [2008] conducted a high-resolution ground-based GPS survey over Lake Vostok over the period 2002–2007. Their results indicated no surface height change that could be attributed to any change of the water level within the lake, a conclusion that is supported by 1-year period repeat laser altimetry from the ICESat satellite [*Shuman et al.*, 2006].

2.4. Seismic Survey

Unlike other methods described above, the analysis of reflected pressure waves allows information to be collected from below the surface of subglacial lakes. Longitudinal seismic or "p" waves have the unique ability to penetrate large distances through ice, rock, and liquid water. This means that by timing reflections from the various subglacial interfaces, seismic studies can be used to infer the depth of water in subglacial lakes and the thickness of any unconsolidated sedimentary layers at their bed.

Seismic investigations, however, are time and labor intensive to carry out and provide only one data point for each "shot" taken. High levels of background noise on the Antarctic plateau can also make interpretation of the returns difficult. During the 1960s, a seismic survey carried out around Vostok Station [*Kapitsa and Sorochtin*, 1965] prior to the drilling of a deep ice core failed to identify the presence of a subglacial lake. The same data set would later provide the first indications of the depth of Lake Vostok after reexamination in the light of knowledge of the lake's surface extent gained from satellite altimetry [*Ridley et al.*, 1993]. The initial data indicated a water depth of around 500 m at a site located a few kilometers from Vostok Station [*Kapitsa et al.*, 1996]. More recent seismic investigations have revealed the maximum depth of the lake to be over 1100 m [*Masolov et al.*, 2006]. A separate basin, with a small surface area, but up to 680 m deep, has also been found to exist immediately below Vostok Station [*Masolov et al.*, 1999].

Lake Vostok was the first, and for many years the only, subglacial lake to have a direct depth measurement by seismic sounding. Recently, however, a subglacial lake situated very near the South Pole [*Peters et al.*, 2008] and another near the Ellsworth Mountains in West Antarctica [*Woodward et al.*, 2010] have been the targets of seismic surveys. At the South Pole Lake, *Peters et al.* [2008] used the amplitude variation with offset technique, which utilizes the observation that seismic reflectivity varies as a function of angle in different ways for different subglacial materials. In this way, they were able to identify liquid water beneath the South Pole independently of the RES data that had originally located the lake. In addition, they used traditional seismic processing to determine the depth of the water column in South Pole Lake to be around 32 m.

As well as characterizing the water depth in subglacial lakes, seismic sounding is currently the only method for investigating the physiography of lake beds and particularly for determining thicknesses of sediments. At Lake Vostok, several of the available seismic records show multiple reflections from around the bed of the lake; these, however, can be interpreted in different ways. Reflections spanning a range of 0.1–0.5 s from a number of different sites above Lake Vostok are thought to represent between 100 and 350 m of sediments with seismic velocity between 1700 and 2100 m s^{-1} [*Masolov et al.*, 1999, 2001]. However, it has also been suggested that these secondary bottom echoes may represent side reflections due to either steep side slopes or high basal roughness at the bed of the lake and, therefore, that the water layer is in direct contact with the seismic basement across the entire lake bottom [*Masolov et al.*, 2006; *Popov et al.*, 2006]. *Filina* [2007] tested these hypotheses and determined that some of the reflections could be due to a sloping lake bed but that, nevertheless, the seismic data were best explained by the presence of a 200- to 300-m thick layer of sediment.

In 2007/2008, seismic measurements were made on Ellsworth Subglacial Lake (hereinafter referred to as Lake Ellsworth) in West Antarctica [*Woodward et al.*, 2010; *Ross et al.*, this volume]. Based on an assessment of sidewall slopes bordering the lake, the water depth was previously hypothesized to be at least several tens of meters [*Siegert et al.*, 2004]. The seismic data upheld this hypothesis and recorded a depth of around 160 m in long location. This result has been used as the basis for the forthcoming exploration of Lake Ellsworth (in 2012/2013), detailed by *Ross et al.* [this volume].

2.5. Characterization of Subglacial Lakes by the Survey of Gravity Anomalies

Geologists use measurements of the free-air gravitational anomaly in Antarctica to map density variations below the surface that are the result of both geological structures and of the topography. If the surface and bed topographies of the ice sheet over a subglacial lake are known (e.g., from RES survey), then the effect of ice sheet geometry can be removed from the free-air gravity anomaly. For a large subglacial lake, such as Lake Vostok, the thickness of the subglacial water column then dominates the remaining gravity signal [*Studinger et al.*, 2004b].

This method has been used to map the bathymetry of Lake Vostok at a higher spatial resolution than is possible with seismic methods alone [*Studinger et al.*, 2004b]. Two separate basins were identified in this way, a larger and deeper southern basin separated by a 40-km-wide bedrock ridge from a smaller and shallower northern one. *Roy et al.* [2005] took the next step by using gravity data to produce a complete model of Lake Vostok including both water column and sediment thickness. Both these studies used aerogeophysical data collected during the 2001/2002 Antarctic summer [*Holt et al.*, 2006b]. Neither model, however, corresponded precisely with the water and sediment depths indicated by more recent seismic studies [e.g., *Masolov et al.*, 2006]. Further work with the original gravity data combined with recent seismic studies has produced a refined bathymetry/sediment model for Lake Vostok [*Filina et al.*, 2008]. This indicates a deep (~1200 m) maximum water depth underlain by ~300 m of unconsolidated sediments in the southern basin and shallower water (~250 m) overlying a thicker sedimentary layer (~350 m) in the northern basin.

The only other subglacial lake to have been characterized at any level of detail using gravity measurements is Lake

Concordia. A University of Texas/Support Office for Aero-geophysics (SOAR) airborne survey covered the region of the lake in 1999/2000 and collected gravity data [*Studinger et al.*, 2004a; *Filina et al.*, 2006]. Inversion modeling of gravity data from the six survey lines crossing the lake was then undertaken by both *Tikku et al.* [2005] and *Filina* [2007]. Gravity results, along with simultaneously acquired RES data, were then used to determine that the lake has a large surface area of between 600 and 800 km^2, but a relatively shallow average water column thickness of ~59 m (maximum depth 126 m), and consequently a volume of just 70 km^3 [*Filina*, 2007; *Thoma et al.*, 2009]. Since the lake is relatively shallow, any underlying sedimentary layer has not been resolved in the gravity data.

The inherent noisiness of airborne gravimetry (due to the sensor being subjected to nongravitational accelerations resulting from aircraft motion) means that extensive low-pass filtering is required to extract that part of the recorded signal, which is of geological origin [*Childers et al.*, 1999]. Analysis of repeat lines and crossovers has determined an accuracy of 1–2 mGal for the surveys of both Lake Vostok and Lake Concordia [*Holt et al.*, 2006b]. This, combined with the thickness of the overlying ice (necessitating that data be acquired at an effective altitude of more than 4 km above the ice base target), currently imposes a limit of around 8–10 km on the resolution of aerogravity surveys over central Antarctica [*Holt et al.*, 2006b]. Modern aerogravity equipment, as used in commercial exploration, has the potential to provide submilligal accuracy to future surveys of the Antarctic subglacial environment [*Studinger et al.*, 2008]; however, as yet, no subglacial lake studies have been published with this kind of resolution. Further errors can potentially be introduced into lake bathymetries determined solely by gravity survey due to the unknown nature of the underlying geology. Unknown or uncertain spatial variation in rock density can be indistinguishable from changes in lake volume in gravity surveys; hence, the use of seismic data is desirable in verifying gravity inversion models [e.g., *Filina et al.*, 2008].

3. GEOGRAPHICAL DISTRIBUTION, DIMENSIONS, AND RELATION TO TOPOGRAPHIC AND GLACIOLOGICAL SETTING

3.1. Inventory of Known Subglacial Lakes

The first published inventory of subglacial lakes is the work of *Oswald and Robin* [1973]. At this time, just 17 lakes had been discovered by the SPRI-NSF-TUD RES program, 14 of which lay beneath the Dome C area of East Antarctica, a region which would later become known as the "Antarctic Lake District."

By the time another systematic catalog of all known lakes was produced [*Siegert et al.*, 1996], the total number had increased to 77. The large majority of these lakes were still to be found under the thick ice of central East Antarctica with two clusters in particular, beneath the Dome C and Ridge B areas, accounting for 77% of the total number [*Siegert et al.*, 1996]. Lake Vostok was included in this inventory. Its large volume, at least two orders of magnitude greater than any other lake, would dominate calculations of water storage beneath the ice sheet [*Dowdeswell and Siegert*, 1999].

The most recent inventory of subglacial lakes [*Siegert et al.*, 2005], while still only incorporating lakes identified by RES, drew upon a larger number of studies including, most significantly, the Italian survey of the Dome C area [*Tabacco et al.*, 2003], the Russian airborne RES of the Dome A and Dome F regions [*Popov and Masolov*, 2003], and the U.S. surveys of the Vostok and Wilkes Basin areas [*Studinger et al.*, 2003a, 2004a]. The tally for this inventory reached 145 separate lakes, even after six individual reflectors from the previous inventory had been reclassified as parts of larger lakes [*Siegert and Ridley*, 1998a].

Since the publication of this inventory, a number of recent studies have further added to our knowledge of subglacial lakes:

1. *Popov and Masolov* [2007] reported 29 new lakes identified during ground-based RES work aimed at mapping the shoreline of Lake Vostok. Most of these are small lakes (0.5- to 10-km-long reflectors) located within valleys several hundred meters deep. Two further small lakes (4 and 26.5 km^2) were detected during an overland traverse between Mirny and Vostok [*Popov and Masolov*, 2007].

2. A systematic categorization of RES reflections from the 1998–2001 SOAR campaigns covering the Pensacola-South Pole transect, the Wilkes Basin Dome C transect, and a survey of the Lake Vostok area by *Carter et al.* [2007] resulted in the identification of 22 new "definite" lakes and a further 58 classed by the authors as "dim lakes" (see section 2.1.4). This reanalysis also determined that several features previously thought to be separate lakes, in fact, belonged to the same water bodies, thereby reducing the total of *Siegert et al.* [2005] by 5. A complete list of all "lake-type" reflectors identified in this study is given by *Blankenship et al.* [2009].

3. As mentioned in section 2.2, *Bell et al.* [2006] contributed revised dimensions for two large lakes at 90°E (~2000 km^2) and at Sovetskaya (~1600 km^2). Similarly, four lakes with surface areas 1500–4500 km^2 have been identified at the head of the Recovery Ice Stream [*Bell et al.*, 2007].

4. Five new lakes were reported by *Cafarella et al.* [2006] as a result of Italian RES work during 2003. Four of these are small (1- to 3-km-long reflectors) and are located within the

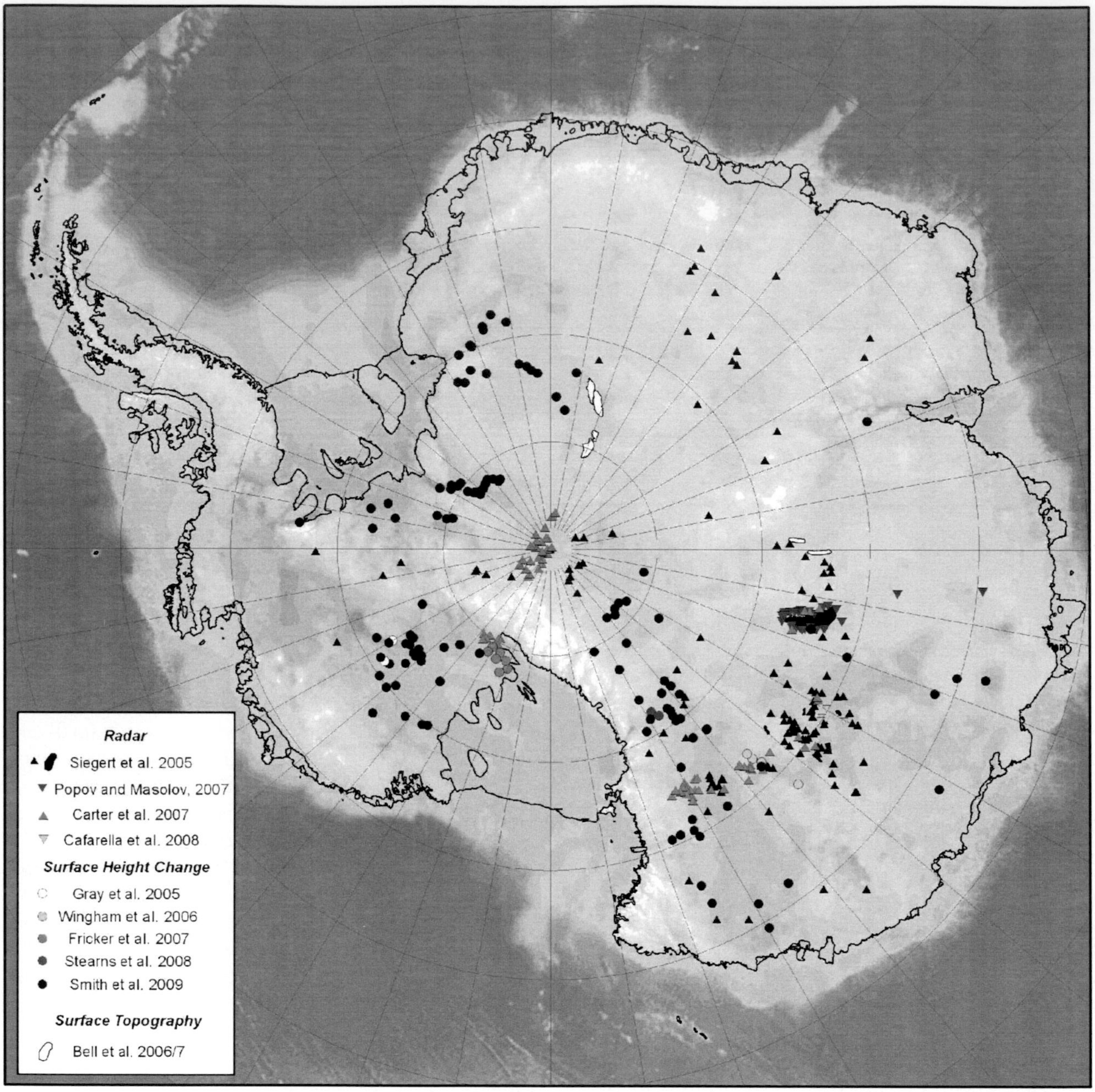

Figure 2. Map of subglacial Antarctica showing the locations of all lakes known from radio echo sounding and all sites of surface height change consistent with lake activity detected by satellite. Larger lakes are shown in outline; smaller lakes and those of unknown surface area are indicated by triangular (RES) and circular markers (satellite). Lakes included in the previous inventory of *Siegert et al.* [2005] are shown in black; lakes discovered since that time are shaded according to the publication in which they were first identified.

Belgica Subglacial Highlands, and one is larger (~10 km) and situated in the southern Aurora Basin.

5. Finally, several active subglacial lakes have been identified in individual studies throughout East and West Antarc-tica [*Gray et al.*, 2005 (2); *Wingham et al.*, 2006 (4); *Fricker et al.*, 2007 (7); *Stearns et al.*, 2008 (2)]. A systematic catalog of all lakes that were active (and crossed by more than one ICESat track) between 2003 and 2008 has been

produced by *Smith et al.* [2009]. This work added 113 previously unrecorded lakes.

The total number of lakes described in the literature as of October 2009, therefore, stands at 387 (see Figure 2).

3.2. The Distribution of Subglacial Lakes

3.2.1. Distribution with ice thickness and with distance from subglacial flow divide. The direction of subglacial water flow at any point on the ice sheet can be shown to depend only upon the slopes of the ice surface and of the bedrock [*Shreve*, 1972]. Thus, given a digital elevation model of the ice sheet surface [e.g., *Bamber et al.*, 2009] and of the bed [*Lythe et al.*, 2001], flow lines can be calculated for subglacial water throughout the ice sheet [e.g., *Wright et al.*, 2008]. The slope of the ice surface has an order of magnitude greater effect on the direction of subglacial water flow than does the topography of the bedrock [*Shreve*, 1972]. The flow lines therefore, largely follow the direction of ice flow. As a consequence of this, an equivalent subglacial watershed is situated approximately beneath the ice divide.

The upstream distance from a lake along the flow path to the subglacial flow divide is a useful parameter in characterizing the distribution of subglacial lakes. Figure 3 shows histograms of this distance to the divide for (a) all lakes, (b) lakes detected by RES fieldwork, and (c) lakes detected by satellite measurements of height change. Figure 3a clearly shows that the densities of all known lakes increase nearly exponentially with approach to the flow divide. Histograms for lakes detected by RES and for active lakes detected from space, however, produce very different results. The total number of lakes from RES surveys (256) is greater than the total for active lakes (128) and the distribution is more heavily skewed; hence, the pattern for radar lakes dominates that for all lakes. (Four lakes beneath the head of the Recovery Ice Stream are known by their surface expression only [see *Bell et al.*, 2007]. One lake (Lake Mercer) is identified by both surface height change and in RES records.) The active lakes, so far, reported do not show a strong relationship to upstream distance to the flow divide. The modal value for this distance is 750 km, which for lakes detected by RES falls within the upper 5% tail of the distribution. This demonstrates that the method of lake

detection, or perhaps the type of lake detected, has a significant effect on the distribution of lakes with regard to the flow divides.

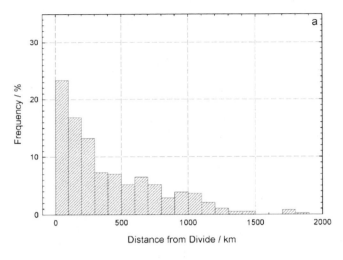

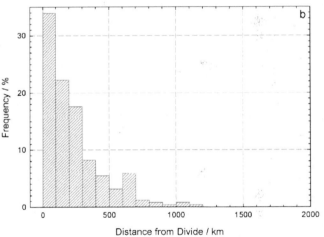

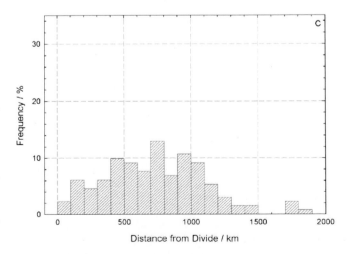

Figure 3. (opposite) Distribution of subglacial lakes in terms of distance along the flow line to a major ice divide. The histograms show (a) all known subglacial lakes, (b) lakes identified by their RES reflection, and (c) lakes identified by satellite measurements of vertical surface movement.

The distribution of subglacial lakes with respect to the thickness of overlying ice is also of interest when categorizing subglacial lake physiography. For lakes identified by RES, ice thickness can be determined from the same data set. For those identified by other means, the ice thicknesses used here have been taken from the BEDMAP digital elevation model, which uses interpolation to fill the gaps between RES and seismic measurements of the ice thickness [*Lythe et al.*, 2001].

Figure 4a shows that the ice thickness distribution for known lakes (bars) is skewed toward thicker ice compared with the distribution of all grounded ice in Antarctica (line). Plotting separate histograms for lakes identified by RES (Figure 4b) and by satellite (Figure 4c) demonstrates that the observed skew [cf. *Dowdeswell and Siegert*, 1999] is due entirely to a skew in the lakes recorded by RES. By comparison, apart from an unobserved peak in ice thickness between 1800 and 2200 m, the distribution of lakes discovered by satellite is in much closer agreement with the overall distribution of ice thickness across the continent (Figure 4c).

The difference in the distribution of overlying ice thickness with measurement technique is likely to be associated with the spatial dependence of the technique. RES has been demonstrated to be very good at detecting subglacial lakes near the center of the ice sheet, but less good at identifying them in the often warmer and more heavily crevassed ice toward the margins. *Dowdeswell and Siegert* [2002] thought this was possibly due to scattering at the ice sheet base that occurs as a consequence of basal sliding, though no evidence for this was available at the time. Conversely, satellite measurements of ice surface changes have revealed many lakes close to the ice margin, but not so many near the divide. This is possibly due to the periodicity of lake drainage that is a function of lake size and the input flux of water. Small lakes toward the ice margin that are filled by high rates of water flow are likely to drain regularly and produce a greater surface height change signal when they do. In comparison, larger lakes nearer to the ice divide, where the input of water is more restricted, are likely to drain less frequently and, for a similar discharge volume, produce a much smaller surface height change [*Siegert et al.*, 2007]. This may then be below the detection threshold for satellite altimetry.

3.2.2. Bias in the distribution due to data collection methods. The SPRI-NSF-TUD survey is currently the only available source of information for around 16% of the known subglacial

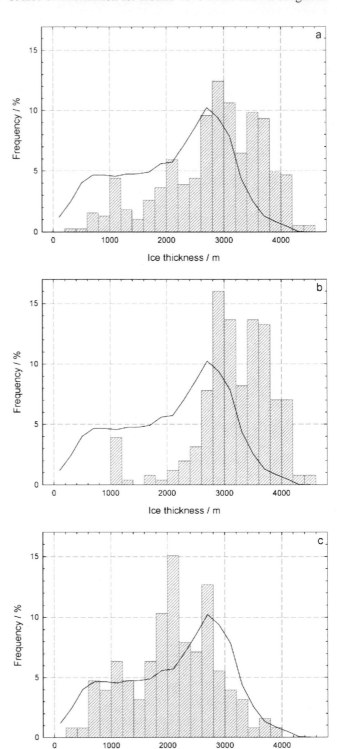

Figure 4. (opposite) Distribution of subglacial lake locations in terms of thickness of overlying ice. The histograms show (a) all known lakes, (b) lakes identified during RES studies, and (c) lakes identified by satellite measurements of vertical surface movement. The distribution of ice thickness for the whole ice sheet (black line) is shown for comparison.

lakes; 32% have been discovered by satellite during the past 5 years, while the remainder have been characterized by RES surveys undertaken for specific study areas since the 1970s. While the SPRI-NSF-TUD survey, together with the Russian surveys in East Antarctica, constitutes an approximately representative sample of the ice sheet [e.g., *Dowdeswell and Siegert*, 1999], recent RES projects, in general, do not. The high density of survey lines flown in the area around Lake Vostok, for example, have identified a large number of small lakes under thick ice and close to the flow divide, thus tending to skew the distributions for RES lakes in Figures 3 and 4.

Satellite techniques also suffer from bias. The separation of ICESat flight tracks decreases toward southerly latitudes (from ~30 km at 70°S to ~5 km at 85°S) to the southern limit of the orbit at 86°S, below which there is no coverage. This leads to an increase in the number of lakes detected together with a decrease in the average size of the lakes as the satellite moves toward this limit, while the area around the South Pole cannot be sampled at all [*Smith et al.*, 2009]. Other forms of geographical bias also apply, such as discrimination against areas that experience high levels of cloud cover and several others recognized by the authors of the inventory.

Very little overlap exists between the inventories of lakes identified by RES studies and those shown to be active through satellite measurements. Satellites have detected lakes (e.g., Engelhardt Subglacial Lake) situated underneath large outlet glaciers, which have not been identifiable in RES records from the same site. The reasons for this are still uncertain, though RES studies have always had difficulties in imaging the bed in very active regions of the ice sheet. In many cases, active lakes occur beneath or in close proximity to the fast flowing ice associated with surface crevassing and basal fluting. The weaker attenuation of the radar signal due to thinner ice cover and the increased likelihood of general basal lubrication in such areas also combine to reduce the contrast in bed reflection power between wet sediment and liquid water lakes at the bed.

3.3. Dimensions and Volumes of Subglacial Lakes

Figure 5 is a histogram showing the distribution of subglacial lake lengths (taking the longest dimension if several records are available). Lake Vostok, 90°E Lake, Sovetskaya Lake, and the Recovery Lakes are left out of the diagram simply because their size skews the graph adversely. The bulk of subglacial lakes are less than 10 km in length with the modal size being just 5 km. Several large lakes exist, with dimensions in excess of 20 km, and these will comprise the bulk of the water currently stored beneath the ice sheet (see below).

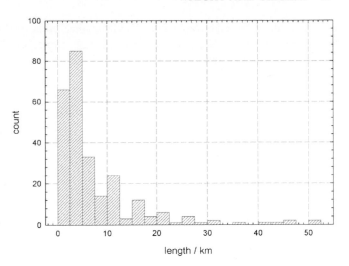

Figure 5. Distribution of known subglacial lake lengths <55 km.

Dowdeswell and Siegert [1999], working from the SPRI-NSF-TUD airborne RES records, identified four distinct regions of subglacial lake occurrence: Dome C, Ridge B, Hercules Dome, and the area around the South Pole. Following the two assumptions that subglacial lakes were scattered randomly throughout each of these areas and that the RES data from each region was organized to cover a representative sample of the ice sheet, they were able to estimate (assuming a circular approximation for the area of lakes crossed by RES records at a random orientation) the total area of subglacial lakes in each region and, hence, the total stored volume of water for a range of mean lake depths. Their calculations led them to predict a total volume for water stored beneath the ice sheet of between 4000 and 12,000 km^3.

Since then, the total number of known lakes has increased, and many lakes have been discovered outside of the regions identified by *Dowdeswell and Siegert* [1999]. The scale of the RES surveys from which these lakes have been discovered is much smaller than the earlier SPRI-NSF-TUD work and can no longer be taken to cover a representative sample of a larger region; rather, detailed surveys have been undertaken of local areas. Therefore, we cannot now update the estimate for total subglacial water storage, as the survey lines are too unevenly distributed. We can, however, estimate, using the same principles, the total known surface area of subglacial lakes and also a range for the total stored volume based on likely mean depths.

Using the circular approximation for lakes crossed by just a single survey line [*Dowdeswell and Siegert*, 1999] and reported surface areas for lakes with multiple crossings or with prominent flat surface features, we estimate the total surface area of known lakes to be ~50,000 km^2. This value, of course, does not take into account undiscovered lakes or those discovered by surface elevation change alone, as length/area measurements for most of these lakes are not yet published. Nevertheless, this value already exceeds that estimated by *Dowdeswell and Siegert* [1999] (~40,000 km^2) for the whole of the ice sheet. The large lakes in the Ridge B area (Vostok, 90°E, Sovetskaya) comprise 35% of this total surface area, while the four lakes proposed at the head of the Recovery Ice Stream contribute a further 27%, leaving only 38% to be accounted for by all the other smaller lakes.

Using estimates of between 50 and 250 m mean water depth based on RES observations of surrounding topography [*Dowdeswell and Siegert*, 1999], and actual volumes for lakes where depth measurements are available, we obtain an envelope for the total stored water volume within known subglacial lakes of 9000–16,000 km^3. Once again, this is an increase from the range estimated by *Dowdeswell and Siegert* [1999] and can be regarded as a minimum for the whole of Antarctica, as many lakes are likely yet to be discovered. Only three lakes (Vostok ~5400 km^3, Concordia ~30 km^3, Ellsworth, 1.8 km^3) have reasonably well-constrained volumes.

4. SUBGLACIAL LAKE CLASSIFICATION, BASAL HYDROLOGY, AND ICE FLOW

Dowdeswell and Siegert [2002] analyzed the first inventory of subglacial lakes [*Siegert et al.*, 1996] to outline four forms of subglacial lake: (1) those where topography is subdued in the middle of large basins, (2) lakes that occupy entire discrete basins, which are often large, (3) lakes perched against the sides of subglacial highlands, and (4) lakes located at or near the onset of major outflow units. Based on the new inventory (of 387 lakes), and following recent discoveries concerning the role of subglacial lakes in basal hydrology, we offer the following revision to the classification.

A necessary condition for the occurrence of a subglacial lake is that the ice sheet base must be at the pressure melting point. This is most likely to be the case where either the ice is thick (>3 km) and, therefore, a high-pressure environment exists at the bed that is well insulated from cold surface temperatures or where vigorous internal deformation and basal sliding are actively generating heat at the ice base through friction [*Wilch and Hughes*, 2000].

Subglacial lakes can be divided into three groups, those associated with thick ice and the trapping of geothermal heat near the ice divides in the interior of the continent, those associated with the onset regions of fast flow features nearer to the margins of the ice sheet, and those situated beneath the trunks of outlet glaciers and ice streams.

4.1. Lakes in the Ice Sheet Interior

The subglacial topography of the Antarctic interior is characterized by several very large, deep, likely sedimentary basins separated by mountainous ridges known as subglacial highlands [*Drewry*, 1983; *Lythe et al.*, 2001]. The summits, or "Domes," of the Antarctic interior are connected by a line of ridges or "ice divides." The ice sheet in the vicinity of these divides is characterized by very low surface slopes and very slow ice flow. The subglacial lakes so far discovered beneath the thick ice of the interior can be divided into three categories based on the characteristics of their surrounding bed topography [*Dowdeswell and Siegert*, 2002; *Tabacco et al.*, 2006].

First, the majority of interior lakes are found where the ice is thickest, and therefore, conditions are favorable for melting at the bed. This most often occurs toward the centers of the deep subglacial basins such as the Aurora, Vincennes, and Wilkes basins. The subglacial topography here is characterized by low relief; *Dowdeswell and Siegert* [1999] found that for more than 60% of the then known lakes, the maximum elevation of the nearest topographic highs was no more than 400 m above the level of the lake, with a maximum gradient of no more than 0.1 adjacent to the lake shore. Many of these lakes consequently have little depth potential for the size of their surface area. Several large lakes fall into this category, e.g., Lake Concordia, Vincennes Lake, and Aurora Lake [*Tabacco et al.*, 2006]. In some cases, features previously identified as lakes in RES records or by flat surface features may be areas of water-saturated sediments, rather than true lakes [*Carter et al.*, 2007].

Second, there are those lakes which are closely related to significant topographic depressions in the bedrock; these are frequently found toward the margins of subglacial basins but still near to one of the major ice divides of the interior. Lake Vostok is an example of this kind of lake, as are the large lakes at 90°E and beneath Sovetskaya Station [*Bell et al.*, 2006]. These lakes inhabit topographic depressions with the elongate, rectilinear morphology characteristic of tectonically controlled features [*Meybeck*, 1995; *Tabacco et al.*, 2006]. Geophysical investigation has confirmed that Lake Vostok is situated within a rift feature forming part of a continental collision zone [*Studinger et al.*, 2003b]. Several smaller examples of this type of lake have been reported including Lake Ellsworth in West Antarctica [*Woodward et al.*, 2010].

The third category is composed of lakes situated on the flanks of subglacial mountain ranges. These are characteristically small (<10-km long) features constrained within steep local topography. Consequently, their presence is not found to leave an imprint on the ice surface morphology.

Several cases have been identified of lakes perched on the stoss face of subglacial mountains [*Dowdeswell and Siegert*, 2002].

Subglacial lakes located near drainage divides in the ice sheet interior are likely to have relatively small catchment areas from which they receive runoff. They are, therefore, likely to be inactive or to have long periodicity between flood discharge events due to the associated low refilling rates [*Evatt et al.*, 2006].

4.2. Lakes Associated With the Onset of Enhanced Ice Flow

A second class of subglacial lakes was postulated by *Siegert and Bamber* [2000], when the association was first made between the locations of some subglacial lakes and the onset regions of fast ice flow features recently identified as extending farther into the interior of Antarctica than previously thought [*Bamber et al.*, 2000]. At least 16 subglacial lakes from the inventory of *Siegert et al.* [1996] could be associated with onset regions of enhanced ice flow [*Dowdeswell and Siegert*, 2002]. These lakes are located relatively large distances (approximately a few hundred kilometers) from the nearest major ice divide, are small (<10 km in length), and probably shallow. *Dowdeswell and Siegert* [2002, p. 234] state that "warm-based fast flowing ice streams provide a possible route by which subglacial lakes, located at the onset of enhanced ice flow, may establish a hydrological connection with the ice-sheet margin."

More recently, four subglacial lakes, each with large (>1000 km^2) surface area have been identified near the onset of fast flow at the head of the Recovery Ice Stream, the fast flow feature that penetrates deepest into the interior of East Antarctica [*Bell et al.*, 2007]. These lakes appear to have more in common with the large lakes in the Ridge B-Vostok region, as they also exhibit signs of tectonic origin, though they appear to be much shallower than Lake Vostok [*Bell et al.*, 2007].

Not all fast ice flow features are associated with subglacial lakes, and the mechanism by which subglacial lakes, where they do occur, influence the initiation of fast ice flow is uncertain. Suggestions include the direct lubrication of the ice/bed interface through the constant supply of water, modification of the basal thermal regime by the release of latent heat during freeze-on of lake water to the underside of the ice sheet and through the creation of subglacial conduits during periodic drainage events [e.g., *Kamb et al.*, 1985; *Bell et al.*, 2007]. Recent work by *Stearns et al.* [2008] has shown that outlet glaciers can speed up as a response to the draining of upstream lakes. That the movement of subglacial water can influence outlet glacier dynamics is an important

observation, however, this relationship has probably operated as long as liquid water has been present at the base of the ice sheet and does not represent a new instability.

4.3. Lakes Beneath the Trunks of Ice Streams

Recent satellite studies, aimed at measuring rapid fluctuations in the ice surface height [e.g., *Smith et al.*, 2009], have been very successful in detecting active subglacial systems beneath the fast flowing trunks of ice streams and glaciers in both East and West Antarctica [*Gray et al.*, 2005; *Fricker et al.*, 2007; *Fricker and Scambos*, 2009].

Due to the nature of the techniques with which they have been detected, all the lakes so far discovered beneath ice streams have been actively filling or discharging. Despite significant developments in the collection and processing techniques for synthetic aperture RES [e.g., *Peters et al.*, 2007], studies using this method, which is capable of detecting static lakes, still experience significant difficulties in imaging the bed of ice streams due to a high degree of scattering from both surface crevasses and from basal fluting. Such surveys have, therefore, not yet been able to detect lakes located in very fast flowing regions of the ice sheet.

This newly discovered class of lakes appears to be generally of smaller size, and the lakes are probably more transient in nature, with large proportions of their water volume filling and draining on an annual or semiannual basis [*Smith et al.*, 2009].

5. DISCUSSION AND SUMMARY

Following the first inventory of 77 subglacial lakes, *Dowdeswell and Siegert* [2002] provided an assessment of lake classifications. In recent years, several new data sets have increased the number of known subglacial lake features to 387 and have greatly improved our understanding of their role in the hydrological system beneath the ice sheet.

The traditional (and earliest) method of subglacial lake detection is from radar sounding (either ground-based or airborne). Radar provides an actual measurement of the ice-water interface and an along-track recording of the extent of this water. Several subglacial lakes have been investigated through a grid of radar transects, detailing with a high level of certainty the lake extent.

Over large lakes, the ice sheet surface slope is noticeably flat as a consequence of ice floatation. In some cases, the outline of lake extent can be mapped from these surface features. The first lake to be outlined in this way was Lake Vostok [*Kapitsa et al.*, 1996], shortly followed by several lakes at Dome C [*Siegert and Ridley*, 1998a], Lake Concordia, two large lakes near Ridge B (90°E and Sovetskaya) [*Bell et al.*, 2006], and a collection of lakes at the head of the Recovery Ice Stream [*Bell et al.*, 2007].

Satellite altimetry time series data have recently been used to determine the location of subglacial lakes that display noticeable outpouring or influx of water; the net change causing the ice sheet surface to fall or rise accordingly and, hence, the outline of the "active" lake detected [e.g., *Wingham et al.*, 2006; *Fricker et al.*, 2007; *Smith et al.*, 2009].

The water depths of subglacial lakes are best measured using seismics, as unlike radio waves, sound can travel well in water. Only a few subglacial lakes have been the subject of seismic surveys (Lake Vostok, South Pole Lake, and Lake Ellsworth). For the latter two cases, seismics have proved beyond doubt the existence of the subglacial lake, as radar data alone was insufficient for some scientists to accept lake detection [e.g., *Price et al.*, 2002].

Water depths can also be modeled using gravity data, although this has been applied only to Lake Vostok and Lake Concordia at present, revealing the former to comprise two distinct basins. As seismic exploration takes considerable time to undertake (Lake Ellsworth, which is only 10-km long, needed a whole season for five lines), gravity data acquired from aircraft provides an efficient means by which the depths of larger lakes can be estimated prior to seismic analysis.

The number of known subglacial lakes has risen from 17 in 1973, to 77 in 1996 and 145 in 2006. The new total, as of January 2010, now stands at 387. Subglacial lakes are widespread beneath the east and west Antarctic ice sheets. Many have been shown to be "active" (i.e., discharging and/or receiving water). Such water flows beneath the ice sheet in an organized hydrological system that connects distinct groups of lakes. Lakes will not be "interconnected," as the flow of water is likely to be one directional. However, there will be a hierarchy of lakes, from source feeder lakes to downstream receiver lakes.

As the basal hydrology in Antarctica is now known to be far more active than had been considered even 5 years ago, the question now is whether the observed movement of water can affect ice flow. *Stearns et al.* [2008] showed that subglacial discharges have indeed affected the flow of Byrd Glacier, albeit for a relatively short time. As a consequence, subglacial lakes at the heads of major ice streams may well have an influence on the flow downstream and may be influential in ice stream changes (e.g., those changes identified in the Siple Coast).

While one can classify subglacial lakes into various categories depending on their size, location, overriding ice thickness, etc, the prevailing view at present is of a common hydrology involving a well-defined hydrological network fed by melting ice and periodic discharges of lake water.

Even beneath the coldest most stable ice sheet of the Cenozoic Era, basal hydrology and subglacial lakes will have represented dynamic physical systems. This has almost certainly always been the case. Our understanding of past, present, and future ice sheet changes requires an awareness of basal hydrology, especially as one considers that the next 10 years may well see the number of subglacial lakes, and their associated dynamics and effects on modern flow processes increase considerably.

Acknowledgments. A.P.W. and M.J.S. acknowledge funding from the U.K. NERC (NE/D003733/1) for the U.S.-U.K.-Australian ICECAP program.

REFERENCES

Bamber, J. L., D. G. Vaughan, and I. Joughin (2000), Widespread complex flow in the interior of the Antarctic ice sheet, *Science*, *287*(5456), 1248–1250.

Bamber, J. L., J. L. Gomez-Dans, and J. A. Griggs (2009), A new 1 km digital elevation model of the Antarctic derived from combined satellite radar and laser data—Part 1: data and methods, *Cryosphere*, *3*, 101–111.

Bell, R. E., M. Studinger, M. A. Fahnestock, and C. A. Shuman (2006), Tectonically controlled subglacial lakes on the flanks of the Gamburtsev Subglacial Mountains, East Antarctica, *Geophys. Res. Lett.*, *33*, L02504, doi:10.1029/2005GL025207.

Bell, R. E., M. Studinger, C. A. Shuman, M. A. Fahnestock, and I. Joughin (2007), Large subglacial lakes in East Antarctica at the onset of fast-flowing ice streams, *Nature*, *445*, 904–907.

Bentley, C. R., N. Lord, and C. Liu (1998), Radar reflections reveal a wet bed beneath stagnant Ice Stream C and a frozen bed beneath ridge BC, West Antarctica, *J. Glaciol.*, *44*(146), 149–156.

Berry, M. V. (1973), The statistical properties of echos diffracted from rough surfaces, *Philos. Trans. R. Soc. London, Ser. A*, *273*(1237), 611–654.

Berry, M. V. (1975), Theory of radio echos from glacier beds, *J. Glaciol.*, *15*, 65–74.

Bianchi, C., A. Forieri, and I. E. Tabacco (2004), Electromagnetic reflecting properties of sub-ice surfaces, *Ann. Glaciol.*, *39*, 9–12.

Bingham, R. G., and M. J. Siegert (2007), Radio-echo sounding over polar ice masses, *J. Environ. Eng. Geophys.*, *12*(1), 47–62.

Blankenship, D. D., D. L. Morse, C. A. Finn, R. E. Bell, M. E. Peters, S. D. Kempf, S. M. Hodge, M. Studinger, J. C. Behrendt, and J. M. Brozena (2001), Geologic controls on the initiation of rapid basal motion for West Antarctic ice streams: A geophysical perspective including new airborne radar sounding and laser altimetry results, in *The West Antarctic Ice Sheet: Behavior and Environment, Antarct. Res. Ser.*, vol. 77, edited by R. A. Alley and R. A. Bindschadler, pp. 105–121, AGU, Washington, D. C.

Blankenship, D. D., S. P. Carter, J. W. Holt, D. L. Morse, M. E. Peters, and D. A. Young (2009), Antarctic subglacial lake classification inventory, Natl. Snow and Ice Data Cent., Digital Media, Boulder, Colo.

Bogorodskiy, V. V., C. R. Bentley, and P. E. Gudmandsen (1985), *Radioglaciology*, D. Riedel, Dordrecht, Netherlands.

Cafarella, L., S. Urbini, C. Bianchi, A. Zirizzotti, I. E. Tabacco, and A. Forieri (2006), Five subglacial lakes and one of Antarctica's thickest ice covers newly determined by radio-echo sounding over the Vostok-Dome C region, *Polar Res.*, *25*(1), 69–73.

Carter, S. P., D. D. Blankenship, M. E. Peters, D. A. Young, J. W. Holt, and D. L. Morse (2007), Radar-based subglacial lake classification in Antarctica, *Geochem. Geophys. Geosyst.*, *8*, Q03016, doi:10.1029/2006GC001408.

Carter, S. P., D. D. Blankenship, D. A. Young, M. E. Peters, J. W. Holt, and M. J. Siegert (2009), Dynamic distributed drainage implied by the flow evolution of the 1996–1998 Adventure Trench subglacial lake discharge, *Earth Planet. Sci. Lett.*, *283*, 24–37.

Catania, G. A., H. B. Conway, A. M. Gades, C. F. Raymond, and H. Engelhardt (2003), Bed reflectivity beneath inactive ice streams in West Antarctica, *Ann. Glaciol.*, *36*, 287–291.

Childers, V. A., R. E. Bell, and J. M. Brozena (1999), Airborne gravimetry: An investigation of filtering, *Geophysics*, *64*, 64–69.

Cudlip, W., and N. F. McIntyre (1987), Seasat altimeter observations of an Antarctic "lake," *Ann. Glaciol.*, *9*, 55–59.

Dowdeswell, J. A., and S. Evans (2004), Investigations of the form and flow of ice sheets and glaciers using radio-echo sounding, *Rep. Prog. Phys.*, *67*, 1821–1861.

Dowdeswell, J. A., and M. J. Siegert (1999), The dimensions and topographic setting of Antarctic subglacial lakes and implications for large-scale water storage beneath continental ice sheets, *Geol. Soc. Am. Bull.*, *111*(2), 254–263.

Dowdeswell, J. A., and M. J. Siegert (2002), The physiography of modern Antarctic subglacial lakes, *Global Planet. Change*, *35*, 221–236.

Drewry, D. J. (1983), *Antarctica: Glaciological and Geophysical Folio*, 9 pp., Cambridge Univ. Press, Cambridge, U. K.

Evatt, G. W., A. C. Fowler, C. D. Clark, and N. R. J. Hulton (2006), Subglacial floods beneath ice sheets, *Philos. Trans. R. Soc. A*, *364*, 1769–1794.

Fatland, D. R., and C. S. Lingle (2002), InSAR observations of the 1993–95 Bering Glacier (Alaska, USA) surge and a surge hypothesis, *J. Glaciol.*, *48*, 439–451.

Filina, I. Y. (2007), Geophysical investigations of subglacial lakes Vostok and Concordia, East Antarctica, Ph.D. thesis, Univ. of Tex., Austin.

Filina, I. Y., D. D. Blankenship, L. Roy, M. Sen, T. Richter, and J. Holt (2006), Inversion of airborne gravity data acquired over subglacial lakes in East Antarctica, in *Antarctica: Contributions to Global Earth Sciences*, edited by D. Fütterer et al., pp. 129–134, Springer, New York.

Filina, I. Y., D. D. Blankenship, M. Thoma, V. V. Lukin, V. N. Masolov, and M. K. Sen (2008), New 3D bathymetry and sediment distribution in Lake Vostok: Implication for pre-glacial origin and numerical modeling of the internal processes within the lake, *Earth Planet. Sci. Lett.*, *276*, 106–114.

Fricker, H. A., and T. Scambos (2009), Connected subglacial lake activity on lower Mercer and Whillans ice streams, West Antarctica, 2003–2008, *J. Glaciol.*, *55*(190), 303–315.

Fricker, H. A., T. Scambos, R. Bindshadler, and L. Padman (2007), An active subglacial water system in West Antarctica mapped from space, *Science*, *315*, 1544–1548.

Fujita, S., H. Maeno, S. Uratsuka, T. Furukawa, S. Mae, Y. Fujii, and O. Watanabe (1999), Nature of radio echo layering in the Antarctic ice sheet detected by a two-frequency experiment, *J. Geophys. Res.*, *104*(B6), 13,013–13,024.

Gades, A. M., C. F. Raymond, H. Conway, and R. W. Jacobel (2000), Bed properties of Siple Dome and adjacent ice streams, West Antarctica, inferred from radio-echo sounding measurements, *J. Glaciol.*, *46*, 88–94.

Glen, J. W., and J. G. Paren (1975), The electrical properties of snow and ice, *J. Glaciol.*, *15*, 15–38.

Gogineni, S., T. Chuah, C. Allen, K. Jezek, and R. K. Moore (1998), An improved coherent radar depth sounder, *J. Glaciol.*, *44*(148), 659–669.

Gorman, M. R., and M. J. Siegert (1999), Penetration of Antarctic subglacial lakes by VHF electromagnetic pulses: Information on the depth and conductivity of subglacial water bodies, *J. Geophys. Res.*, *104*(B12), 29,311–29,320.

Gray, L., I. Joughin, S. Tulaczyk, V. B. Spikes, R. Bindshadler, and K. Jezek (2005), Evidence for subglacial water transport in the West Antarctic Ice Sheet through three-dimensional satellite radar interferometry, *Geophys. Res. Lett.*, *32*, L03501, doi:10.1029/2004GL021387.

Gudmundsson, G. H. (2003), Transmission of basal variability to a glacier surface, *J. Geophys. Res.*, *108*(B5), 2253, doi:10.1029/2002JB002107.

Holt, J. W., D. D. Blankenship, D. L. Morse, D. A. Young, M. E. Peters, S. D. Kempf, T. G. Richter, D. G. Vaughan, and H. F. J. Corr (2006a), New boundary conditions for the West Antarctic Ice Sheet: Subglacial topography of the Thwaites and Smith glacier catchments, *Geophys. Res. Lett.*, *33*, L09502, doi:10.1029/2005GL025561.

Holt, J. W., T. G. Richter, S. D. Kempf, D. L. Morse, and D. D. Blankenship (2006b), Airborne gravity over Lake Vostok and adjacent highlands of East Antarctica, *Geochem. Geophys. Geosyst.*, *7*, Q11012, doi:10.1029/2005GC001177.

Iken, A., H. Röthlisberger, A. Flotron, and W. Haeberli (1983), The uplift of Unteraargletscher at the beginning of the melt season—A consequence of water storage at the bed?, *J. Glaciol.*, *29*, 28–47.

Kamb, B., C. F. Raymond, W. D. Harrison, H. Engelhardt, K. A. Echelmayer, N. Humphrey, M. M. Brugman, and W. T. Pfeffer (1985), Glacier surge mechanism—1982–1983 surge of Variegated Glacier, Alaska, *Science*, *227*(4686), 469–479.

Kapitsa, A., J. K. Ridley, G. D. Robin, M. J. Siegert, and I. Zotikov (1996), A large deep freshwater lake beneath the ice of central East Antarctica, *Nature*, *381*, 684–686.

Kapitsa, A. P., and O. G. Sorochtin (1965), Ice thickness measurements on the Vostok-Molodezhnaya traverse, *Sov. Antarct. Exped. Inf. Bull., Engl. Transl.*, *5*(5), 299–302.

Le Brocq, A. M., A. Hubbard, M. J. Bentley, and J. L. Bamber (2008), Subglacial topography inferred from ice surface terrain analysis reveals a large un-surveyed basin below sea level in East Antarctica, *Geophys. Res. Lett.*, *35*, L16503, doi:10.1029/2008GL034728.

Lythe, M. B., and D. G. Vaughan, and the BEDMAP Consortium (2001), BEDMAP: A new ice thickness and subglacial topographic model of Antarctica, *J. Geophys. Res.*, *106*(B6), 11,335–11,351.

Masolov, V. N., G. A. Kudryavtzev, and G. L. Leitchenkov (1999), Earth science studies in the Lake Vostok region: Existing data and proposal for future research, paper presented at the SCAR International Workshop on Subglacial Lake Exploration, Cambridge, U. K.

Masolov, V. N., V. V. Lukin, A. N. Sheremetyev, and S. V. Popov (2001), Geophysical investigations of the subglacial Lake Vostok in Eastern Antarctica, *Dokl. Earth Sci.*, *279a*(6), 734–738.

Masolov, V. N., S. Popov, V. Lukin, A. Sheremetyev, and A. Popkov (2006), Russian geophysical studies of Lake Vostok, central East Antarctica, in *Proceedings of the 9th ISAES*, edited by D. Fütterer et al., pp. 135–140, Springer, Berlin.

Meybeck, M. (1995), Global distribution of lakes, in *Lakes: Chemistry, Geology, Physics*, edited by A. Lerman, pp. 1–35, Springer, New York.

Neal, C. S. (1976), Radio-echo power profiling, *J. Glaciol.*, *17*, 527–530.

Oswald, G. K. A. (1975), Investigation of sub-ice bedrock characteristics by radio-echo sounding, *J. Glaciol.*, *15*(73), 75–87.

Oswald, G. K. A., and G. D. Robin (1973), Lakes beneath the Antarctic ice sheet, *Nature*, *245*, 251–254.

Pattyn, F., B. de Smedt, and R. Souchez (2004), Influence of subglacial Vostok Lake on the regional ice dynamics of the Antarctic ice sheet: A model study, *J. Glaciol.*, *50*(171), 583–589.

Peters, L. E., S. Anandakrishnan, C. W. Holland, H. J. Horgan, D. D. Blankenship, and D. E. Voigt (2008), Seismic detection of a subglacial lake near the South Pole, Antarctica, *Geophys. Res. Lett.*, *35*, L23501, doi:10.1029/2008GL035704.

Peters, M. E., D. D. Blankenship, and D. L. Morse (2005), Analysis techniques for coherent airborne radar sounding: Application to West Antarctic ice streams, *J. Geophys. Res.*, *110*, B06303, doi:10.1029/2004JB003222.

Peters, M. E., D. D. Blankenship, S. P. Carter, S. D. Kempf, D. A. Young, and J. W. Holt (2007), Along-track focusing of airborne radar sounding data from West Antarctica for improving basal reflection analysis and layer detection, *IEEE Trans. Geosci. Remote Sens.*, *45*(9), 2725–2736.

Plewes, L. A., and B. Hubbard (2001), A review of the use of radio-echo sounding in glaciology, *Prog. Phys. Geogr.*, *25*(2), 203–236.

Popov, S. V., and V. N. Masolov (2003), Novye danye o podlednikovih ozerah tsentral'noy chasty Vostochnoy Antarktidy, [New data on subglacial lakes in central part of Eastern Antarctica] (in Russian with English summary), *Mater. Glatsiol. Issled.*, *95*, 161–167.

Popov, S. V., and V. N. Masolov (2007), Forty-seven new subglacial lakes in the 0–110°E sector of East Antarctica, *J. Glaciol.*, *53* (181), 289–297.

Popov, S. V., I. Y. Filina, O. B. Soboleva, V. N. Masolov, and N. I. Khlupin (2002), Small-scale airborne radio-echo sounding investigations in Central East Antarctica, in *Proceedings XVI-XIX All-Russian Symposium on Radio-Echo Sounding of the Natural Environments*, vol. 2, pp. 84–86.

Popov, S. V., A. N. Sheremetyev, V. N. Masolov, V. V. Lukin, A. V. Mironov, and V. S. Luchininov (2003), Velocity of radio-wave propagation in ice at Vostok Station, Antarctica, *J. Glaciol.*, *49* (165), 179–183.

Popov, S. V., A. N. Lastochkin, V. N. Masolov, and A. M. Popkov (2006), Morphology of the subglacial bed relief of Lake Vostok basin area, in *Proceedings of the 9th ISAES*, edited by D. Fütterer et al., pp. 141–146, Springer, Berlin.

Price, P. B., O. V. Nagornov, R. Bay, D. Chrikin, Y. He, P. Miocinovic, A. Richards, K. Woschnagg, B. Koci, and V. Zagorodnov (2002), Temperature profile for glacial ice at the South Pole: Implications for life in a nearby subglacial lake, *Proc. Natl. Acad. Sci. U. S. A.*, *99*, 7844–7847.

Rémy, F., and I. E. Tabacco (2000), Bedrock features and ice flow near the EPICA ice core site (Dome C, Antarctica), *Geophys. Res. Lett.*, *27*, 405–408.

Richter, A., S. V. Popov, R. Dietrich, V. V. Lukin, M. Fritsche, V. Y. Lipenkov, A. Y. Matveev, J. Wendt, A. V. Yuskevich, and V. N. Masolov (2008), Observational evidence on the hydro-glaciological regime of subglacial Lake Vostok, *Geophys. Res. Lett.*, *35*, L11502, doi:10.1029/2008GL033397.

Ridley, J. K., W. Cudlip, and S. W. Laxon (1993), Identification of subglacial lakes using the ERS-1 radar altimeter, *J. Glaciol.*, *39*, 625–634.

Rippin, D. M., J. L. Bamber, M. J. Siegert, D. G. Vaughan, and H. F. J. Corr (2003), Basal topography and ice flow in the Bailey/Slessor region of East Antarctica, *J. Geophys. Res.*, *108*(F1), 6008, doi:10.1029/2003JF000039.

Robin, G. D., C. W. M. Swithinbank, and B. M. E. Smith (1970), Radio echo exploration of the Antarctic ice sheet, *Int. Assoc. Sci. Hydrol. Publ.*, *86*, 97–115.

Robin, G. D., D. J. Drewry, and D. T. Meldrum (1977), International studies of ice sheet and bedrock, *Philos. Trans. R. Soc. London, Ser. A*, *279*, 185–196.

Robinson, R. V. (1960), Experiment in visual orientation during flights in the Antarctic, *Sov. Antarct. Exped. Inf. Bull.*, *18*, 28–29.

Roemer, S., B. Legrésy, M. Horwath, and R. Dietrich (2007), Refined analysis of radar altimetry data applied to the region of the subglacial Lake Vostok, Antarctica, *Remote Sens. Environ.*, *106*, 269–284.

Ross, N., et al. (2011), Ellsworth Subglacial Lake, West Antarctica: A review of its history and recent field campaigns, in *Antarctic Subglacial Aquatic Environments*, Geophys. Monogr. Ser., doi:10.1029/2010GM000936, this volume.

Roy, L., M. Sen, D. D. Blankenship, P. Stoffa, and T. G. Richter (2005), Inversion and uncertainty estimation of gravity data using simulated annealing: An application over Lake Vostok, East Antarctica, *Geophysics*, *70*(1), J1–J12.

Sergienko, O. V., D. R. MacAyeal, and R. A. Bindschadler (2007), Causes of sudden, short-term changes in ice-stream surface elevation, *Geophys. Res. Lett.*, *34*, L22503, doi:10.1029/2007GL031775.

Shabtaie, S., I. M. Whillans, and C. R. Bentley (1987), The morphology of ice streams A, B and C, West Antarctica and their environs, *J. Geophys. Res.*, *92*, 8865–8883.

Shoemaker, E. M. (1990), The ice topography over sub-glacial lakes, *Cold Reg. Sci. Technol.*, *18*, 323–329.

Shreve, R. L. (1972), Movement of water in glaciers, *J. Glaciol.*, *11*(62), 205–214.

Shuman, C. A., H. J. Zwally, B. E. Schutz, A. C. Brenner, J. P. DiMarzio, V. P. Suchdeo, and H. A. Fricker (2006), ICESat Antarctic elevation data: Preliminary precision and accuracy assessment, *Geophys. Res. Lett.*, *33*, L07501, doi:10.1029/2005GL025227.

Siegert, M. J. (2000), Antarctic subglacial lakes, *Earth Sci. Rev.*, *50*, 29–50.

Siegert, M. J., and J. L. Bamber (2000), Subglacial water at the heads of Antarctic ice stream tributaries, *J. Glaciol.*, *46*(155), 702–703.

Siegert, M. J., and J. K. Ridley (1998a), Determining basal ice-sheet conditions at Dome C, central East Antarctica, using satellite radar altimetry and airborne radio-echo sounding information, *J. Glaciol.*, *44*, 1–8.

Siegert, M. J., and J. K. Ridley (1998b), An analysis of the ice-sheet surface and subsurface topography above the Vostok Station subglacial lake, central East Antarctica, *J. Geophys. Res.*, *103*(B5), 10,195–10,207.

Siegert, M. J., J. A. Dowdeswell, M. R. Gorman, and N. F. McIntyre (1996), An inventory of Antarctic subglacial lakes, *Antarct. Sci.*, *8*, 281–286.

Siegert, M. J., R. Hindmarsh, H. Corr, A. Smith, J. Woodward, E. C. King, A. J. Payne, and I. Joughin (2004), Subglacial Lake Ellsworth: A candidate for *in situ* exploration in West Antarctica, *Geophys. Res. Lett.*, *31*, L23403, doi:10.1029/2004GL021477.

Siegert, M. J., S. Carter, I. Tabacco, S. Popov, and D. D. Blankenship (2005), A revised inventory of Antarctic subglacial lakes, *Antarct. Sci.*, *17*(3), 453–460.

Siegert, M. J., A. LeBrocq, and A. J. Payne (2007), Hydrological connections between Antarctic subglacial lakes, the flow of water beneath the East Antarctic Ice Sheet and implications for sedimentary processes, in *Glacial Sedimentary Processes and Pro-*

ducts, edited by M. J. Hambrey et al., *Spec. Publ. Int. Assoc. Sedimentol.*, *39*, 3–10.

Smith, B. E., H. A. Fricker, I. R. Joughin, and S. Tulaczyk (2009), An inventory of active subglacial lakes in Antarctica detected by ICESat (2003-2008), *J. Glaciol.*, *55*(129), 573–595.

Stearns, L. A., B. E. Smith, and G. S. Hamilton (2008), Increased flow speed on a large East Antarctic outlet glacier caused by subglacial floods, *Nat. Geosci.*, *1*, 827–831.

Studinger, M., et al. (2003a), Ice cover, landscape setting and geological framework of Lake Vostok, East Antarctica, *Earth Planet. Sci. Lett.*, *205*, 195–210.

Studinger, M., G. D. Karner, R. E. Bell, V. Levin, C. A. Raymond, and A. A. Tikku (2003b), Geophysical models for the tectonic framework of the Lake Vostok region, East Antarctica, *Earth Planet. Sci. Lett.*, *216*, 663–677.

Studinger, M., R. E. Bell, W. R. Buck, G. D. Karner, and D. D. Blankenship (2004a), Sub-ice geology inland of the Transantarctic Mountains in light of new aerogeophysical data, *Earth Planet. Sci. Lett.*, *220*, 391–408.

Studinger, M., R. E. Bell, and A. A. Tikku (2004b), Estimating the depth and shape of Lake Vostok's water cavity from aerogravity data, *Geophys. Res. Lett.*, *31*, L12401, doi:10.1029/2004GL019801.

Studinger, M., R. E. Bell, and N. Frearson (2008), Comparison of AIRGrav and GT-1A airborne gravimeters for research applications, *Geophysics*, *73*(6), 151–161.

Tabacco, I. E., C. Bianchi, M. Chiappini, A. Zirizzotti, and E. Zuccheretti (2000), Analysis of the bottom morphology of the David Glacier-Drygalski Ice Tongue, East Antarctica, *Ann. Glaciol.*, *30*, 47–51.

Tabacco, I. E., A. Forieri, A. D. Vedova, A. Zirizzotti, C. Bianchi, P. De Michelis, and A. Passerini (2003), Evidence of 14 new subglacial lakes in the Dome C-Vostok area, *Terra Antarct. Rep.*, *8*, 175–179.

Tabacco, I. E., P. Cianfarra, A. Forieri, F. Salvini, and A. Zirizzotti (2006), Physiography and tectonic setting of the subglacial lake district between Vostok and Belgica subglacial highlands (Antarctica), *Geophys. J. Int.*, *165*, 1029–1040.

Thoma, M., K. Grosfeld, I. Y. Filina, and C. Mayer (2009), Modelling flow and accreted ice in subglacial Lake Concordia, Antarctica, *Earth Planet. Sci. Lett.*, *286*, 278–284.

Tikku, A. A., R. E. Bell, M. Studinger, G. K. C. Clarke, I. Tabacco, and F. Ferraccioli (2005), The influx of meltwater to subglacial Lake Concordia, East Antarctica, *J. Glaciol.*, *51*(172), 96–104.

Vaughan, D. G., H. F. J. Corr, F. Ferraccioli, N. Frearson, A. O'Hare, D. Mach, J. W. Holt, D. D. Blankenship, D. L. Morse, and D. A. Young (2006), New boundary conditions for the West Antarctic Ice Sheet: Subglacial topography beneath Pine Island glacier, *Geophys. Res. Lett.*, *33*, L09501, doi:10.1029/2005GL025588.

Wilch, E., and T. J. Hughes (2000), Calculating basal thermal zones beneath the Antarctic ice sheet, *J. Glaciol.*, *46*(153), 297–310.

Wingham, D. J., M. J. Siegert, A. Shepherd, and A. S. Muir (2006), Rapid discharge connects Antarctic subglacial lakes, *Nature*, *440*, 1033–1036.

Woodward, J., A. M. Smith, N. Ross, M. Thoma, H. F. J. Corr, E. C. King, M. A. King, K. Grosfeld, M. Tranter, and M. J. Siegert (2010), Location for direct access to subglacial Lake Ellsworth: An assessment of geophysical data and modelling, *Geophys. Res. Lett.*, *37*, L11501, doi:10.1029/2010GL042884.

Wright, A. P., M. J. Siegert, A. M. Le Brocq, and D. B. Gore (2008), High sensitivity of subglacial hydrological pathways in Antarctica to small ice sheet changes, *Geophys. Res. Lett.*, *35*, L17504, doi:10.1029/2008GL034937.

_____

M. J. Siegert and A. Wright, Grant Institute, School of Geo-Sciences, University of Edinburgh, West Mains Road, Edinburgh EH9 3JW, UK. (A.P.Wright@ed.ac.uk)

Antarctic Subglacial Lake Discharges

Frank Pattyn

Laboratoire de Glaciologie, Département des Sciences de la Terre et de l'Environnement
Université Libre de Bruxelles, Brussels, Belgium

Antarctic subglacial lakes were long time supposed to be relatively closed and stable environments with long residence times and slow circulations. This view has recently been challenged with evidence of active subglacial lake discharge underneath the Antarctic ice sheet. Satellite altimetry observations witnessed rapid changes in surface elevation across subglacial lakes over periods ranging from several months to more than a year, which were interpreted as subglacial lake discharge and subsequent lake filling, and which seem to be a common and widespread feature. Such discharges are comparable to jökulhlaups and can be modeled that way using the Nye-Röthlisberger theory. Considering the ice at the base of the ice sheet at pressure melting point, subglacial conduits are sustainable over periods of more than a year and over distances of several hundreds of kilometers. Coupling of an ice sheet model to a subglacial lake system demonstrated that small changes in surface slope are sufficient to start and sustain episodic subglacial drainage events on decadal time scales. Therefore, lake discharge may well be a common feature of the subglacial hydrological system, influencing the behavior of large ice sheets, especially when subglacial lakes are perched at or near the onset of large outlet glaciers and ice streams. While most of the observed discharge events are relatively small (10^1–10^2 m^3 s^{-1}), evidence for larger subglacial discharges is found in ice free areas bordering Antarctica, and witnessing subglacial floods of more than 10^6 m^3 s^{-1} that occurred during the middle Miocene.

How still,
How strangely still
The water is today,
It is not good
For water
To be so still that way.

Hughes [2001, p. 48]

1. INTRODUCTION

Subglacial lakes are omnipresent underneath the Antarctic ice sheet. A recent inventory [*Smith et al.*, 2009a] brings the total to more than 270 (Figure 1), i.e., 145 from an inventory by *Siegert et al.* [2005] and more than 130 added since [*Bell et al.*, 2006, 2007; *Carter et al.*, 2007; *Popov and Masolov*, 2007; *Smith et al.*, 2009a]. Subglacial lakes are usually identified from radio echo sounding and characterized by a strong basal reflector and a constant echo strength (corroborating a smooth surface). They are brighter than their surroundings by at least 2 dB (relatively bright) and both are consistently reflective (specular). They are therefore called "definite" lakes [*Carter et al.*, 2007]. The larger ones are characterized by a flat surface compared to the surroundings with a slope around one tenth, and in the opposite direction to, the lake surface slope [*Siegert et al.*, 2005], hence the ice column above such a subglacial lake is in hydrostatic equilibrium. "Fuzzy" lakes are defined by high absolute and relative reflection coefficients, but are not specular [*Carter et al.*,

Antarctic Subglacial Aquatic Environments
Geophysical Monograph Series 192
Copyright 2011 by the American Geophysical Union
10.1029/2010GM000935

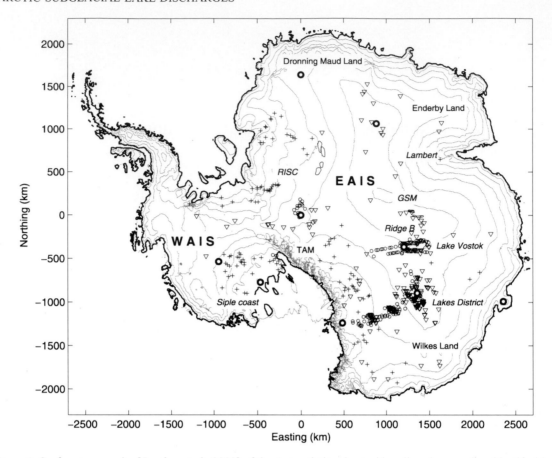

Figure 1. Surface topography [*Bamber et al.*, 2009] of the Antarctic ice sheet with radio echo sounding-identified large subglacial lakes (black lines), "definite" (inverted triangles) subglacial lakes [*Siegert et al.*, 2005; *Popov and Masolov*, 2007; *Carter et al.*, 2007], "fuzzy" (small circles) lakes [*Carter et al.*, 2007] and "active" (plus signs) lakes [*Smith et al.*, 2009a]. The bold circles show the position of deep ice core drill sites. Abbreviations are WAIS, West Antarctic Ice Sheet; EAIS, East Antarctic Ice Sheet; TAM, Transantarctic Mountains; RISC, Recovery Ice Stream Catchment; GSM, Gamburtsev Subglacial Mountains.

2007]. They are interpreted as corresponding to saturated basal sediments [*Peters et al.*, 2005; *Carter et al.*, 2007]. Most lakes lie under a thick ice cover of >3500 m and are therefore situated close to ice divides. Although the majority of subglacial lakes are small (<20 km in length), the largest lake is Subglacial Lake Vostok, containing 5400 ± 1600 km^3 of water [*Studinger et al.*, 2004]. Assuming an average water depth of 1000 m for large lakes and 100 m for shallow lakes, the volume of water in known Antarctic subglacial lakes is ~22,000 km^3, or approximately 25% of the water worldwide in surface lakes [*SALE Workshop Report*, 2007]. This is equivalent to a uniform sheet of water ~1-m thick if spread out underneath the whole Antarctic ice sheet.

Subglacial lakes were long time supposed to be relatively closed and stable environments with long residence times and slow circulations [*Siegert et al.*, 2001; *Kapitsa et al.*, 1996; *Bell et al.*, 2002]. As long as the huge amount of water they contain

is not moving around through a hydrological network, they pose no threat to the stability of the ice sheet. However, the idea of closed subglacial environments has recently been challenged with evidence of subglacial lake discharge. First, from ERS laser altimetry data and later on from NASA's Ice, Cloud and land Elevation Satellite (ICESat) laser altimetry data, more than 120 "active" subglacial lakes have been determined [*Wingham et al.*, 2006a; *Fricker et al.*, 2007; *Smith et al.*, 2009a]. These are spots where a rapid lowering or rising of the ice surface is detected, a phenomenon that has been interpreted as either a sudden drainage (over a period of several months) or a rapid lake infill with subglacial water drained from an upstream subglacial lake [*Fricker et al.*, 2007; *Fricker and Scambos*, 2009].

In this chapter, we will give an overview of subglacial lake drainage, how it is observed, what the possible mechanisms are and whether such events may influence the stability of the ice sheet.

2. OBSERVATIONS OF SUBGLACIAL LAKE DISCHARGE

2.1. Jökulhlaups

Subglacial lake drainage is probably the most common feature in active volcanic regions, where they are called jökulhlaups. They originally refer to the well-known subglacial outburst floods from Vatnajökull, Iceland, and are often triggered by volcanic subglacial eruptions. More generally, jökulhlaups describe any large and abrupt release of water from a subglacial or supraglacial lake. Well-documented jökulhlaups are those from the subglacial caldera Grímsvötn, beneath Vatnajökull [*Björnsson*, 2002]: high rates of geothermal heat flux cause enhanced subglacial melting, and subglacial hydraulic gradients direct this meltwater to Grímsvötn where it collects. Jökulhlaups occur when the lake level reaches within tens of meters of the hydraulic seal, from where water is released and flows subglacially to the ice cap margin [*Björnsson*, 1988]. During this phase, ice velocities may well increase up to threefold over an area up to 8 km wide around the subglacial flood path [*Magnusson et al.*, 2007]. Owing to the large quantity of subglacial meltwater produced by high geothermal heat rates, jökulhlaups lead to a significant impact on ice dynamics. Not surprisingly, this type of jökulhlaups is unlikely to occur in Antarctica due to low geothermal heat flux. Geothermal heat flow measurements in the Lakes District (East Antarctica) point to values as low as 40 mW m^{-2} [*Shapiro and Ritzwoller*, 2004], which is supported by both the temperature gradient in the Vostok ice core as by inverse modeling [*Siegert*, 2000]. Nevertheless, these low values are sufficient to guarantee bottom melting due to the insulating ice cover. Moreover, since large quantities of subglacial water are stored underneath the ice sheet (e.g., 5400 km^3 for Lake Vostok), there is a substantial potential for subglacial lake discharge.

Despite low geothermal heating, evidence from Antarctic jökulhlaups is at hand: a first documented outburst occurred near Casey Station, Law Dome, Antarctica [*Goodwin*, 1988]. The discharge event started in March 1985 and lasted 6 months, with occasional outbursts during the austral autumn and winter of 1986. An oxygen-isotope and solute analysis of the spilled water revealed that its origin was basal meltwater, originating from an ice-marginal subglacial lake. However, the extent and depth of the reservoir remained unknown.

2.2. Satellite Observations

With the advent of satellite image interferometry, it became possible to observe and monitor small vertical changes at the surface of the Earth. Such changes measured in the direction of the satellite can be mapped onto both a vertical and horizontal component. The technique revealed very powerful to map horizontal flow speeds of Antarctic and Greenland outlet glaciers and ice streams using a speckle-tracking technique [*Joughin*, 2002]. *Gray et al.* [2005] analyzed RADARSAT data from the 1997 Antarctic Mapping Mission and used them interferometrically to solve for the three-dimensional surface ice motion in the interior of the West Antarctic Ice Sheet. They found an area of ~125 km^2 in a tributary of the Kamb Ice Stream displaced vertically downward by up to 50 cm between 26 September and 18 October 1997. Similar upward and downward surface displacements were also noted in the Bindschadler Ice Stream. Both sites seemed to correspond to areas where basal water is apparently ponding (hydraulic potential well). The authors therefore suggested transient movement of pockets of subglacial water as the most likely cause for the vertical surface displacements.

A well-documented Antarctic subglacial lake discharge is reported in the vicinity of the Lakes District, central East Antarctica. *Wingham et al.* [2006a] observed ice sheet

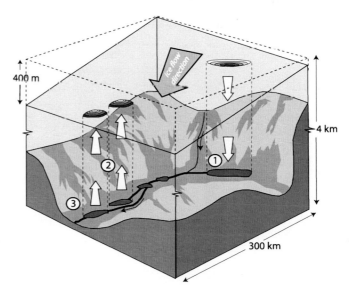

Figure 2. Rapid drainage and hydraulic connection of Antarctic subglacial lakes, inferred from satellite altimetric measurements of the ice sheet surface over the Adventure subglacial trench, East Antarctica between 1996 and 2003. Circled numbers indicate the following: 1, lake drainage results in ice sheet surface lowering of ~3 m over ~16 months; 2, ice sheet surface uplift occurs over a series of known subglacial lakes ~290 km from the initial lake drainage; 3, once surface lowering over the upstream lakes ceases (implying a reduction in the supply of water), ice surface lowering over at least one downstream lake occurs. Adapted from *Wingham et al.* [2006a].

surface elevation changes that were interpreted to represent rapid discharge from a subglacial lake. The altimeter survey by the satellite ERS-2 revealed two anomalies of ice sheet surface elevation change in the vicinity of the Adventure subglacial trench. One anomaly, at the northern (upstream) end of the trench, shows an abrupt fall in ice-surface elevation (Figure 2 and curve L1 in Figure 3). Some 290 km distant from L1, a corresponding abrupt rise occurred at the southern (downstream) end of the trench (Figure 2 and U1 and U2 in Figure 3). The only mechanism that explains these observations is a rapid transfer of basal water from a subglacial lake beneath the region of surface lowering to lakes beneath the regions of uplift. *Wingham et al.* [2006a] estimated the magnitude and rate of subglacial drainage of 1.8 km^3 of water transferred at a peak discharge of 50 m^3 s^{-1} during a period of 16 months.

High-resolution evidence of subglacial lake discharges was obtained with ICESat laser altimetry [*Fricker et al.*, 2007]. Satellite laser altimeter elevation profiles from 2003 to 2006 collected over the lower parts of Whillans and Mercer ice streams, West Antarctica, revealed 14 regions of temporally varying elevation, which were interpreted as the surface expression of subglacial water movement. Contrary to the Adventure trench lakes, perched near the ice divide of

the East Antarctic ice sheet in a region of relatively slow ice movement, the lakes in West Antarctica were detected in an area of fast ice flow, close to the grounding line.

Subglacial Lake Engelhardt is one of the larger subglacial lakes in the area, situated just upstream of the grounding line and flanking Engelhardt Ice Ridge. Image differencing presented by *Fricker et al.* [2007] and *Fricker and Scambos* [2009] revealed a spatial extent of the drawdown of this lake of 339 km$^2 \pm 10\%$. This area is sufficiently large that the ice above the lake is in hydrostatic equilibrium, corroborated by the flatness of profiles across it (Figure 4). Draining and filling rates were estimated from the 2003–2006 flood and the total water loss over that period estimated as 2.0 km^3 [*Fricker and Scambos*, 2009], a value comparable to the Adventure trench discharge. Since Lake Engelhardt is situated ~7 km upstream from the grounding line, the floodwater is almost certainly discharged directly into the subglacial lake [*Fricker and Scambos*, 2009]. Since June 2006, Lake Engelhardt is steadily filling at a rate of 0.14 km^3 a^{-1} (Figure 4).

The analysis of *Fricker and Scambos* [2009] was further extended to the whole Antarctic ice sheet north of 86.6°S, based on 4.5 years (2003–2008) ICESat laser altimeter data [*Smith et al.*, 2009a]. This analysis detected 124 "active" lakes. Most of these lakes are situated in the coastal regions

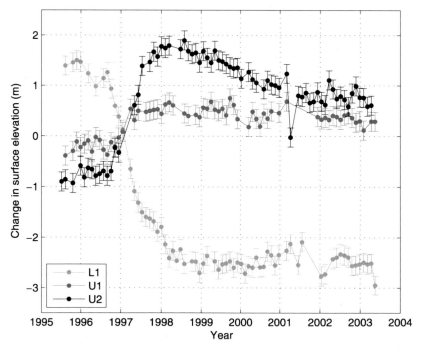

Figure 3. ERS-2 altimetric data from three sites, L1, U1, and U2, in the Adventure trench area (East Antarctica). Lake U3 has the same signature of U2 and is not shown here for clarity. The error bars are 1-sigma errors, which are determined empirically and determined as ±0.18 m. Adapted from *Wingham et al.* [2006a].

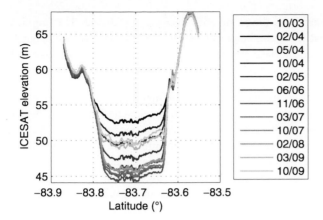

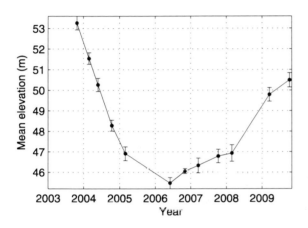

Figure 4. (top) Evolution of surface evolution across Engelhardt Lake (West Antarctica). (bottom) Mean elevation across the lake surface [*Fricker and Scambos*, 2009; II. A. Fricker, personal communication, 2009].

of the ice sheets and present underneath ice streams (Figure 1). They may, therefore, form reservoirs that may contribute pulses of water to produce rapid temporal changes in glacier speeds [*Smith et al.*, 2009a].

However, one should remain careful in interpreting rapid surface changes detected through satellite laser altimetry as sudden drainage of subglacial lakes. The interpretation of small-scale surface displacement to recover subglacial water movement is complicated by the fact that other subglacial processes can also result in surface deformation [*Smith et al.*, 2009a]. Ice-flow models show that local changes both in basal friction and in basal topography can produce changes in surface elevation [*Gudmundsson*, 2003; *Pattyn*, 2003, 2004; *Sergienko et al.*, 2007]. Furthermore, upward vertical changes may not always imply lake filling with water coming from upstream [*Pattyn*, 2008]. This point will be discussed later.

Besides satellite altimetry, present-day as well as past subglacial lake dynamics may be detected through other remote sensing techniques. Both melting and refreezing above subglacial lakes may be estimated by ice-penetrating radar by measuring along-flow changes in the thickness of the basal internal layers where thinning of the basal ice is indicative of melting and thickening points to freezing [*Bell et al.*, 2002; *Siegert et al.*, 2000; *Tikku et al.*, 2004]. Large distortions in basal internal layers may alternatively be indicative for subglacial lake discharge. *Siegert et al.* [2004] calculated from the convergence of a number of internal layers across the West Antarctic ice sheet that observed distortions would require basal melt rates of up to 6 cm a^{-1}, which are an order of magnitude greater than those calculated for the Siple Coast region. In view of the recent evidence regarding subglacial lake discharge in this area [*Fricker and Scambos*, 2009; *Smith et al.*, 2009a], large melting rates may witness such events as well.

2.3. Ice Flow Acceleration

Water plays a crucial role in ice sheet stability and the onset of ice streams. Since water may move between subglacial lakes by a rapid drainage, enhanced lubrication of ice streams and glaciers is expected, resulting in a speedup of ice flow. Many subglacial lakes sit at the onset of ice streams, hence have the potential to enhance the ice flow further downstream [*Siegert and Bamber*, 2000]. *Bell et al.* [2007] detected a number of large subglacial lakes at the onset region of the Recovery Glacier Ice Stream (RISC in Figure 1), where ice is moving at rates of 20 to 30 m a^{-1}. *Stearns et al.* [2008] report an observed acceleration of ice velocity on Byrd Glacier, East Antarctica, of about 10% of the original speed between December 2005 and February 2007. The acceleration extended along the entire 75-km glacier trunk, and its onset coincided with the discharge of about 1.7 km^3 of water from two large subglacial lakes located about 200 km upstream of the grounding line. Deceleration of the ice flow coincided with the termination of the flood. These findings provide direct evidence that an active lake drainage system can cause large and rapid changes in glacier dynamics. More spectacular interactions between floods and glacier surges are reported from Icelandic jökulhlaups [e.g., *Björnsson*, 1998].

Subglacial lakes may also play a crucial role in the redistribution of subglacial meltwater. Water piracy has recently been found a suitable mechanism in explaining the on and off switching of streaming flow, as is the case for the Siple Coast ice streams [*Anandakrishnan and Alley*, 1997] or Rutford Ice Stream [*Vaughan et al.*, 2008], a process that may be controlled by periodical subglacial lake discharge.

3. MECHANISMS OF SUBGLACIAL LAKE DISCHARGE

3.1. Hydraulic Geometry

Water is almost ubiquitously present underneath the Antarctic ice sheet, either stored in subglacial lakes or part of a subglacial drainage network that transports this water to the edge of the ice sheet. Free water can exist at the ice-bed contact and interstitially in subglacial sediment. The pressure p_w of subglacial water is an independent variable that varies temporally and spatially in a complicated manner that is determined by the balance between influx and outflux of water, the geometry of the subglacial water system, the physical properties of the glacier substrate, thermodynamic conditions near the ice-bed interface, and the ice overburden pressure [Clarke, 2005].

Glaciers are through buoyancy supported by the subglacial water pressure, and the magnitude of this support stems from comparing the water pressure p_w to the ice overburden pressure $p_i = \rho_i g h$, where ρ_i is the ice density, g the gravity acceleration, and H is the ice thickness. The effective pressure $p_e = p_i - p_w$ is a common measure for the importance of buoyancy [Clarke, 2005]. For subglacial lakes, where the overlying ice is in hydrostatic equilibrium with the underlying water column, it follows that $p_e = 0$. Subglacial water movement is controlled by gradients in the fluid potential, the latter defined by

$$\phi_w = p_w + \rho_w g z, \tag{1}$$

where ρ_w is the water density and z is the elevation ($z = z_b$ at the ice/bed contact, where z_b is the bed elevation). For a subglacial lake that obeys the condition $p_w \equiv p_i$, differentiating equation (1) leads to

$$\nabla\phi_w = \rho_i g \nabla z_s + (\rho_w - \rho_i) g \nabla z_b, \tag{2}$$

where ∇ is the gradient operator and z_s is the surface elevation [Shreve, 1972]. For $\rho_i = 910$ kg m^{-3} and $\rho_i = 1000$ kg m^{-3}, the first term on the right-hand side of equation (2) is roughly an order of magnitude larger than the second term. This means that subglacial water flow is mainly driven by the surface topography of the glacier and, to a lesser extent, by the bed topography. Water ponding in a subglacial lake implies that $\nabla\phi_w = 0$. This ponding threshold can therefore also be expressed as a simple relationship between surface and bed slopes, i.e.,

$$\frac{dz_s}{dx} = \frac{\rho_i - \rho_w}{\rho_i}\frac{dz_b}{dx}, \tag{3}$$

where x is the along-path distance coordinate. This implies that water can be pushed out of a subglacial cavity whenever the ice surface slope is larger than one tenth the adverse bed slope, or $dz_s/dx > -\frac{1}{10} dz_b/dx$.

Based on the hydraulic potential gradient, Siegert et al. [2007] and Wright et al. [2008] calculated subglacial water flow patterns. The steepest downslope gradient of the hydraulic potential indicates the direction of the water flow, from which flow paths can be calculated that predict the drainage network [Wright et al., 2008]. An example of such a reconstruction is given in Figure 5, based on an initial distribution of local meltwater production using a thermomechanical ice sheet model with appropriate boundary conditions of surface temperature and geothermal heat flux, the latter corrected for the presence of subglacial lakes [Pattyn, 2010]. The input geometry is based on a 5-km resolution updated BEDMAP database [Lythe and Vaughan, 2001; Pattyn, 2010] and a resampled surface topography [Bamber et al., 2009]. Figure 5 clearly shows that major drainage paths converge in the large outlet glaciers and ice streams, dominated by the surface topographic slopes.

Water transport is largely depending on the hydraulic geometry, whether water flows through an aquifer (or a sheet) or through a pipe. Numerous idealized drainage structures have been proposed to describe water flow at the base of a glacier, including ice-walled conduits [Röthlisberger, 1972; Shreve, 1972], bedrock channels [Nye, 1976], water films and sheets [Weertman, 1972; Walder, 1982; Weertman and Birchfield, 1983], linked cavities [Walder, 1986; Kamb, 1987], soft sediment canals [Walder and Fowler, 1994], or porous sediment sheets [Clarke, 1996]. For glaciers or ice sheets resting on a hard bed, there are essentially three possibilities for subglacial water flow, either through a system of channels, a linked cavity system, or drainage in a water film. Observed jökulhlaups generally drain through conduits that connect the subglacial lake to the edge of the ice cap. A similar theory can be applied to connecting Antarctic subglacial lakes.

Röthlisberger [1972] based his theory of water flow through a subglacial conduit on the Gauckler-Manning-Stricker formula for the mean velocity of turbulent flow, i.e.,

$$\mathbf{v}_w = \frac{Q}{S} = m_n^{-1} R^{\frac{2}{3}} \left(-\frac{1}{\rho_w g}\frac{\partial\phi_w}{\partial x}\right)^{\frac{1}{2}}, \tag{4}$$

where $m_n = 0.08$ m$^{-1/3}$s is the Manning coefficient determining the roughness of the conduit, Q is the water discharge, and R is the hydraulic radius, which is defined as the cross-sectional area over a wetted perimeter. For a conduit of semicircular cross-section, $S = \pi r^2/2$, or $r = \sqrt{2S/\pi}$, the wetted perimeter is defined as:

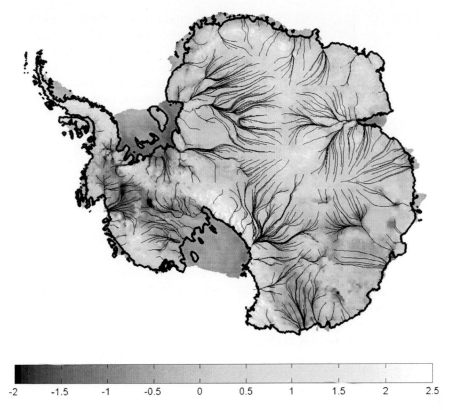

Figure 5. Calculated pattern of subglacial Antarctic drainage, superimposed on the subglacial topography (grey scale, in kilometers above sea level). Contrary to *Wright et al.* [2008], subglacial hollows where water is stored are not smoothed out, with the result that not all water drainage reaches the grounding line.

$$R = \frac{S}{\pi r + 2r} = \frac{S}{\pi \left(\frac{2S}{\pi}\right)^{\frac{1}{2}} + 2 \left(\frac{2S}{\pi}\right)^{\frac{1}{2}}}. \qquad (5)$$

Substituting equation (5) in equation (4) and rearranging terms, leads to an equation relating the size of the subglacial conduit to subglacial water discharge [*Röthlisberger*, 1972; *Peters et al.*, 2009]

$$S = \left[Qm_n \left(\frac{\sqrt{2}(\pi + 2)}{\sqrt{\pi}}\right)^{\frac{2}{3}} \left(-\frac{1}{\rho_w g} \frac{\partial \phi_w}{\partial x}\right)^{\frac{1}{2}} \right]^{\frac{3}{4}}. \qquad (6)$$

A similar and easier-to-derive expression can be found for a circular conduit, i.e.,

$$S = \left[Qm_n (2\sqrt{\pi})^{\frac{2}{3}} \left(-\frac{1}{\rho_w g} \frac{\partial \phi_w}{\partial x}\right)^{\frac{1}{2}} \right]^{\frac{3}{4}}. \qquad (7)$$

Peters et al. [2009] used equation (6) to derive the size of the conduit through which water would flow connecting two lakes along the hydraulic flow path in the Adventure trench basin, mentioned earlier. A similar approach was followed by *Wingham et al.* [2006a]. The geometry used is sketched in Figure 6: water from Lake L is drained via a conduit of $l = 290$ km in length over a period of 16 months. Owing to the discharge, the surface of the lake has lowered by $\Delta h_L = 3$ m. Since the lake size is estimated as $S_L = 600$ km^2 and assuming a cylindrical geometry, this leads to a water volume of $V = 1.8$ km^3 at a mean discharge of $Q = 43$ m^3 s^{-1}. The peak discharge of the lake has been estimated at 50 m^3 s^{-1} [*Wingham et al.*, 2006a]. Given a difference in ice thickness above both lakes of $\Delta H = -450$ m and the difference in height of the lakes (bedrock elevation difference) of $\Delta z_b = 260$ m, the hydraulic potential gradient can be approximated using equation (2), so that $\partial \phi_w/\partial x \approx -6.6$ Pa m^{-1}. Considering the system in equilibrium, the conduit size that sustains this discharge is obtained from equation (6), leading to a cross-sectional area of $S = 78$ m^2. This represents a semicircular conduit with a diameter of 14 m.

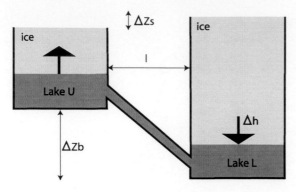

Figure 6. A simple model of the exchange between two lakes. An uphill flow of water is forced by the larger ice overburden at the lower Lake L. Both lakes are considered to have the same water depth. Adapted from the works of *Wingham et al.* [2006a] and *Peters et al.* [2009].

The question whether a conduit with that size can be formed between two lakes underneath the Antarctic ice sheet depends on the amount of energy that is available to form the conduit in the first place. Assuming again a cylindrical geometry for both lakes, the total energy release E_T is the energy due to the lowering of lake L (E_L) minus the energy needed to (1) raise lake U (E_U), (2) raise the water through the conduit by Δz_b (E_H), and (3) to maintain the water at pressure melting point (E_W). Assuming that the volume of water is conserved, i.e., $S_L \Delta h_L = S_U \Delta h_U$, it follows that

$$E_L - E_U = S_L \Delta h_L \rho_i g |\Delta H|, \qquad (8)$$

$$E_H = V \rho_w g \Delta z_b, \qquad (9)$$

$$E_W = -c_v V \frac{T_0}{L} \rho |\Delta H| \left(\frac{1}{\rho_w} - \frac{1}{\rho_i} \right), \qquad (10)$$

where $T_0 = 273.15$ K, $c_v = 4.2$ kJ kg^{-1} is the specific heat capacity of water, and $L = 3.3 \times 10^5$ J kg^{-1} is the latent heat for melting of ice. The total amount of energy equals $E_T = E_L - E_U - E_H - E_W$. Applied to the Adventure lake system, this leads to $E_T = 7.2 \times 10^{15} - 4.5 \times 10^{15} - 2.5 \times 10^{11} = 2.6 \times 10^{15}$ J. The energy required to melt a conduit of cross-sectional area S is $E_M = Sl\rho_i L$. As such, the available energy for melting a conduit over a distance l ($E_T = E_M$) is sufficient to sustain a conduit with a cross-sectional area $S = 30$ m^2 [*Peters et al.*, 2009]. This is about half the size of conduit that can be formed by the given discharge rate according to equation (6). However, in the former derivation, a relatively high Manning coefficient was used, typical for a rough conduit. Lower values of the Manning coefficient, e.g., 0.02 instead of 0.08 m$^{-1/3}$ s, corresponding to a smoother conduit, will give rise to a cross-sectional area of

≈ 30 m^2 [*Peters et al.*, 2009], as can be seen from Figure 7. Note also that the shape of the conduit (circular or semicircular) does not have a significant effect on the final conduit size corresponding to a given value for the Manning coefficient.

These calculations show that when considering the ice at the bed at pressure melting point, subglacial lake discharges can occur, and water can be transported over substantial distances of several hundreds of kilometers through a conduit with a diameter of 5 to 10 m.

3.2. The Nye-Röthlisberger Model

Rapid discharge of a subglacial lake is one of the different types of recognized jökulhlaups [*Roberts*, 2005]. The physical understanding of its time-dependent dynamics is based on empirical data from a rather limited number of events. Typical jökulhlaups from Grímsvötn (Iceland) increase toward a peak and fall rapidly. The periodicity is between 1 and 10 years, with peak discharges of 600 to 4–5 × 10^4 m^3 s^{-1} at the glacier margin, a duration of 2 days to 4 weeks, and a total water volume of 0.5–4.0 km^3 [*Björnsson*, 2002]. The difference with Antarctic subglacial lake discharge is striking: peak discharges are several orders of magnitude smaller and discharge duration generally more than a year. However, the periodicity may well be of the same order of magnitude, but the lack of longer time series prevents a proper evaluation.

The basic theory that governs this type of jökulhlaups is due to *Nye* [1976], extending the hydraulic theory of *Röthlisberger* [1972]. The theory assumes a single, straight subglacial conduit linking a meltwater reservoir directly to the terminus (or to another reservoir). For water to flow from one

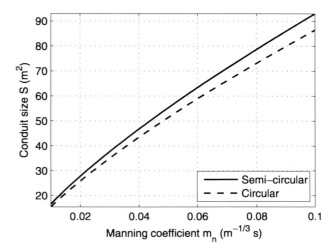

Figure 7. Conduit size S as a function of the Manning coefficient m_n for both a circular (7) and a semicircular (6) channel. Both curves are produced using the discharge geometry corresponding to the Adventure trench lakes.

to the other, a hydraulic gradient must exist. The size of the conduit will be due to a competition between the enlargement of the conduit and the processes that tend to shrink it. Energy is dissipated by flowing water, and some of the energy is transferred to the conduit walls causing ice to melt. Creep closure of the tunnel is due to ice deformation [*Clarke et al.*, 2004]. Nye derived partial differential equations describing the nonsteady flow in a conduit, accounting for the geometry and flow of ice, continuity of water, flow of water and energy, and heat transfer [*Nye*, 1976; *Fowler*, 2009]. *Peters et al.* [2009] applied the Nye-Röthlisberger model in its simplified form to the Adventure trench lakes, for which a hydrograph proxy exists (Figure 3).

Contrary to Icelandic jökulhlaups, temperatures of the lake water and the ice are likely to be similar at or close to pressure melting point, so that the energy equation and heat transfer equation can be simplified [*Peters et al.*, 2009]. The equations for the geometry and flow of ice, continuity and flow of water are

$$\frac{\partial S}{\partial t} = \frac{m_r}{\rho_i} - KS(p_i - p_w)^n, \tag{11}$$

$$\frac{\partial S}{\partial t} + \frac{\partial Q}{\partial x} = \frac{m_r}{\rho_w}, \tag{12}$$

$$\rho_w g \nabla z_b - \frac{\partial p_w}{\partial x} = f \rho_w g \frac{Q|Q|}{S^{8/3}}, \tag{13}$$

where K is a creep closure coefficient due to the nonlinear flow law $\dot{\varepsilon} = A\tau^n$, where $\dot{\varepsilon}$ is the strain rate and τ is the stress. More specifically, $K = 2A/n^n$ [*Fowler*, 2009]. The term m_r/ρ_i is the volumetric melt of the side-walls of the conduit, and f is a friction factor. *Wingham et al.* [2006a] used equation (11) to determine whether the subglacial conduit can remain open for a sustained period (at least more than a year). A fourth equation for energy (equation (19) in *Nye* [1976]), can be greatly reduced by removing all temperature dependencies and keeping the internal energy constant [*Peters et al.*, 2009]. However, as shown by *Spring and Hutter* [1982], this may lead to an overestimation of the peak discharge. Assuming steady state conditions (constant conduit size in time), it is possible to determine the effective pressure ($p_e = p_i - p_w$) to balance growth of the tunnel through melting with closure through ice flow. Neglecting sensible heat advection, the mean melt rate of the conduit is defined by

$$m_r = \frac{Q}{\rho_i L}\left(-\frac{\partial \phi}{\partial x}\right), \tag{14}$$

so that, by making use of the previously determined parameters, the melt rate of the Adventure lake conduit is estimated

as $m_r = 8.3 \times 10^{-7}$ m^2 s^{-1}. Using $A = 2.5 \times 10^{-24}$ Pa^{-3} s^{-1} (the value of the flow parameter at pressure melting point) and a conduit size of 30 m^2, and rearranging equation (11) leads to $p_e = 690$ kPa. This is a very large value compared to the change in pressure at lake L by the discharge event ($\Delta p_L = \rho_i g \Delta h_L = 27$ kPa). This led *Wingham et al.* [2006a] to conclude that the water flow was not stopped by the closure of the tunnel and that the lake could well have been emptied.

Peters et al. [2009] modeled the time-dependent evolution of a two-lake system, solving equations (11)–(13). They found that the Manning coefficient influences the peak discharge, but also the time for the flood to initiate as well as the lifetime of the flood. Furthermore, comparison with a circular conduit showed that the latter results in a faster onset and higher peak discharge. Considering more than two lakes shows the limits of the Nye-Röthlisberger model. It results in multipeak discharges, which were not observed along the Adventure trench [*Peters et al.*, 2009]. This may be due to the fact that the drainage system is not along a single conduit, but forms a distributed system based on broad shallow canals, as advocated by *Carter et al.* [2009]. Their study of the Adventure trench system is based on satellite altimetry and radio echo sounding data along the whole flow path and shows that the volume release from the source lake L exceeded the volume of the other lakes by >1 km^3, implying water loss of the system. This downstream release continued until at least 2003, when nearly 75% of the initial water release had traveled downstream from the filling lakes. Such discharge would only be sustained effectively by a broad shallow water system. A distributed system is also consistent with the 3-month delay between water release at the source lake and water arrival at the destination lake [*Carter et al.*, 2009]. Neither a porous aquifer or thin water layer at the ice bed interface are capable of transporting even a fraction of the inferred discharge. However, recent work by *Creyts and Schoof* [2009] suggests that distributed water sheets can be stable to much greater depth than previously quantified, as the presence of protrusions that bridge the ice-bed gap can stabilize them.

However, little remains known about how vertical deformations at the base of an ice sheet are transmitted to the surface, which is what is measured in the first place. Factors that influence this are, for instance, (1) the extent of the basal deformation anomaly compared to the ice thickness, (2) the time over which such deformation occurs, and (3) ice viscosity. A number of these effects will be explored in the next section.

4. EFFECT OF SUBGLACIAL LAKES AND LAKE DISCHARGE ON ICE FLOW

Observations by *Wingham et al.* [2006a], *Fricker et al.* [2007], and *Stearns et al.* [2008] suggest the rapid transfer of subglacial lake water and periodically flushing of subglacial lakes connected with other lakes that consequently fill through a hydrological network. These observations point to similar events of similar magnitude. However, they stem from rapid changes at the surface from which the drainage events are inferred. With the exception of the Law Dome jökulhlaup [*Goodwin*, 1988], direct observation of such subglacial events is lacking and, therefore, knowledge on the mechanisms that trigger them as well. A major assumption in the previous analysis is that surface elevation changes are directly related to changes in subglacial water volume. This would be correct if ice deformation would not occur. One way, however, to investigate the effect of ice sheet dynamics on sudden changes in subglacial lake volume is to use a numerical ice sheet model. While most models applied to large-scale flow of ice sheets neglect longitudinal components, this is not applicable on the scale of subglacial lakes. Ice flow across a subglacial lake experiences no (basal) friction at the ice/water interface [*Pattyn*, 2003; *Sergienko et al.*, 2007]. As such, the ice column behaves as an "embedded" or "captured" ice shelf within the grounded ice sheet [*Pattyn*, 2003; *Pattyn et al.*, 2004; *Erlingsson*, 2006]. This is why the ice sheet surface across large subglacial lakes, such as Subglacial Lake Vostok, is relatively flat and featureless, consistent with the surface of an ice shelf. Since ice shelf deformation is governed by longitudinal or membrane stresses, effects of stretching and compression should be taken into account.

4.1. Subglacial Lakes and Ice Dynamics

In general, numerical ice sheet models are based on balance laws of mass and momentum, extended with a constitutive equation. Solving the complete momentum balance leads to a so-called full Stokes model of ice flow [*Martin et al.*, 2003; *Zwinger et al.*, 2007; *Pattyn*, 2008]. Further simplifications can be made to this system of equations. A common approach is the higher-order approximation, where it is assumed that the full vertical stress is balanced by the hydrostatic pressure [*Blatter*, 1995; *Pattyn*, 2003]. Further approximations to the Stokes flow, still including longitudinal stress gradients, are governed by the shallow-shelf approximation [*MacAyeal*, 1989; *Sergienko et al.*, 2007]. Not only is hydrostatic equilibrium in the vertical applied, but all vertical dependence, such as vertical shearing, is omitted. The whole system is integrated over the vertical, so that the Stokes problem is simplified to the two plane directions [*Hindmarsh*, 2004]. However, its applicability is reduced to areas of low basal frictions, such as ice shelves (subglacial lakes) and ice streams, as shown by *Sergienko et al.* [2007].

The simplest way of mimicking subglacial lake presence in an ice sheet model is to introduce a slippery spot within an area of high(er) basal friction. For this purpose, consider a uniform slab of ice of 80 by 80 km in size and $H = 1600$ m thick, lying on gently sloping bed ($\alpha = 0.115°$). The basal boundary condition is written as $\tau_b - \mathbf{v}_b\beta^2$, where τ_b is the basal drag, $\mathbf{v}_b$ is the basal velocity vector and β^2 is a friction coefficient. For large β^2, $\mathbf{v}_b$ is small or zero (ice is frozen to the bedrock); for $\beta^2 = 0$, ice experiences no friction at the base (slippery spot) as is the case for an ice shelf. In this experiment, the basal friction coefficient β^2 is defined by a sine function ranging between 0 and 20 kPa a m^{-1} (Figure 8d). Periodic lateral boundary conditions were applied, and the

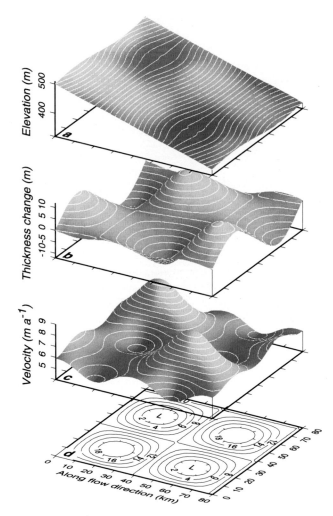

Figure 8. Effect of a slippery spot (subglacial lake) on ice sheet geometry and velocity field: (a) predicted steady state surface topography; (b) predicted change in ice thickness compared to initial uniform slab of 1600 m; (c) predicted horizontal surface velocity magnitude; (d) basal friction field β^2 varying between 0 (subglacial lake marked by L) and 20 kPa a m^{-1}. Ice flow is from left to right. From *Pattyn* [2004].

model was run to steady state. The effect of a slippery spot on the ice slab is shown by a local increase in ice velocity where friction is low (Figure 8c) as well as a flattening of the ice surface above this spot (Figure 8a). This flattening is due to a thinning of the ice upstream from the slippery spot and thickening of the ice downstream (Figure 8b) and is a direct result of the lack of basal shear across the slippery spot. *Gudmundsson* [2003] found a similar behavior for a linear viscous medium.

The effect of an embedded ice shelf is not only limited to a surface flattening, also the velocity field is influenced as the direction of the flow is not forced to follow the steepest surface gradient (as is the case when longitudinal stress gradients are neglected). This particularly applies to Subglacial Lake Vostok, where the observed surface flow is not along the steepest surface slope, since the latter is governed by the embedded ice shelf being in hydrostatic equilibrium with the underlying water body and the thickest ice, hence, highest surface elevation is found in the northern part of the lake. The effect of treating Lake Vostok as a slippery spot on the ice velocity is shown in Figure 9: the lack of a lake makes ice flow from north to south (right to left) or circumvents the lake, and no ice flow goes across Lake Vostok. Appropriate ice flow is obtained by treating Lake Vostok as a slippery spot, where ice flows from West to East (top to bottom), in

agreement with observations [*Kwok et al.*, 2000; *Bell et al.*, 2002; *Tikku et al.*, 2004].

4.2. Effect of Lake Discharge on Ice Flow

As shown in Figures 3 and 4, subglacial lake discharge results in a sudden drop of surface elevation. The effect of such a sudden drainage can easily be mimicked by lowering the bedrock topography across a slippery spot that represents a subglacial lake. Using a numerical ice stream model applied to an idealized ice stream geometry, *Sergienko et al.* [2007] show that this effect significantly modulates the surface-elevation expression and that the observed surface elevation changes do not directly translate the basal elevation changes, due to the viscoplastic behavior of the ice when it flows across the lake. A sudden drop in ice surface will be filled in gradually due to the ice flow and its deformation. Therefore, subglacial water volume change is not directly proportional to the area integral of surface-elevation changes [*Sergienko et al.*, 2007].

In reality, the relation between subglacial lakes and the overlying ice sheet is even more complicated as the ice on top of larger subglacial lakes is in hydrostatic equilibrium. This is expressed by the fact that on most lakes, it holds that $\nabla z_s \approx -\frac{1}{10}\nabla z_b$. Whenever water is trapped in a subglacial cavity,

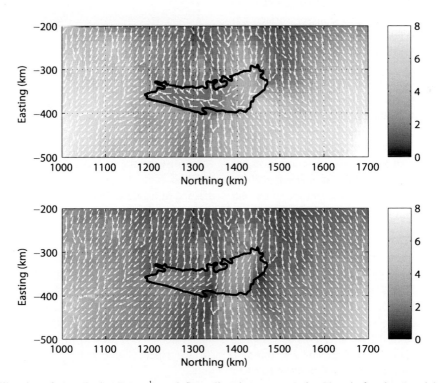

Figure 9. Predicted surface velocity (m a^{-1}) and flow direction across Lake Vostok for the "no lake" and "lake" experiment. The position of Lake Vostok is given by the black line. Adapted from *Pattyn et al.* [2004].

the effect will be transmitted to the surface by this relation. *Pattyn* [2008] implemented this effect in a full-Stokes ice sheet model of ice flow over a subglacial lake with a finite water volume. Full basal hydraulics, as those treated in the previous section, are not implemented in the model as such. A basal hydrological algorithm is used to check whether the lake "seal" breaks or not, based on the hydraulic potential gradient displayed in Figure 5. The hydrostatic seal is, thus, broken when water originating from the lake potentially arrives at the downstream edge of the domain. Nevertheless, as the lake empties, the marginal ice could collapse or fracture to form a cauldron, a common feature of jökulhlaups. The rate of subsidence is directly related to the rate of lake discharge, but requires an underpressure in the lake to draw down the overlying ice [*Evatt et al.*, 2006; *Evatt and Fowler*, 2007]. However, as small floods creating small surface depressions are considered here, only a flotation criterion for the ice over the lake is considered, as in the work of *Nye* [1976].

Conservation of water volume in the subglacial lake implies the definition of a buoyancy level (for a floating ice shelf this is sea level, but for a subglacial lake this is defined as $z_b + H\rho_i/\rho_w$). In an iterative procedure, the buoyancy level is determined for the local ice thickness to be in hydrostatic equilibrium on top of the lake for a given water volume. This procedure determines the position of both the upper and lower surfaces of the ice sheet across the subglacial lake in the Cartesian coordinate system. The contact surface between the ice sheet and the lake is then set to $\beta^2 = 0$.

The basic model setup is an idealized subglacial lake underneath an idealized ice sheet (Figure 10). The lake is defined as a Gaussian cavity. Ice thickness is $H_0 = 3500$ m, and the depth of the water cavity C_0 is taken as 400 m. Initially, the lake cavity was filled with 40×10^9 m^3 of water. The horizontal domain is L by L, where $L = 80$ km, and a grid size of 2 km was used (order of magnitude of ice thickness). The general characteristics of the steady state ice sheet geometry are typical for those of a slippery spot [*Pattyn*, 2003, 2004; *Pattyn et al.*, 2004], i.e., a flattened surface of the ice/air interface across the lake and the tilted lake ceiling in the opposite direction of the surface slope, due to hydrostatic equilibrium (Figure 10). The tilt of this surface in the direction of the ice flow will determine the stability of the lake, since the hydraulic gradient is dominated by surface slopes and therefore the flatter this air/ice surface the easier water is kept inside the lake cavity.

A perturbation experiment was carried out in which water is added to the lake at a rate of 0.1 m^3 s^{-1}. This corresponds to melting at the ice/lake ceiling at a rate of 2.5 mm a^{-1}, which is of the order of magnitude of basal melting to occur under ice sheets. It is regarded as a common gradual change in the glacial environment that on the time scales of decades

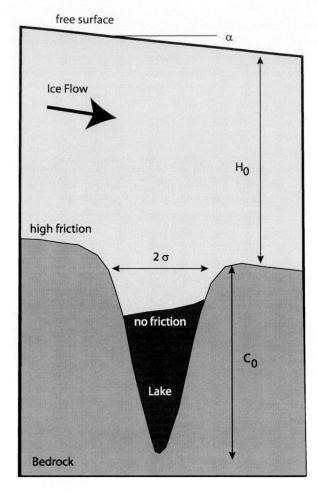

Figure 10. Model geometry of ice flow across an idealized subglacial lake. After *Pattyn* [2008]. Reprinted from the *Journal of Glaciology* with permission of the International Glaciological Society.

are hardly noticeable, hence, forming part of the natural variability of the system. Once the drainage condition is fulfilled, i.e., the hydrostatic seal is broken, the lake is drained at a rate of 50 m^3 s^{-1} for a period of 16 months, a value given by *Wingham et al.* [2006a] and which is of the same order of magnitude as the drainage rate of Lake Engelhardt [*Fricker et al.*, 2007]. This essentially means that enough energy is released to keep the subglacial tunnel open for a while and a siphon effect can take place. It is not clear whether this effect is only temporary or ends when the lake is completely emptied as suggested by *Wingham et al.* [2006a]. However, the drainage event of Lake Engelhardt suggests that water is still present in the lake after the event, as the surface aspect (flatness) has not changed over this period of time.

The time-dependent response of the lake system to such small perturbations is shown in Figure 11. When drainage

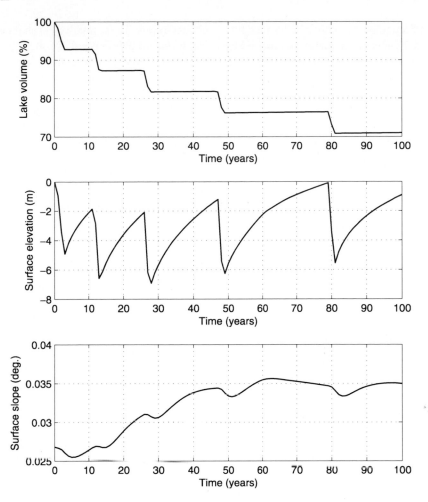

Figure 11. Time series of the perturbation experiment: (top) time evolution of lake water volume, (middle) ice surface elevation on top of the lake, and (bottom) ice surface slope across the lake. Adapted from *Pattyn* [2008].

occurs, episodic events take place, even though the initial geometric conditions are not met. The perturbation is, therefore, only a minor trigger to get the episodic lake drainage going. These events occur at a higher frequency at the beginning, i.e., when more water is present in the lake than later on and occur with frequencies of less than a decade. The episodic drainage results in variations in the surface elevation across the lake of rapid lowering, followed by a gradual increase until the initial level is more or less reached. This surface rise is clearly nonlinear, and the rate of uplift decreases with time. The frequency decreases with time as well as less water is present in the system, hence, hampering drainage. Interesting to note, however, is that the surface rise is not caused by an influx of subglacial water entering from upstream. On the contrary, less water is subsequently present in the whole subglacial system. Surface increase is due to an increased ice flux, filling up the surface depression created by the sudden lake drainage. Evidently, the maximum speed

of surface increase is significantly lower than the surface lowering due to drainage.

As shown before, Nye's hydraulic theory offers an appropriate description of the drainage of subglacial lakes, especially where the lake is not connected to the ice surface, and the ice over the lake responds dynamically to lake level fluctuations [*Nye*, 1976]. According to that model, the magnitude and duration of floods appear to be controlled by the channel hydraulics. *Evatt et al.* [2006] show with a reduced hydraulic model based on Nye theory that the peak discharge is essentially related to lake volume, i.e., big lakes produce big floods. Therefore, small floods apparently observed by *Wingham et al.* [2006a] could be associated with drainage of small lakes or through channels at low effective pressure [*Evatt and Fowler*, 2007]. *Pattyn* [2008] supposed that once the hydrostatic seal is broken, a subglacial channel is established in which the water pressure is lower than the ice pressure. However, in order to keep the channel from closing through viscous creep of ice, closure should be balanced by the melting of the channel walls. This energy is supplied by viscous dissipation of the turbulent water in the channel. As long as creep closure is balanced by the dissipated heat, the channel will remain open and discharge can take place. *Evatt et al.* [2006] have shown that lake drainage events as described by *Wingham et al.* [2006a] can occur over sustained periods of tens of months without the complete emptying of the lake. Furthermore, they also corroborate the fact that lake drainage is a common mechanism underneath the Antarctic ice sheet.

5. STABILITY OF SUBGLACIAL LAKES

The above-described mechanisms demonstrate that the stability of subglacial lakes is given by their nature of keeping the ice surface slope across the lake as small as possible due to the vanishing traction at the ice/water interface. The ice above Subglacial Lake Vostok, for instance, has a surface gradient of 0.0002 from north to south, which corresponds to 50 m of surface elevation change [*Siegert*, 2005]. If the slope of the grounded ice across the lake's western margin were changed, so too must the ice surface slope over the lake. Thus, the reason that a lake exists within the Vostok trough, and that ice-shelf flow is subsequently permitted, is due primarily to the flow direction of grounded ice upstream of the trough [*Siegert*, 2005]. Subglacial water will flow "uphill" if the ice surface slopes exceed 1/10 of the basal slope [*Shreve*, 1972]. Nevertheless, the ice surface gradient above Lake Vostok is 100 times less than the minimum basal slope of the head wall of the Vostok trough [*Studinger et al.*, 2003], which makes the lake stable unless the surface gradient increases tenfold, the value required to force the water out of the trough to the south [*Siegert and Ridley*, 1998]. However, the water level in Subglacial Lake Vostok, reaches close to a hydrostatic seal situated at the southeastern part of the lake, nearby Vostok Station [*Erlingsson*, 2006]. Water circulation modeling of Lake Vostok [*Thoma et al.*, 2007] points to an imbalance in the water mass balance of the lake which indicates either a constant growth of the lake or its continuous (or periodical) discharge into a subglacial drainage system. Even if periodic subglacial discharges occur of the order of magnitude as those observed by *Wingham et al.* [2006a], this would still remain unnoticeable for precise satellite altimetry due to the immense size of the lake.

Although the Antarctic ice sheet remained more or less stable and in its present configuration for at least the last 14 million years [*Kennett*, 1977; *Denton et al.*, 1993; *Huybrechts*, 1994], ice flow across the lake could also change drastically over less long time scales, such as glacial/inter-

glacial cycles. These involve (1) changes in ice thickness and ice surface elevation and (2) migration of ice divides and alteration in the grounded flow direction. Surface elevation variations of the order of +50 m during interglacials and −100 m during glacials may well have occurred [*Ritz et al.*, 2001; *Siegert*, 2005]. Such changes may induce a migration of the ice divide, which traverses at present Lake Vostok from east to west [*Pattyn et al.*, 2004], thereby changing the ice thickness distribution. Since a change in ice thickness of >50 m at either side of the lake is sufficient to reverse the surface gradient across the lake, such an event will change the direction and magnitude of the ice/water slope. In certain cases, this may lead to a large unstable situation that potentially leads to a large drainage of Subglacial Lake Vostok.

6. EVIDENCE OF FORMER SUBGLACIAL OUTBURSTS

Evidence from catastrophic drainage of glacial and subglacial lakes stems from geomorphological evidence. The Channeled Scablands, for instance, cover an area of approximately 40,000 km^2 in Washington State (United States of America). They are the result of a catastrophic drainage of Glacial Lake Missoula, commonly known as the Missoula floods, at the end of the last glaciation. The floods are witnessed by a large anastomosis of flood channels and recessional gorges [*Baker et al.*, 1987]. *Clarke et al.* [1984] computed the water discharge from the ice-dammed lake through the ice dam, based on a modified Nye theory to account for effects of lake temperature and reservoir geometry [*Clarke*, 1982]. Maximum discharge was estimated as 2.7–13.7 × 10^6 m^3 s^{-1} [*Clarke et al.*, 1984].

Another major flood stems from the catastrophic discharge of Lake Agassiz, a proglacial lake formed at the southern margin of the Laurentide ice sheet during deglaciation of the latter and drained through a subglacial system [*Clarke et al.*, 2004]. Flood hydrographs for floods that originate in subglacial drainage conduits were simulated using the Spring-Hutter theory [*Clarke*, 2003], leading to a flood magnitude of ~5 × 10^6 m^3 s^{-1} and a duration of half a year [*Clarke et al.*, 2004].

Spectacular meltwater features associated with subglacial outburst floods are also reported in the Transantarctic Mountains in southern Victoria Land, Antarctica. Since the Antarctic ice sheet has been relatively stable over the last 14 million years, they must predate Pleistocene glacial periods. Fortunately, and because of the long-term hyperarid polar climate, the outburst features are well preserved and suggest Miocene ice sheet overriding of the Transantarctic Mountains [*Denton and Sugden*, 2005]. These features consist of channels associated with areal scouring, scablands with scallops, potholes, and plunge holes cut in sandstone and dolerite [*Denton and Sugden*, 2005]. Channels systems and canyons of as much as 600 m wide and 250 m deep can, for instance, be witnessed in the northern part of Wright Valley (Figure 12). Some of the observed potholes are >35 m deep [*Lewis et al.*, 2006]. These features are consistent with incision from subglacial meltwater, and the estimate discharge is of the order of $1.6–2.2 \times 10^6 \, m^3 \, s^{-1}$ [*Lewis et al.*, 2006]. Their analysis also shows that the major channel incision predates 12.4 Ma, and that the last subglacial flood occurred sometime between 14.4 and 12.4 Ma ago. The source for such catastrophic discharge event should be a large subglacial lake. In view of the subglacial drainage flow paths (Figure 5) and the likelihood that the ice sheet was slightly bigger in size at that time, Subglacial Lake Vostok is a likely candidate. *Lewis et al.* [2006] advocate that the number and total volume of subglacial lakes beneath the East Antarctic ice sheet during the Middle Miocene could have been considerably greater than today due to the warmer basal conditions and larger ice sheet size. However, the present-day ice sheet has a predominant wet bed (reaching pressure-melting point) in the central parts where the ice is thickest, so that Miocene conditions could well be similar to present-day ones.

Besides the possible ice sheet instability due to the large quantity of subglacial water release, the huge freshwater discharge may also have an impact on deep-water formation in the Ross Embayment. Moreover, such type of subglacial floods could have formed a trigger for changes in middle Miocene climate [*Zachos et al.*, 2001; *Lewis et al.*, 2006].

Figure 12. View of the Labyrinth, Wright Valley, Antarctica. The channelized system is formed by subglacial water routing due to a major Miocene subglacial drainage. From Photo Library, U.S. Antarctic Program. Photograph by P. Rejcek, National Science Foundation (2007).

Subglacial meltwater channel systems have also been detected offshore on the continental shelf of West Antarctica in the western Amundsen Sea Embayment [*Lowe and Anderson*, 2002, 2003; *Smith et al.*, 2009b]. They lie in the alignment of large outlet glaciers and ice streams, such as Pine Island and Thwaites Glaciers, both characterized by an important present-day dynamic ice loss [*Wingham et al.*, 2006b; *Shepherd and Wingham*, 2007; *Rignot et al.*, 2008]. The offshore meltwater systems relate to periods when the ice sheet was expanded and/or exhibiting a different dynamic behavior. While *Lewis et al.* [2006] investigate the morphology of subglacial channel systems, *Smith et al.* [2009b] focus on the sediment infill of the channels, since they have the potential to reveal important information on channel genesis and drainage processes. The presence of deformed till, for instance, at one core site and the absence of typical meltwater deposits (such as sorted sands and gravels) at other cores suggest that the channel incision predates the overriding by fast flowing ice streams during the last glacial period [*Smith et al.*, 2009b]. The channels were, therefore, likely formed over multiple glaciations, possible since Miocene as well [*Smith et al.*, 2009b].

7. CONCLUSIONS AND OUTLOOK

Recent observations on rapid discharge of water from subglacial lakes, as well as modeling of subglacial water drainage and interactions with the ice sheet, confirm that rapid subglacial discharge is a common feature of the Antarctic ice sheet. Although discharge rates seem rather small compared to observed jökulhlaups of Vatnajökull, evidence of former catastrophic outbursts exists, and they probably predate the Pleistocene epoch. These events are of a similar order of magnitude as those witnessed in North America during the deglaciation of the Laurentide ice sheet (e.g., Missoula floods).

Theory of subglacial lake drainage, developed several decades ago, has been improved over recent years and has been applied to Antarctic subglacial lake discharges. It shows that sustained drainage between linked lakes over distances of several hundreds of kilometers is possible for conduits of tens of meters in diameter. Linking discharge events with viscoplastic ice sheet models exhibits a nonlinear response of the ice surface to sudden discharge events, hence, complicating the interpretation of satellite altimetry observations.

Future challenges therefore lie in a better understanding of the dynamics of Antarctic subglacial hydrology, its relation to overall ice sheet dynamics, and its possible influence on the general climate evolution. Evidence presented here supports the idea that subglacial outbursts are a major component of the ice dynamical system. Including subglacial discharge of lakes into large-scale numerical ice sheet models for Antarctic ice sheet evolution (paleo as well as future predictions) will become necessary.

The relation between ice dynamics, subglacial lake dynamics, and circulation is complicated and influences the mass balance at the interface between both systems. A first step in fully coupling both subglacial lake circulation and ice flow has been realized [*Thoma et al.*, 2010]. However, the processes involved are similar to those that govern any interaction between the ice sheet and a water body, such as grounding line migration and marine ice sheet stability. Therefore, subglacial lake discharge research will be crucial in future developments and improvements of cryospheric dynamics.

Acknowledgments. This paper forms a contribution to the Belgian Research Programme on the Antarctic (Belgian Federal Science Policy Office), Project nr. SD/CA/02 "Antarctic Subglacial Processes and Interactions: Role of transition zones in ice sheet stability (ASPI)" and the ARC research project "IceCube-Dyn."

REFERENCES

Anandakrishnan, S., and R. B. Alley (1997), Stagnation of Ice Stream C, West Antarctica by water piracy, *Geophys. Res. Lett.*, *24*(3), 265–268.

Baker, V. R., R. Greeley, P. D. Komar, D. A. Swanson, and R. B. Waitt (1987), Columbia and Snake River plains, in *Geomorphic Systems of North America*, edited by W. L. Graf, *Geol. Soc. Am. Centennial Spec.*, vol. 2, pp. 403–468, Geol. Soc. of Am., Boulder, Colo.

Bamber, J. L., J. L. Gomez-Dans, and J. A. Griggs (2009), A new 1 km digital elevation model of the antarctic derived from combined satellite radar and laser data — part 1: Data and methods, *The Cryosphere*, *3*(1), 101–111.

Bell, R. E., M. Studinger, A. A. Tikku, G. C. K. Clarke, M. M. Gutner, and C. Meertens (2002), Origin and fate of Lake Vostok water frozen to the base of the East Antarctic ice sheet, *Nature*, *416*, 307–310.

Bell, R. E., M. Studinger, M. A. Fahnestock, and C. A. Shuman (2006), Tectonically controlled subglacial lakes on the flanks of the Gamburtsev Subglacial Mountains, East Antarctica, *Geophys. Res. Lett.*, *33*, L02504, doi:10.1029/2005GL025207.

Bell, R. E., M. Studinger, C. A. S. M. A. Fahnestock, and I. Joughin (2007), Large subglacial lakes in East Antarctica at the onset of fast-flowing ice streams, *Nature*, *445*, 904–907, doi:10.1038/nature05554.

Björnsson, H. (1988), *Hydrology of Ice Caps in Volcanic Regions*, Soc. Sci. Islandica, Univ. of Iceland, Reykjavik, Iceland.

Björnsson, H. (1998), Hydrological characteristics of the drainage system beneath a surging glacier, *Nature*, *395*, 771–774.

Björnsson, H. (2002), Subglacial lakes and jökulhlaups in Iceland, *Global Planet. Change*, *35*(3–4), 255–271.

Blatter, H. (1995), Velocity and stress fields in grounded glaciers: A simple algorithm for including deviatoric stress gradients, *J. Glaciol.*, *41*(138), 333–344.

Carter, S. P., D. D. Blankenship, M. F. Peters, D. A. Young, J. W. Holt, and D. L. Morse (2007), Radar-based subglacial lake classification in Antarctica, *Geochem. Geophys. Geosyst.*, *8*, Q03016, doi:10.1029/2006GC001408.

Carter, S. P., D. D. Blankenship, D. A. Young, M. F. Peters, J. W. Holt, and M. J. Siegert (2009), Dynamic distributed drainage implied by the flow evolution of the 1996–1998 Adventure Trench subglacial lake drainage, *Earth Planet. Sci. Lett.*, *283*, 24–37.

Clarke, G. K. C. (1982), Glacier outburst floods from "Hazard Lake", Yukon Territory, and the problem of flood magnitude prediction, *J. Glaciol.*, *28*(98), 3–21.

Clarke, G. K. C. (1996), Lumped-element analysis of subglacial hydraulic circuits, *J. Geophys. Res.*, *101*, 17,547–17,599.

Clarke, G. K. C. (2003), Hydraulics of subglacial outburst floods: New insights from the Spring–Hutter formulation, *J. Glaciol.*, *49*(165), 299–313.

Clarke, G. K. C. (2005), Subglacial processes, *Annu. Rev. Earth Planet. Sci.*, *33*, 247–276.

Clarke, G. K. C., W. H. Mathews, and R. T. Pack (1984), Outburst floods from glacial Lake Missoula, *Quat. Res.*, *22*(3), 289–299.

Clarke, G. K. C., D. W. Leverington, J. T. Teller, and A. S. Dyke (2004), Paleohydraulics of the last outburst flood from glacial Lake Agassiz and the 8200 BP cold event, *Quat. Sci. Rev.*, *23*, 389–407.

Creyts, T. T., and C. G. Schoof (2009), Drainage through subglacial water sheets, *J. Geophys. Res.*, *114*, F04008, doi:10.1029/2008JF001215.

Denton, G. H., and D. E. Sugden (2005), Meltwater features that suggest Miocene ice-sheet overriding of the Transantarctic Mountains in Victoria Land, Antarctica, *Geogr. Ann. Ser. A*, *87*(1), 67–85.

Denton, G. H., D. E. Sugden, D. R. Marchant, B. L. Hall, and T. I. Wilch (1993), East Antarctic ice sheet sensitivity to Pliocene climatic change from a Dry Valleys perspective, *Geogr. Ann. Ser. A*, *75*(4), 155–204.

Erlingsson, U. (2006), Lake Vostok behaves like a 'captured lake' and may be near to creating an Antarctic jökulhlaup, *Geogr. Ann., Ser. A*, *88*(1), 1–7.

Evatt, G. W., and A. C. Fowler (2007), Cauldron subsidence and subglacial floods, *Ann. Glaciol.*, *45*, 163–168.

Evatt, G. W., A. C. Fowler, C. D. Clark, and N. R. J. Hulton (2006), Subglacial floods beneath ice sheets, *Philos. Trans. R. Soc. A*, *364*, 1769–1794.

Fowler, A. C. (2009), Dynamics of subglacial floods, *Proc. R. Soc. A*, *465*, 1809–1828.

Fricker, H. A., and T. Scambos (2009), Connected subglacial lake activity on lower Mercer and Whillans ice streams, West Antarctica, 2003–2008, *J. Glaciol.*, *55*(190), 303–315.

Fricker, H. A., T. Scambos, R. Bindschadler, and L. Padman (2007), An active subglacial water system in West Antarctica mapped from space, *Science*, *315*, (1544), doi:10.1126/science.1136897.

Goodwin, I. D. (1988), The nature and origin of a jökulhlaup near Casey Station, Antarctica, *J. Glaciol.*, *34*(116), 95–101.

Gray, L., I. Joughin, S. Tulaczyk, V. B. Spikes, R. Bindschadler, and K. Jezek (2005), Evidence for subglacial water transport in the West Antarctic ice sheet through three-dimensional satellite radar interferometry, *Geophys. Res. Lett.*, *32*, L03501, doi:10.1029/2004GL021387.

Gudmundsson, G. H. (2003), Transmission of basal variability to a glacier surface, *J. Geophys. Res.*, *108*(B5), 2253, doi:10.1029/2002JB002107.

Hindmarsh, R. C. A. (2004), A numerical comparison of approximations to the Stokes equations used in ice sheet and glacier modeling, *J. Geophys. Res.*, *109*, F01012, doi:10.1029/2003JF000065.

Hughes, L. (2001), *The Collected Works of Langston Hughes*, vol. 1, *The Poems: 1921–1940*, Univ. of Mo. Press, Columbia.

Huybrechts, P. (1994), Formation and desintegration of the Antarctic ice sheet, *Ann. Glaciol.*, *20*, 336–340.

Joughin, I. (2002), Ice-sheet velocity mapping: A combined interferometric and speckle-tracking approach, *Ann. Glaciol.*, *34*, 195–201.

Kamb, B. (1987), Glacier surge mechanism based on linked cavity configuration of the basal water conduit system, *J. Geophys. Res.*, *92*, 9083–9100.

Kapitsa, A., J. K. Ridley, G. Robin, M. J. Siegert, and I. Zotikov (1996), A large deep freshwater lake beneath the ice of central East Antarctica, *Nature*, *381*, 684–686.

Kennett, J. P. (1977), Cainozoic evolution of Antarctic glaciation, the circum-Antarctic ocean, and their impact on global paleoceanography, *J. Geophys. Res.*, *82*, 3843–3860.

Kwok, R., M. J. Siegert, and F. D. Carsey (2000), Ice motion over Lake Vostok, Antarctica: Constraints on inferences regarding the accreted ice, *J. Glaciol.*, *46*(155), 689–694.

Lewis, A. R., D. R. Marchant, D. E. Kowalewski, S. L. Baldwin, and L. E. Webb (2006), The age and origin of the Labyrinth, western Dry Valleys, Antarctica: Evidence for extensive middle Miocene subglacial floods and freshwater discharge to the Southern Ocean, *Geology*, *34*(7), 513–516.

Lowe, A. L., and J. B. Anderson (2002), Reconstruction of the West Antarctic ice sheet in Pine Island Bay during the Last Glacial Maximum and its subsequent retreat history, *Quat. Sci. Rev.*, *21* (16–17), 1879–1897.

Lowe, A. L., and J. B. Anderson (2003), Evidence for abundant subglacial meltwater beneath the paleo-ice sheet in Pine Island Bay, antarctica, *J. Glaciol.*, *49*, 125–138.

Lythe, M. B., and D. G. Vaughan (2001), BEDMAP: A new ice thickness and subglacial topographic model of Antarctica, *J. Geophys. Res.*, *106*(B6), 11,335–11,351.

MacAyeal, D. R. (1989), Large-scale ice flow over a viscous basal sediment: Theory and application to Ice Stream B, Antarctica, *J. Geophys. Res.*, *94*(B4), 4071–4087.

Magnusson, E., H. Rott, H. Björnsson, and F. Palsson (2007), The impact of jökulhlaups on basal sliding observed by SAR interferometry on Vatnajökull, Iceland, *J. Glaciol.*, *53*(181), 232–240.

Martin, C., F. Navarro, J. Otero, M. L. Cuadrado, and M. Corcuera (2003), Three-dimensional modelling of the dynamics of Johnson Glacier (Livingstone Island, Antarctica), *Ann. Glaciol.*, *39*, 1–8.

Nye, J. F. (1976), Water flow in glaciers: Jökulhlaups, tunnels and veins, *J. Glaciol.*, *17*(76), 181–207.

Pattyn, F. (2003), A new three-dimensional higher-order thermomechanical ice sheet model: Basic sensitivity, ice stream development, and ice flow across subglacial lakes, *J. Geophys. Res.*, *108*(B8), 2382, doi:10.1029/2002JB002329.

Pattyn, F. (2004), Comment on the comment by M. J. Siegert on "A numerical model for an alternative origin of Lake Vostok and its exobiological implications for Mars" by N. S. Duxbury et al., *J. Geophys. Res.*, *109*, E11004, doi:10.1029/2004JE002329.

Pattyn, F. (2008), Investigating the stability of subglacial lakes with a full Stokes ice-sheet model, *J. Glaciol.*, *54*(185), 353–361.

Pattyn, F. (2010), Antarctic subglacial conditions inferred from a hybrid ice sheet/ice stream model, *Earth Planet. Sci. Lett.*, *295*, 451–461.

Pattyn, F., B. De Smedt, and R. Souchez (2004), Influence of subglacial Lake Vostok on the regional ice dynamics of the Antarctic ice sheet: a model study, *J. Glaciol.*, *50*(171), 583–589.

Peters, M. E., D. D. Blankenship, and D. L. Morse (2005), Analysis techniques for coherent airborne radar sounding: Application to West Antarctic ice streams, *J. Geophys. Res.*, *110*, B06303, doi:10.1029/2004JB003222.

Peters, N. J., I. C. Willis, and N. S. Arnold (2009), Numerical analysis of rapid water transfer beneath Antarctica, *J. Glaciol.*, *55*(192), 640–650.

Popov, S. V., and V. N. Masolov (2007), Forty-seven new subglacial lakes in the 0 110° sector of East Antarctica, *J. Glaciol.*, *53*(181), 289–297.

Rignot, E. J., J. L. Bamber, M. R. van den Broeke, C. Davis, Y. Li, W. J. van de Berg, and E. van Meijgaard (2008), Recent Antarctic ice mass loss from radar interferometry and regional climate modelling, *Nat. Geosci.*, *1*, 106–110.

Ritz, C., V. Rommelaere, and C. Dumas (2001), Modeling the evolution of the Antarctic ice sheet over the last 420000 years: Implications for altitude changes in the Vostok region, *J. Geophys. Res.*, *106*(D23), 31,943–31,964.

Roberts, M. J. (2005), Jökulhlaups: A reassessment of floodwater flow through glaciers, *Rev. Geophys.*, *43*, RG1002, doi:10.1029/2003RG000147.

Röthlisberger, H. (1972), Water pressure in intra- and subglacial channels, *J. Glaciol.*, *11*(62), 177–203.

SALE Workshop Report (2007), Subglacial Antarctic Lake Environments (SALE) in the International Polar Year 2007–2008, in *Advanced Science and Technology Planning Workshop, Grenoble France, 24–26 April 2006*, 47 pp.

Sergienko, O. V., D. R. MacAyeal, and R. A. Bindschadler (2007), Causes of sudden, short-term changes in ice-stream surface elevation, *Geophys. Res. Lett.*, *34*, L22503, doi:10.1029/2007GL031775.

Shapiro, N. M., and M. H. Ritzwoller (2004), Inferring surface heat flux distributions guided by a global seismic model: Particular application to Antarctica, *Earth Planet. Sci. Lett.*, *223*, 213–224.

Shepherd, A., and D. Wingham (2007), Recent sea-level contributions of the Antarctic and Greenland ice sheets, *Science*, *315*, 1529–1532.

Shreve, R. L. (1972), Movement of water in glaciers, *J. Glaciol.*, *11*(62), 205–214.

Siegert, M. J. (2000), Antarctic subglacial lakes, *Earth Sci. Rev.*, *50*, 29–50.

Siegert, M. J. (2005), Reviewing the origin of subglacial Lake Vostok and its sensitivity to ice sheet changes, *Prog. Phys. Geogr.*, *29*(2), 156–170.

Siegert, M. J., and J. L. Bamber (2000), Subglacial water at the heads of Antarctic ice-stream tributaries, *J. Glaciol.*, *46*(155), 702–703.

Siegert, M. J., and J. K. Ridley (1998), An analysis of the ice-sheet surface and subsurface topography above the Vostok Station subglacial lake, central East Antarctica, *J. Geophys. Res.*, *103* (B5), 10,195–10,207.

Siegert, M. J., R. Kwok, C. Mayer, and B. Hubbard (2000), Water exchange between the subglacial Lake Vostok and the overlying ice sheet, *Nature*, *403*, 643–646.

Siegert, M. J., J. C. Ellis-Evans, M. Tranter, C. Mayer, J. R. Petit, A. Salamatin, and J. C. Priscu (2001), Physical, chemical and biological processes in Lake Vostok and other Antarctic subglacial lakes, *Nature*, *404*, 603–609.

Siegert, M. J., B. Welch, D. Morse, A. Vieli, D. D. Blankenship, I. Joughin, E. C. King, G. J.-M. C. Leysinger Vieli, A. J. Payne, and R. Jacobel (2004), Ice flow direction change in interior West Antarctica, *Science*, *305*, 1948–1951.

Siegert, M. J., S. Carter, I. Tobacco, S. Popov, and D. Blankenship (2005), A revised inventory of Antarctic subglacial lakes, *Antarct. Sci.*, *17*(3), 453–460.

Siegert, M. J., A. Le Brocq, and A. J. Payne (2007), Hydrological connections between Antarctic subglacial lakes, the flow of water beneath the East Antarctic ice sheet and implications of sedimentary processes, in *Glacial Sedimentary Processes and Products*, edited by M. J. Hambrey et al., *Spec. Publ. Int. Assoc. Sedimentol.*, *39*, 3–10.

Smith, B. E., H. A. Fricker, I. R. Joughin, and S. Tulaczyk (2009a), An inventory of active subglacial lakes in Antarctica detected by ICESat (2003–2008), *J. Glaciol.*, *55*(192), 573–595.

Smith, J. A., C. D. Hillenbrand, R. D. Larter, A. G. C. Graham, and G. Kuhn (2009b), The sediment infill of subglacial meltwater channels on the West Antarctic continental shelf, *Quat. Res.*, *71*, 190–200.

Spring, U., and K. Hutter (1982), Conduit flow of a fluid through its solid phase and its application to intraglacial channel flow, *Int. J. Eng. Sci.*, *20*(2), 327–363.

Stearns, L. A., B. A. Smith, and G. S. Hamilton (2008), Increased flow speed on a large East Antarctic outlet glacier caused by subglacial floods, *Nat. Geosci.*, *1*, 827–831.

Studinger, M. et al. (2003), Ice cover, landscape setting, and geological framework of Lake Vostok, East Antarctica, *Earth Planet. Sci. Lett.*, *205*, 195–210.

Studinger, M., R. E. Bell, and A. A. Tikku (2004), Estimating the depth and shape of subglacial Lake Vostok's water cavity from aerogravity data, *Geophys. Res. Lett.*, *31*, L12401, doi:10.1029/2004GL019801.

Thoma, M., K. Grosfeld, and C. Mayer (2007), Modelling mixing and circulation in subglacial Lake Vostok, antarctica, *Ocean Dyn.*, *57*, 531–540.

Thoma, M., K. Grosfeld, C. Mayer, and F. Pattyn (2010), Interaction between ice sheet dynamics and subglacial lake circulation: A coupled modelling approach, *The Cryosphere*, *4*(1), 1–12.

Tikku, A. A., R. E. Bell, M. Studinger, and G. K. C. Clarke (2004), Ice flow field over Lake Vostok, East Antarctica inferred by structure tracking, *Earth Planet. Sci. Lett.*, *227*(3–4), 249–261.

Vaughan, D. G., H. F. Corr, A. M. Smith, H. D. Pritchard, and A. Shepherd (2008), Flow-switching and water piracy between Rutford Ice Stream and Carlson Inlet, West Antarctica, *J. Glaciol.*, *54*(184), 41–48.

Walder, J. S. (1982), Stability of sheet flow of water beneath temperate glaciers and implications for glacier surging, *J. Glaciol.*, *28*, 273–293.

Walder, J. S. (1986), Hydraulics of subglacial cavities, *J. Glaciol.*, *32*, 439–445.

Walder, J. S., and A. Fowler (1994), Channelized subglacial drainage over a deformable bed, *J. Glaciol.*, *40*(134), 3–15.

Weertman, J. (1972), General theory of water flow at the base of a glacier or ice sheet, *Rev. Geophys.*, *10*, 287–333.

Weertman, J., and G. E. Birchfield (1983), Basal water film, basal water pressure, and velocity of travelling waves on glaciers, *J. Glaciol.*, *29*, 20–27.

Wingham, D. J., M. J. Siegert, A. P. Shepherd, and A. S. Muir (2006a), Rapid discharge connects Antarctic subglacial lakes, *Nature*, *440*, 1033–1036.

Wingham, D. J., A. Shepherd, A. Muir, and G. J. Marshall (2006b), Mass balance of the Antarctic ice sheet, *Philos. Trans. R. Soc. A*, *364*, 1627–1635.

Wright, A. P., M. J. Siegert, A. M. Le Brocq, and D. B. Gore (2008), High sensitivity of subglacial hydrological pathways in Antarctica to small ice-sheet changes, *Geophys. Res. Lett.*, *35*, L17504, doi:10.1029/2008GL034937.

Zachos, J. C., M. Pegani, L. Stone, E. Thomas, and K. Billups (2001), Trends, rhythms, and aberrations in global climates 65 Ma to present, *Science*, *292*, 686–693.

Zwinger, T., R. Greve, O. Gagliardini, T. Shiraiwa, and M. Lyly (2007), A Full Stokes-flow thermo-mechanical model for firn and ice applied to the Gorshkov crater glacier, Kamchatka, *Ann. Glaciol.*, *45*, 29–37.

F. Pattyn, Laboratoire de Glaciologie, Département des Sciences de la Terre et de l'Environnement, Université Libre de Bruxelles, Av. F.D. Roosevelt 50, B-1050 Brussels, Belgium. (fpattyn@ulb.ac.be)

Vostok Subglacial Lake: A Review of Geophysical Data Regarding Its Discovery and Topographic Setting

Martin J. Siegert

School of GeoSciences, University of Edinburgh, Edinburgh, UK

Sergey Popov

Polar Marine Geosurvey Expedition, St. Petersburg, Russia

Michael Studinger[1]

Lamont-Doherty Earth Observatory of Columbia University, Palisades, New York, USA

Vostok Subglacial Lake is the largest and best known sub-ice lake in Antarctica. The establishment of its water depth (>500 m) led to an appreciation that such environments may be habitats for life and could contain ancient records of ice sheet change, which catalyzed plans for exploration and research. Here we discuss geophysical data used to identify the lake and the likely physical, chemical, and biological processes that occur in it. The lake is more than 250 km long and around 80 km wide in one place. It lies beneath 4.2 to 3.7 km of ice and exists because background levels of geothermal heating are sufficient to warm the ice base to the pressure melting value. Seismic and gravity measurements show the lake has two distinct basins. The Vostok ice core extracted >200 m of ice accreted from the lake to the ice sheet base. Analysis of this ice has given valuable insights into the lake's biological and chemical setting. The inclination of the ice-water interface leads to differential basal melting in the north versus freezing in the south, which excites circulation and potential mixing of the water. The exact nature of circulation depends on hydrochemical properties, which are not known at this stage. The age of the subglacial lake is likely to be as old as the ice sheet (~14 Ma). The age of the water within the lake will be related to the age of the ice melting into it and the level of mixing. Rough estimates put that combined age as ~1 Ma.

1. INTRODUCTION

The concept of liquid water beneath the ice sheets of Antarctica is, to those unfamiliar with glacial processes, somewhat incongruous [*Siegert*, 2005]. The surface air temperatures in central East Antarctica often reach below −60°C, and the coldest official temperature ever recorded on Earth, −89.2°C (−128.6°F), occurred at Vostok Station on 21 July 1983. Yet, a little less than 4 km below the ice

---

[1]Now at Cryospheric Sciences Branch, Goddard Earth Sciences and Technology Center, NASA Goddard Space Flight Center, Greenbelt, Maryland, USA.

Antarctic Subglacial Aquatic Environments
Geophysical Monograph Series 192
Copyright 2011 by the American Geophysical Union
10.1029/2010GM000934

45

surface at this Russian base, a huge body of water named Vostok Subglacial Lake exists.

Lakes beneath the Antarctic Ice Sheet were first reported from airborne radio echo sounding (RES) records in the late 1960s and early 1970s [*Robin et al.*, 1970; *Oswald and Robin*, 1973], and Vostok Subglacial Lake was first measured using this technique by *Robin et al.* [1977] on Christmas Eve in 1974. This lake is the largest of >380 known lakes that lay under the East and West Antarctic Ice Sheets [*Wright and Siegert*, this volume].

Temperatures can attain the melting value beneath an ice sheet because of three factors. First, the pressure beneath an ice sheet (i.e., the weight of ice) causes a reduction in the melting point, which beneath 4 km of ice is around −3.2°C. Second, the ice sheet insulates the base from the ultracold temperatures at the surface. Third, heat is transmitted to the base from deep within the lithosphere (geothermal heat). For an ice sheet 4 km thick in central East Antarctica, the heat required to melt basal ice is about 50 mW m^{-2}, which is roughly the background geothermal value [*Siegert and Dowdeswell*, 1996]. Thus, subglacial water, and lakes, can occur beneath the center of a large ice sheet without the need for unusual glacial or geothermal conditions.

Water flow beneath an ice sheet is controlled by the hydraulic potential (a combination of gravity and ice overburden). In simple terms, water may flow "uphill" if the slope of the ice surface exceeds about one tenth of the opposing slope at the ice sheet base. In other cases, subglacial water simply flows downhill. The production and flow of water at the ice sheet bed, through what glaciologists expect to be an organized subglacial drainage network [*Siegert et al.*, 2007], lead to its accumulation within topographic hollows and hence the formation of subglacial lakes.

There has been a huge level of scientific and media interest in Vostok Subglacial Lake (and subglacial lakes in general) following the discovery that the water depth of the lake was more than 500 m [*Kapitsa et al.*, 1996]. Discussion about whether to make in situ measurements of the lake has been driven by two scientific hypotheses. The first is that unique microorganisms inhabit the lake. The second is that a complete record of ice sheet history is available from the sediments that lie across the lake floor. Future exploration of Vostok Subglacial Lake, and other subglacial lakes, will be focused on testing these hypotheses [*Lukin and Bulat*, this volume; *Fricker et al.*, this volume; *Ross et al.*, this volume]. If the hypotheses are correct, future investigations of subglacial lakes could enable valuable insights into the history of Antarctica, detailing its response to and control on climate change, and our understanding of biological functioning within extreme environments.

This chapter presents an overview of geophysical and glaciological investigations of Vostok subglacial lake as an introduction to the lake and the processes taking place within it and other Antarctic subglacial lake environments.

2. DISCOVERY OF VOSTOK SUBGLACIAL LAKE, 1960s–1990s

The history of the discovery of Vostok Subglacial Lake spans 30 years between the 1960s and 1990s, involving scientists and research organizations from the United Kingdom, Russia, mainland Europe, and the United States (see *Zotikov* [2006]). The first person to mention lakes within central Antarctica was R. V. Robinson, the senior navigator of the fourth Soviet Antarctic Expedition, in 1959. *Robinson* [1961] explained, in a recorded radio broadcast from Mirny Station, that navigation of intracontinental flights utilized various ice-surface features, including discrete flat surfaces that appeared as "lakes" with apparent "shorelines." Unwittingly, Robinson may have been reporting the ice-surface manifestation of a subglacial lake several kilometers below. Subglacial lakes, especially large ones, are associated with an extremely flat and smooth ice sheet surface above them. This is because there is effectively zero friction between the ice sheet and the lake. However, a significant amount of friction occurs between ice and bedrock outside the lake margin. Such friction will lead to surface roughness and will be related to a noticeable slope of the ice sheet. Therefore, areas of low basal friction, such as Vostok Subglacial Lake, can be effectively "mapped" through detailed surveying of the ice surface. Of course, Robinson did not know this in 1961, and so no attempt was made to establish a relationship between Vostok Subglacial Lake and the ice-surface geomorphology until the 1990s and the advent of accurate satellite radar altimetry.

Russian and British scientists debated whether water might exist beneath the surface of the Antarctic Ice Sheet, using quite simple thermodynamic considerations [e.g., *Robin*, 1955; *Zotikov*, 1963]. While they predicted that basal water production in central regions of East Antarctica was likely, no specific mention of subglacial lakes was made.

In 1964, a team of Russian scientists, led by Andrei Kapitsa (Moscow State University), performed a seismic experiment on the ice surface over what is now known to be Vostok Subglacial Lake. Seismic data are useful in the examination of subglacial lakes because the compressional sound wave (P wave) will propagate as well through water as ice and reflect off boundaries where there is a significant density contrast. Records of the "two-way travel time" of the first-arrival P wave, multiplied by the velocity of sound waves in the propagating medium, yield the distance traveled. Thus, seismic sounding can provide information on

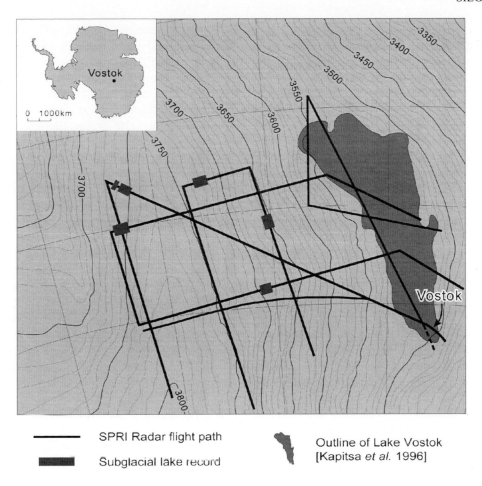

Figure 1. ERS-1 ice sheet surface elevation over, and upstream, of Vostok Subglacial Lake, with 1970s radio echo sounding (RES) transects illustrated, showing evidence of the lake itself, and a series of small lakes in the Ridge B locale.

(1) the ice thickness, (2) the depth of water within the lake, and (3) the thickness of any sediment at the base of the lake. In 1964, Kapitsa's team was interested in recording the thickness of ice and not aware of the lake, so the data collected from the lake itself were not interpreted until much later. After the field work, these seismic data were stored in Dr. Kapitsa's garden shed in Moscow, luckily surviving a fire that subsequently destroyed it. The tightly rolled seismic paper resulted in only the edges of the data being charred, a feature that is clearly visible in the original record.

A problem for using seismic techniques to establish the ice thickness around Antarctica is that the upper ~100 m of the ice sheet consists of snow that is not yet compressed enough to be classified as glacier ice. Sound waves do not travel well through snow, which acts like a sound-proofing blanket, and one has to drill down to at least 40 m into the dense glacier ice, where the seismic explosion and seismometers can be set. Thus, seismic investigations of ice sheets take a long time to perform and are not well suited to surveying at a continental scale.

During the 1970s, a United Kingdom-United States-Danish consortium led by Gordon Robin (Scott Polar Research Institute in Cambridge) performed a systematic airborne radar sounding survey of the ice sheet base. Radar works by the emission of a very high frequency VHF radio wave that is transmitted down into the ice sheet and is reflected off layers of contrasting electrical properties. Such boundaries are found between air and ice at the surface, internal ice sheet layers, and between ice and bedrock or water at the base of the ice sheet. As in seismic sounding, radar can be used to measure the thickness of ice by multiplying the two-way travel time by the velocity of radio waves in ice. By stacking the reflections from individual pulses of VHF radio wave reflections in a time-dependent manner as one traverses across the ice sheet, a pseudo-cross-section can be recorded. Since the equipment is mounted on an aircraft, information about the ice base is recorded along a flight line at a speed of ~ 300 km h^{-1}. Thus, airborne radar is a very efficient means of surveying the Antarctic ice base.

The radar reflections off a subglacial lake surface are easily distinguishable from those of the ice-bedrock interface [see *Wright and Siegert*, this volume]. Specifically, subglacial lake radar records are characterized by (1) very bright radar returns due to the high reflectivity of radio waves at an ice-water interface, (2) extremely constant reflected radar signal strength over horizontal distances (because the lake surface is very smooth, and therefore, the radio waves are not subject to scattering associated with a rough surface), and (3) a virtually straight and horizontal ice-water interface observed in the radar-derived cross-section. In contrast, ice-bedrock interfaces yield generally weak, variable radar reflections, which are observed in cross section to undulate beneath the ice sheet. Details of a number of small Antarctic subglacial lakes were first reported from analysis of airborne radar in 1973 by Oswald and Robin. Soon after, a large expanse of water was noticed in radar data close to Vostok Station and the existence of Vostok Subglacial Lake discovered [*Robin et al.*, 1977]. Scientific interest in Vostok Subglacial Lake did not follow its discovery, however. In fact, evidence of Vostok Subglacial Lake was lost by many within the glaciological literature for over a decade. For example, the Vostok ice core's depth-age chronology was first modeled assuming the ice rested on the bed, rather than in water; an issue that was corrected only in the late 1990s [*Petit et al.*, 1999].

Accurate mapping of the ice sheet surface through the altimeter of the European Space Agency's ERS-1 was undertaken in the early 1990s. Satellite altimetry provides very accurate measurements of the Earth's surface and is therefore useful for mapping the flat ice sheet regions that occur above subglacial lakes. *Ridley et al.* [1993] were able to use such data to establish the general shape of a very flat region above Vostok Subglacial Lake. The margins of the flat ice-surface corresponded extremely well with the edge of the lake identified from radar measurements. Ridley and other's investigation of satellite altimetry of the ice surface led to a reexamination of Kapitsa's seismic data at Vostok Station, the studying of Robin's radar data across Vostok Subglacial Lake, and, following this, an international conference on Vostok Subglacial Lake to discuss these data sets (held in Cambridge 1994), which in turn led to a seminal publication by *Kapitsa et al.* [1996]. From these investigations, the lake was shown to be 230 km in length, 50 km wide, and about 14,000 km^2 in surface area (Figure 1). Vostok Station was shown to lie over the southern extreme of the lake where the depth of the water is 510 m.

The water within Vostok Subglacial Lake has been shown to be extremely fresh. VHF radio wave penetration through even 10 m of water, as demonstrated by *Gorman and Siegert* [1999] in the shallow northern end of the lake, can be achieved only if the conductivity of the water is unusually low (i.e., very little salt content). Further, the manner in which the ice sheet floats in the lake water is also indicative of a water density consistent with fresh water (1000 kg m^{-3}) compared with salty water (density of sea water is 1025 kg m^{-3}) [*Kapitsa et al.*, 1996]. Small levels of salinity, especially at depth, cannot be ruled out from analysis of the radar data, however.

The age of Vostok Subglacial Lake can be thought about in three ways: the age of the subglacial lake itself, the age of any (preglacial) lake at the same site, and the age of the water within the lake. The age of the subglacial lake is likely to be the same as the age of the ice sheet, which could be as much as 14 million years [*Sugden*, 1992]. It is conceivable that a preglacial lake, occupying the same basin, originates from well before this time. The age of the lake water will be related to the age of the ice that melts into it. The base of the Antarctic ice sheet at Vostok Station is around 700,000 years old [*Petit et al.*, 1997]. Therefore, the age of water within the lake cannot be less than 700,000 years old. The mean age of the lake water will be controlled by how efficient the lake system is at replacing old water with new water. Several processes may occur that are important to this issue. First is the melting rate of basal ice into the lake. Second is the drainage of water out of the lake. Third is the rate of ice freezing to the base of the ice sheet from the lake water. Fourth is the flow of water from upstream regions into the lake. None of these processes, bar the location and degree of basal refreezing, are known with any certainty, however. We are left to estimate the minimum age of the lake, based on the calculated age of the ice sheet base, to be around 1 Ma [*Siegert et al.*, 2001].

3. RECENT GEOPHYSICAL CAMPAIGNS

In recent years, concerted efforts have been made to map the size and extent of Vostok Subglacial Lake. The results (discussed in Section 4) have allowed an enhanced appreciation of the lake's physiography and the physical, biological, and chemical processes that are likely to take place within it.

3.1. Italian Airborne Geophysics Campaign 1999–2000

In response to the enhanced scientific interest in Vostok Subglacial Lake generated by the discovery of its water depth [*Kapitsa et al.*, 1996], Italian geophysicists undertook the first radar sounding measurements of the lake and its surrounding since the Scott Polar Research Institute RES measurements in the 1970s [*Tabacco et al.*, 2002]. In the Austral summer of 1999–2000, 12 new RES transects were

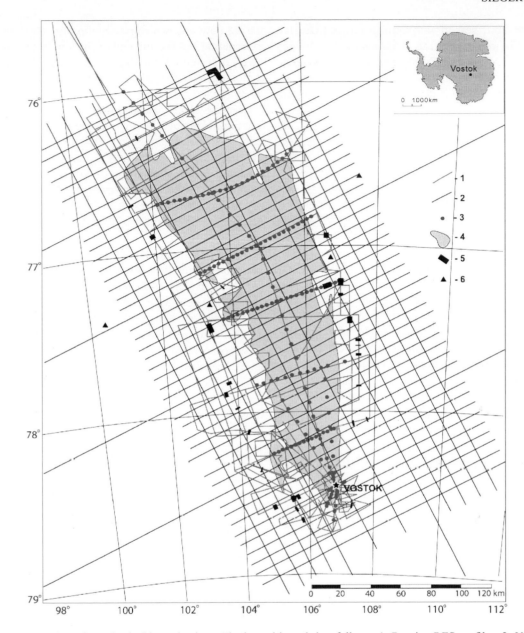

Figure 2. Location of geophysical investigations. The legend is coded as follows: 1, Russian RES profiles; 2, US RES flights; 3, Russian reflection seismic shots; 4, outline of the lake, based on Russian RES data; 5, subglacial lakes as given by *Siegert et al.* [2005]; 6, subglacial lakes as given by *Popov and Masolov* [2007].

collected over the lake, including one continuous flight across the long axis of the lake. The data were used to confirm the findings of *Kapitsa et al.* [1996] and to better define the aerial extent of the lake.

3.2. US Airborne Geophysics Campaign, 2000–2001

In order to further understand the physiographic setting, geological framework, and ice dynamics of the Vostok Sub-

glacial Lake area, scientists from the U.S. Antarctic Program carried out an aerogeophysical survey during the 2000–2001 Antarctic summer (Figure 2). The survey was designed to simultaneously acquire laser altimeter, ice-penetrating radar, gravity, and magnetic data and was complemented by ground geophysical measurements of ice-surface velocities and seismic monitoring. More than 20,000 line-km of aerogeophysical data were acquired by the U.S. National Science Foundation's Support Office for Aerogeophysical Research during the 2000–2001 season with an instrumented De Havilland DHC-6 Twin Otter aircraft [*Studinger et al.*, 2003a; *Richter et al.*, 2001; *Holt et al.*, 2006]. The main grid covered an area of 157.5 × 330 km and was augmented by 12 regional lines, extending outside of the main grid between 180 and 440 km.

3.3. Russian Over-Snow-Based Geophysical Investigations, 1995–2008

Russian researchers began systematic investigation of Vostok Subglacial Lake almost immediately after its outline had been discovered [*Ridley et al.*, 1993]. During the 1995/1996 field season, the Polar Marine Geosurvey Expedition began to study the Vostok Subglacial Lake area by conducting reflection seismic sounding within the framework and under the auspices of the Russian Antarctic Expedition. Since 1998, this work has continued together with ground-based RES [*Masolov et al.*, 2001, 2006]. In 2008, an important stage of the Russian investigations was completed; seismic and radio echo studies aimed at mapping the lake bottom and its environment were finished. The resulting data set (Figure 2) was used to determine an appreciation of the bedrock landscape in the Vostok Subglacial Lake area. In total, 318 seismic reflection soundings were carried out, and 5190 km of RES were acquired.

4. GEOPHYSICS RESULTS

4.1. Ice-Penetrating Radar

The ice thickness above Vostok Subglacial Lake has been determined with unprecedented accuracy by combining radar data in the two major recent surveys (565,735 and 710,448 radar data points from the Russian and U.S. programs, respectively; Plate 1). By integrating information on ice thickness with that of surface elevation, an enhanced map of the bed elevation was also established. The result of this data amalgamation is the most complete depiction of the surface of Vostok Subglacial Lake and the surrounding topography, to date.

The extent of the lake is 15,500 km^2 (excluding 70 km^2 of "islands"), and the elevation of the lake surface varies between 800 m below sea level (bsl) in the south and 200 m bsl in the north. The coastline of the lake is 1030 km long and is complicated by numerous bays and peninsulas. Importantly, two islands were found in the south-western part of the lake. One is situated in 8 km south-westward from Vostok Station. Its size is about 15 × 3 km. The other one is situated 48 km away from the station and is about 11 × 4 km. The latter island deserves special attention as it is located directly on

the ice flow line, which passes through borehole 5G-1. One explanation for the mineral inclusions found in the lower levels of the core is that they were captured by the glacier as it flowed over the island [*Jouzel et al.*, 1999] (see also Section 5.4).

Over the lake, the ice is thicker in the north (up to 4300 m) and thins to about 3700 m in the south. The 600 m change in thickness of the floating ice is associated with a 60 m change in ice-surface topography. On the southwestern shore of Vostok Subglacial Lake, a 10 km wide and 30 km long region with thicker ice and strong bright reflectors indicative of subglacial water has been identified as an embayment separated from the main lake by a narrow bedrock ridge.

The ice flowing into Vostok Subglacial Lake from Ridge B in the west (Figure 1) is characterized by large thickness variations along the western shoreline. In the north between 77°S and 76.5°S, the flank of the bounding topography is at an average elevation of around 200 m bsl, 500 m above the lake surface at 700 m bsl. In the south between 78°S and 77.5°S, a steep shoulder on the western side with average elevations around 400 m above sea level rests almost 1000 m above the lake surface at 550 m bsl.

The differences in surrounding topography and lake surface elevation result in variations to the thickness of ice flowing onto Vostok Subglacial Lake. This dominates the melting and freezing pattern within the lake. Melting occurs in regions with thicker ice, while accretion dominates in regions with thinner ice [*Tikku et al.*, 2004]. The thicker ice in the north enters the lake through a depression in the subglacial topography, while the thinner ice in the south flows over a region with elevated topography. The bounding topography on the eastern side is generally much steeper than on the western side. The eastern side forms a straight shoreline in contrast to the rugged western shoreline. The subglacial topography on the west is very rugged with large differences in elevation over short distances compared with the relatively smooth topography east of the lake. This change in roughness in the topography reflects a change in subglacial geology [*Studinger et al.*, 2003b]. The continuity of the eastern shoreline as a straight segment over more than 200 km indicates that the bounding topography is fault controlled.

4.2. Gravity

The free-air gravity anomaly reflects both the major geologic and topographic structures and changes in the water depth of Vostok Subglacial Lake. A pronounced north-south trending gradient dominates the free-air gravity field (Plate 2a). This gradient parallels the eastern shore of Vostok Subglacial Lake. The steep gradient separates an area of positive

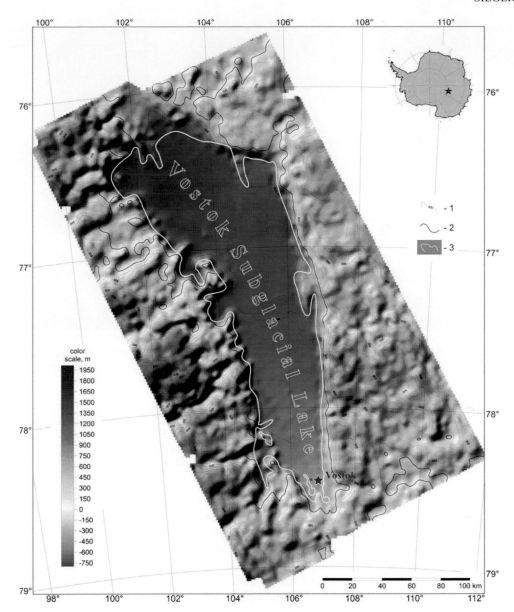

Plate 1. Detailed map of the lake surface and surrounding topography from a combination of Russian and U.S. geophysical data. The legend is coded as follows: 1, ice base, contours in meters; 2, sea level (WGS-84 surface); 3, outline of the lake, based on Russian RES data.

free-air anomalies (up to 50 mGal) in the east from negative values (less than −120 mGal) over the lake. A gravity low in the northern part of the lake (−90 mGal) is separated by a saddle (−70 mGal) from the main gravity low over the southern part of the lake (−120 mGal). West of the lake, the free-air gravity field comprises positive anomalies ranging from −40 to +20 mGal. To remove the gravitational effect of the ice-bedrock density contrast, a complete Bouguer anomaly has been calculated using the ice surface and subglacial

topography grids. East of the lake, the amplitudes reach −170 to −140 mGal, while west of the lake, the Bouguer gravity shows amplitudes around −200 mGal in the north and −220 mGal in the south. This step in Bouguer gravity over a short distance of less than 100 km reflects a significant change in crustal structure east and west of the lake that has been interpreted as a thrust sheet emplacement onto an earlier passive continental margin [*Studinger et al.*, 2003b]. Minor normal reactivation of the thrust sheet offers a simple

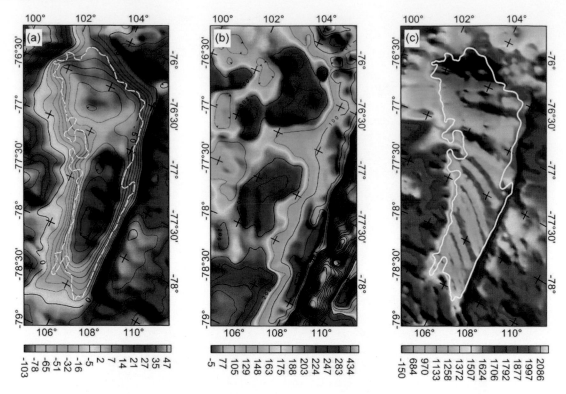

Plate 2. Gravity and magnetic fields, and internal layer structures, over Vostok Subglacial Lake. (a) Free-air gravity in mGal. Contour interval is 5 mGal (thin lines) and 10 mGal (thick, annotated lines). The data have been reduced from the flight elevation down to sea level (free-air correction), and the predicted gravity for the latitude on the Geodetic Reference System 1980 ellipsoid has been subtracted. (b) Total field magnetic anomaly (nT) in 3960 m elevation. Contour interval is 100 nT (thick, annotated lines) and 50 nT. (c) Depth of an internal layer in meters above sea level.

mechanism to explain the formation of the Vostok Subglacial Lake basin.

4.3. Magnetics

The magnetic anomaly map and the regional lines show a wide variety of characteristic wavelengths and amplitudes indicating changes in subglacial geology (Plate 2b). In the main grid, a pronounced north-south striking 900 nT anomaly dominates the magnetic field in the southeastern portion. This linear anomaly is about 30 km wide and can be traced outside the main grid on the two southern and four eastern regional lines. This anomaly is bounded on its western side by a linear low that follows closely the positive magnetic anomaly up to 76.5°S. North of 76.5°S, the trend of the

magnetic low is oriented more to the west. The pair of linear highs and lows parallels the eastern shore line of Vostok Subglacial Lake. West of this structure, the magnetic field is very smooth and comprises long-wavelength anomalies on the order of 50 to 70 km and up to 350 nT with almost entirely positive amplitudes. The distinct change in magnetic character over Vostok Subglacial Lake from a short-wavelength, high-amplitude field in the east to a long-wavelength smooth field over the lake and the west is related to a change in subglacial geology and is not an artifact of the change in subglacial elevation. The long-wavelength anomalies in the western part of the grid reflect sources located between 10 km depth and the Curie Point isotherm, while the high-amplitude, short-wavelength anomalies in the eastern part are likely to be dominated by near-surface sources.

Plate 3. (opposite) Water depth of Vostok Subglacial Lake. (a) Water depth from seismic studies. The legend is coded as follows: 1, isobaths in meters; 2, outline of the lake, based on Russian RES data. (b) Bathymetry of Vostok Subglacial Lake. Hypsographic curves are depicted in the inserts. The legend is as follows: 1, bedrock contours in meters; 2, sea level (WGS-84 surface); 3, outline of the lake, based on Russian RES data. (c) Water depth of the cavity in meters determined from inversion of aerogravity data. White regions mark locations where grounding is observed in the ice-penetrating radar data.

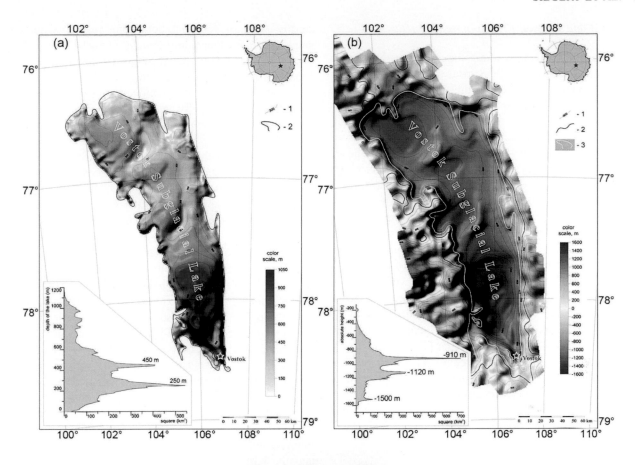

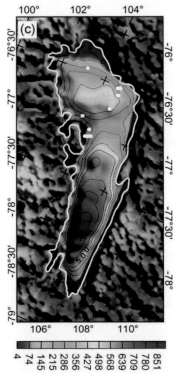

4.4. Flow Field Derived From Internal Structures

The complex topography along the upstream shoreline is preserved in the deep internal layers as the ice sheet traverses the lake. Topographic peaks are preserved as ridges in the internal layers, and topographic depressions are preserved as troughs (Plate 2c). As ice flows over a topographic peak along the shoreline, the silhouette of this peak is preserved as vaults in the topography of the lake surface because the lee side of such obstacles is filled with accreted lake ice, which prevents ice flowing from the sides of the peak to fill this cavity [*Tikku et al.*, 2004]. These structures are resolved in three distinct internal layers at depths between 900 and 3750 m. The flow field derived by structure tracking for Vostok Subglacial Lake displays a large but gradual rotation in the flow direction, from W-E in the northern end to NNW-SSE in the southern end (Plate 2c). The observed rotation is consistent with the general divergence associated with the ice divide over the lake. Accretion ice, lake water frozen to the bottom of the ice sheet, is preferentially imaged along flow lines emanating from topographic ridges [*Tikku et al.*, 2004]. The coincidence of the accretion ice reflector with the flow lines both provides independent support for the flow field and suggests focused accretion along the western shoreline. It also testifies that the flow of ice over the lake has remained largely unchanged for ~20,000 years (the approximate time taken for ice to cross the lake).

4.5. Bathymetry

As mentioned in section 2, the principal problem with reflection seismics in the interior of Antarctica concerns the thick snow-firn layer where acoustic wave attenuation occurs. During the 1995–1997 period, a new technique, using five to six 75 m lines of a detonating cord as a simple alternative to drilling a shot hole, was developed that ensured both efficiency and reliable data acquisition [*Popkov et al.*, 1998].

The seismic data reveal the average depth of Vostok Subglacial Lake to be about 410 m, and the volume of the water body is about 6350 km^3 (Plate 3a). Plate 3b shows the bathymetry is divided into two different-sized basins. The first (southern) part is the deepest (around 800 m). The second (northern) part is relatively shallow (at around 300 m) [*Masolov et al.*,, 2008].

While seismic information provides accurate measurements of the lake water depth, the results across the entire lake area remain subject to interpolations between data-free zones. The broad lake water depth can, however, be established well by inverting the dense aerogravity data [*Studinger et al.*, 2004]. The free-air gravity anomaly field reflects density variations related to both major geological and topographic structures including changes in the lake water depth. The influence of the regional subglacial topography and the geometry of the overlying ice sheet is well constrained from ice-penetrating radar measurements and can be removed from the observed gravity anomaly. The unknown parameter that dominates the remaining gravity anomaly is the lake water depth.

The estimated bathymetry of the gravity inversion is consistent with the seismic results; Vostok Subglacial Lake consists of two sub-basins (Plate 3c). The southern sub-basin is much deeper and approximately twice the spatial area of the smaller northern sub-basin. The two sub-basins are separated by a ridge with very shallow (~200 m) water depths.

The distribution of melting and freezing at the base of the ice sheet appears to be intimately linked to the two-basin structure. The regions with basal melting and freezing have been estimated from thickness changes between internal layers along ice flow. The same pattern is visible in characteristic signatures in the ice-penetrating radar data. The thin layer of accreted ice at the base of the ice sheet is imaged as a weak reflector in the radar data [*Bell et al.*, 2002; *Tikku et al.*, 2004]. Regions of melting correlate with a fuzziness of the ice-water interface in the radar data. Over the northern basin, basal melting is dominant, while over the southern basin, basal freezing characterizes the lake/ice interaction. The intimate link between the regions of melting and freezing and the bathymetric structure of the lake has important ramifications for the water circulation. If the lake water is fresh, basal meltwater in the northern basin would sink to the bottom (section 5.4) [*Siegert et al.*, 2001], and water exchange between the two basins will be limited. The two separate basins may, therefore, have different chemical and biological compositions. Furthermore, sediments released by basal melting are likely to accumulate in the northern basin, while preglacial sediments are more likely to be found in the southern, deeper basin. The sampling strategy for the future recovery of sediments from the lake bottom depends on the type of sediments targeted.

5. PHYSICAL PROCESSES IN THE LAKE

Having established Vostok Subglacial Lake's broad physiographical setting, we are now able to consider the likely chemical and physical processes within the lake. Vostok Subglacial Lake has two obvious advantages for evaluating such processes. First, it is a very large subglacial lake. Because of this, large-scale processes within it are likely to be more obvious and identifiable than in small subglacial lakes. For example, there have been several models of water circulation within Vostok Subglacial Lake, and these have been developed from large-scale ocean models, which have a resolution of the order of kilometers. Second, by chance, the Vostok ice core is located above the southern end of Vostok Subglacial Lake, and this contains at its base some ice that is refrozen from the lake water.

5.1. Details of Accretion Ice Acquisition

Several deep ice cores have been extracted from the ice sheet at Vostok Station (at the southern end of Vostok Subglacial Lake) since drilling began in the mid-1960s (the first 500 m deep dry borehole was extracted in 1965), providing important information about the climate during the last glacial cycle. The most recent and deepest (3623 m) ice core terminated ~130 m from the base of the ice sheet [Masolov et al., 2001]. The upper 3310 m of the ice core provides a detailed paleoclimate record spanning the past 420,000 years [Petit et al., 1997, 1999]. In addition, microbiological analysis of the ice core has revealed a range of microbiota, including bacteria, fungi, and algae, some of which have been reported to be culturable in the laboratory [Abyzov et al., 1998; Priscu et al., 1999; Karl et al., 1999].

Typical glacier ice contains a record of gases and isotopes from which paleoclimate information is inferred. In the Vostok ice core, this type of ice exists to a depth of 3310 m. Lower layers of ice, between depths of 3310 and 3538 m, are reported to have been reworked, making the extraction of paleoclimate information difficult to establish. The basal 84 m of the ice core, from 3539 to 3623 m, has a chemistry and crystallography that are distinctly different from the "normal" glacier ice above. The basal ice has an extremely low conductivity, huge (up to 1 m) crystal sizes, and sediment-particle inclusions (in the upper half) [Jouzel et al., 1999]. The mineral composition of ice-bound sediments below 3539 m is dominated by micas and is clearly different than typical crustal composition and particles within the overlying glacial ice [Priscu et al., 1999]. Its isotopic composition, distinct from the "meteoric" ice above, suggests that it formed by the refreezing of lake water to the underside of the ice sheet. Thus, there is ~210 m of accreted Vostok Subglacial Lake ice beneath Vostok Station [Jouzel et al., 1999].

5.2. Microbiology

The accreted ice offered the first opportunity for aquatic biologists and geochemists to investigate material derived from a subglacial lake. Two recent independent studies of accreted ice subsampled from different depths (3590 and 3603 m) near the base of the Vostok ice core (maximum depth, 3623 m) have shown that these samples contain both low numbers and low diversity of bacteria [Karl et al., 1999;

Priscu et al., 1999]. The low diversity (seven phylotypes) may reflect the small sample size analyzed (~250 mL of melt) and should be considered as a lower limit. Low concentrations of "growth nutrients" and evidence of mineralization of ^{14}C-labeled organic substrates were also found, although activity was measured under potentially more benign laboratory conditions of +3°C and 1 atm pressure [Karl et al., 1999]. Since the accreted ice has been frozen from Vostok Subglacial Lake water, the inference is that these microbes were present in the lake water, at some point, and viable prior to freezing. Priscu et al. [1999], using ice-water partitioning coefficients from the permanently ice-covered lakes in the McMurdo Dry Valleys, estimated that the bacterial density within Vostok Subglacial Lake's water column could be on the order of 10^6 mL^{-1}. Microbiological analyses from the Vostok accreted ice are discussed in more detail by Skidmore [this volume].

5.3. Geochemistry

Solutes are added to the lake water during ice melt and via chemical weathering of debris in and around the base of the lake. The average chemistry of the meltwater entering Vostok Subglacial Lake can be inferred from Legrand et al. [1988], assuming that ice from glacial periods makes up 85% of the melt and that from interglacials makes up 15%. The average initial meltwater is equivalent to a very dilute mix of marine-derived aerosol, Ca-rich dust, and strong acids (i.e., HNO_3 and H_2SO_4). Solutes are rejected from the ice lattice during refreezing [Killawee et al., 1998]; hence, there should be an accumulation of nutrients, gases, and solutes in the lake water over time. The isotopic and major ion composition of Vostok Subglacial Lake has been inferred from the composition of the accreted basal ice in the Vostok ice core. The accreted ice is enriched in ^{18}O and ^{2}H compared to the Vostok precipitation line [Jouzel et al., 1999; Priscu et al., 1999]. This is because there is isotopic fractionation during water freezing, but none during melting [Souchez et al., 1988, 2000]. The accreted ice has values of δ^{18}O and δD that differ from the time-averaged melting ice by only 60% of the theoretical fractionation, and it has been suggested that 30% to 58% of unfractionated lake water is entrained in the accreted ice during freezing, so helping to maintain less extreme values of δD and δ^{18}O [Souchez et al., 2000].

Royston-Bishop et al. [2004] inspected the δD and δ^{18}O data to explore whether Vostok Subglacial Lake is in isotopic steady state. A simple box model showed that the lake is likely to be in steady state on timescales on the order of 10^4 to 10^5 years (three to four residence times of the water in the lake), given our current knowledge of north-south and east-

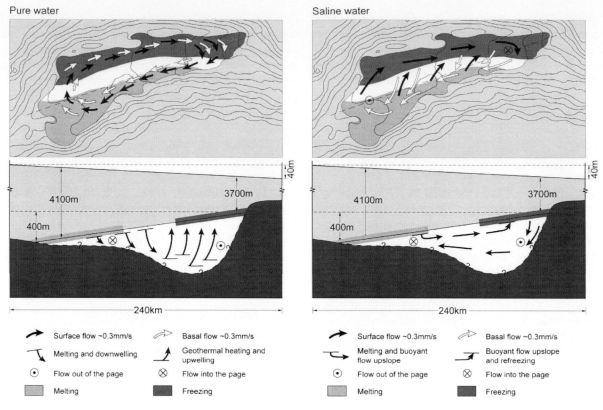

Figure 3. Water circulation within Vostok Subglacial Lake. (a) Circulation assuming that the water is pure. The white arrows show the bottom water circulation, and the black arrows denote the higher level circulation close to the ice base. Dots refer to upwelling of lake water; crosses denote downwelling. Dark gray shading refers to predicted zones of subglacial freezing; light gray shading indicates subglacial melting. (b) Circulation of Vostok Subglacial Lake thought to occur as a result of saline conditions (i.e., 1.2‰–0.4‰). Adapted from *Siegert et al.* [2001].

west gradients in the stable isotopic composition of precipitation in the vicinity of Vostok Station and Ridge B (where the deepest ice originates from the surface). This suggests that the lake has not been subject to any recent major perturbations, such as volume changes. However, they also showed that the lake may not be in perfect steady state, depending on the precise location of the melting area, which determines the source region of inflowing ice, and on the magnitude of the east-west gradient in isotopic compositions in the vicinity of Vostok Station and Ridge B.

5.4. Water Circulation and Water Balance

Borehole temperature measurements along the full length of the Vostok ice core have been used to establish the energy balance between the ice sheet and the lake [*Salamatin et al.,* 1998; *Salamatin,* 2000]. The mean basal temperature gradient is ~0.02°C m^{-1}, which relates to a heat flux through the ice from the lake ceiling of 46 mW m^{-2}, indicating that the rates of subglacial freezing above Vostok Subglacial Lake

are most likely to be ~4 mm yr^{-1} [*Salamatin et al.,* 1998]. In the extreme case where ice at −10°C flows over the western lake margin, rates of melting and freezing beneath Vostok Station will probably not be higher than about 11 mm yr^{-1} [*Salamatin,* 2000].

The spatial distribution of subglacial melting and freezing can be estimated theoretically from isochronous internal radar layering, by observing the loss or gain of basal ice along a flow line. Using this technique, it has been shown that subglacial melting occurs in the north of Vostok Subglacial Lake [*Siegert et al.,* 2000], and freezing (accretion) takes place in the south [*Bell et al.,* 2002]. Rates of melting and freezing calculated from radar layering have been much higher (of the order of centimeters) than those from the ice core's temperature record. It is possible that heat used for melting can be taken from the lake water, but this requires a dynamic water circulation system.

The zones of subglacial melting in the north and freezing in the south of Vostok Subglacial Lake are thought to be controlled by the slope of the ice-water interface, since the thickness of ice dictates the pressure melting temperature and the density of meltwater. Melting and freezing induce circulation in the lake, which itself will be influenced heavily by the lake hydrochemistry.

There are two possible ways in which water within Vostok Subglacial Lake could circulate (Figure 3). One is if the lake contains pure water; the other is if the lake water is slightly saline. These two end-member possibilities are detailed below. In the first instance, circulation of pure water is discussed.

Since the surface of Vostok Subglacial Lake is inclined, the pressure melting point in the south will be slightly (~0.3°C) less than that in the north. The circulation of pure (nonsaline) water in Vostok Subglacial Lake will be driven by the differences between the density of meltwater and lake water. Geothermal heating will warm the bottom water to a temperature higher than that of the upper layers. The water density will decrease with increasing temperature because Vostok Subglacial Lake is in a high-pressure environment, resulting in an unstable water column [*Wüest and Carmack*, 2000]. This leads to convective circulation conditions in the lake in which cold meltwater sinks down the water column and water warmed by geothermal heat ascends up the water column. However, a pool of slightly warmer and stratified water may occur below the ice roof in the south, where the ice sheet is thinner and subglacial freezing takes place [*Wüest and Carmack*, 2000]. Here, the water would not be involved in convective motion as heat is transferred from the ice toward the lake (i.e., the temperature will decrease with depth). There have been four models from which the circulation of pure water in Vostok Subglacial Lake can be evaluated [*Mayer et al.*, 2003; *Wüest and Carmack*, 2000; *Williams*, 2001; *Thoma et al.*, 2010]. The models indicate that meltwater will be colder and denser in the northern area of Vostok Subglacial Lake, where the ice is thickest, than both the surrounding lake water and meltwater in areas with thinner ice cover. It appears therefore that this region is the main zone of downwelling of pure water. However, the circulation is complicated by the geometry of the lake cavity and the Coriolis force. This means that circulation in Vostok Subglacial Lake will include horizontal transfer and vertical overturning. The models agree that northern meltwater will sink and be transported horizontally to the south, via a clockwise circulation system, to a region where the pressure melting point is higher, allowing refreezing to occur.

An alternate point of view is that the lake is saline to a small extent [*Souchez et al.*, 2000]. The fresh glacier meltwater will, therefore, be buoyant compared with the resident, more saline lake water. The northern meltwater likely spreads southward and upward, traveling into regions of progressively lower pressure and displacing lake water in the south if the horizontal salinity gradient (north-south) is high enough to compensate for geothermal warming. The possibility of such a regime is controlled by (1) the melting-freezing rates, (2) the rates of mixing between the fresh ascending meltwater layer and the underlying saline water, and (3) vertical free convection driven by the geothermal heating of water at the lake bottom. The cold northern water will eventually enter a region where its temperature is at, or below, the pressure melting point, if the heat flux from the basal water is not sufficiently high. The water will then refreeze back onto the ice sheet base some distance away from where it was first melted into the lake. In this case, a conveyor of fresh cool meltwater is established, which migrates from north to south immediately beneath the ice sheet, which causes displacement of warmer dense lake water from south to north. In contrast, if the bulk salinity is not high enough, a stable stratification will develop in the upper water layers below the tilted lake ceiling, with more saline warmer water in the south and fresher, cooler water in the north [*Wüest and Carmack*, 2000]. The deep-water stratum will be subject to vertical thermal convection because, for any reasonable level of salinity, the temperature at the lake bottom will be high enough to start the convection.

Royston-Bishop et al. [2005] studied the size-frequency distribution of the microscopic particulates in accreted ice from the Vostok ice core. They demonstrated that the particles have similar distributions of major axis lengths, surface areas and shape factors (aspect ratio and compactness) irrespective of ice core depth, suggesting a common single process is responsible for their incorporation in the ice. In addition, *Royston-Bishop et al.* [2005] calculated Stokes settling velocities for particulates of various sizes and showed that 98% could float to the ice-water interface with upward water velocities of only 0.0003 m s^{-1}. This is well within the range of water flow speeds predicted by circulation models of Vostok Subglacial Lake [*Mayer et al.*, 2003]. The presence of larger particles in the ice (2%) and the uneven distribution of observed particulates in the core suggest that periodic perturbations to the lake's circulation, involving increased velocities, may have occurred in the past and are likely now.

Early models of water circulation in Vostok Subglacial Lake have been mostly conceptual in nature owing to a lack of observation to constrain the models. The models aimed at understanding the basic physics of water circulation and interaction between the lake water and the overlying ice sheet. The definition of the lake cavity and distribution of melting and freezing from geophysical soundings have enabled the next generation of numerical models with realistic constraints on bathymetry, melting and freezing rates, and other critical parameters. These models are coupled ice sheet and water circulation models and attempt to realistically predict not only the water circulation in the lake, but also the pattern of melting and freezing, the rates of melting and freezing, and the interaction between ice flow and water circulation in the lake (see *Thoma et al.* [2010], and references therein).

6. SUMMARY

Vostok Subglacial Lake is the largest of more than 380 subglacial lakes in Antarctica, being more than 240 km in length and 80 km wide. While its discovery, through airborne RES in the 1970s, was forgotten by many for more than a decade, scientific interest was suddenly generated following the discovery, in 1996, of its water depth being more than 500 m. This led microbiologists to hypothesize that Vostok Subglacial Lake is a viable habitat for life and that such life may have developed in isolation from the rest of the planet for as much as 14 million years [see *Skidmore*, this volume]. This hypothesis received widespread media and public interest, as well as further interest from scientists aiming to undertake the geophysical and, ultimately, direct exploration.

The first stage in the process of exploration was to conduct a comprehensive survey of the lake, given that the data collected in the 1970s comprised a mere handful of radio echo transects. The first airborne geophysical reconnaissance of Vostok Subglacial Lake since the 1970s was undertaken by the Italian Antarctic Research Programme. Several RES lines were flown, revealing the limits of the lake shoreline and enhanced definition of the lake surface.

In the Austral summer of 2000–2001, U.S. geophysicists acquired more than 20,000 line-km of data. Radio echo sounding revealed detailed information about the lake surface and its surrounding topography. These data were also used to show that large-scale melting occurs in the north of the lake and to reconfirm that freezing takes place in the south, resulting in more than 200 m of ice accreted to the underside of the ice sheet (as discovered in the ice core extraction of accreted ice a year or two earlier). Magnetic data revealed important information regarding the tectonic structure in which the lake lies, and gravity data were used to model the bathymetry of the lake, showing it to comprise two discrete basins.

The distinction between northern melting and southern freezing leads to circulation of the lake water, in a manner dependent on the salinity of the water. For pure water, the circulation behaves as in the very deep ocean, with cold water descending and rising up the column after heating from geothermal sources. For slightly saline water, melted ice will be lighter than the lake water and so will rise up the inclined ice-water interface. Both cases can account for the large-scale freezing that takes place in the south, as cold water will be transferred from north to south regardless of salinity. There has been much speculation about the hydrochemistry of Vostok Subglacial Lake. Clearly, direct access and sampling of the lake water will resolve this speculation.

While airborne geophysical data are essential for defining the ice sheet above the lake and its surface characteristics, its bathymetry can be measured only by ground-based seismic measurements. Russian scientists have been collecting such data since the mid-1990s. These data were used in conjunction with the U.S. bathymetry data to first reveal, and later to better define, the bathymetry of the lake. In addition, Russian geophysicists have acquired more than 5000 km of radio echo data of the ice sheet and lake surface. In combination with U.S. data, we now have an excellent understanding of Vostok Subglacial Lake's physiographical setting and of the physical processes operating within it.

As a consequence of the geophysical data collected to date, the direct measurement and sampling of Vostok Subglacial Lake can now be contemplated.

Acknowledgements. M. J. Siegert's contribution is supported by U.K. NERC grants NE/D008638/1, NE/D003733/1, and NE/G00465x/1. S. Popov's contribution is supported by RFBR grant 10-05-91330-NNIO_a. M. Studinger's contribution is supported by NSF awards ANT 05-37752 and ANT 06-36584 to Lamont-Doherty Earth Observatory.

REFERENCES

Abyzov, S. S., I. N. Mitskevich, and M. N. Poglazova (1998), Microflora of the deep glacier horizons of central Antarctica, *Microbiology*, 67, 66–73.

Bell, R. E., M. Studinger, A. A. Tikku, G. K. C. Clarke, M. M. Gutner, and C. Meertens (2002), Origin and fate of Lake Vostok water frozen to the base of the East Antarctic ice sheet, *Nature*, 416, 307–310.

Fricker, H. A., et al. (2011), Siple Coast subglacial aquatic environments: The Whillans Ice Stream Subglacial Access Research Drilling project, in *Antarctic Subglacial AquaticEnvironments*, *Geophys. Monogr. Ser.*, doi: 10.1029/2010GM000932, this volume.

Gorman, M. R., and M. J. Siegert (1999), Penetration of Antarctic subglacial water masses by VHF electromagnetic pulses: Estimates of minimum water depth and conductivity of basal water bodies, *J. Geophys. Res.*, 104(B12), 29,311–29,320, doi:10.1029/1999JB900271.

Holt, J. W., et al. (2006), Airborne gravity over Lake Vostok and adjacent highlands of East Antarctica, *Geochem. Geophys. Geosyst.*, 7, Q11012, doi:10.1029/2005GC001177.

Jouzel, J., J. R. Petit, R. Souchez, N. I. Barkov, V. Y. Lipenkov, D. Raynaud, M. Stievenard, N. I. Vassiliev, V. Verbeke, and F.

Vimeux (1999), More than 200 meters of lake ice above subglacial Lake Vostok, Antarctica, *Science*, *286*, 2138–2141.

Kapitsa, A. P., J. K. Ridley, G. D. Robin, M. J. Siegert, and I. Zotikov (1996), A large deep freshwater lake beneath the ice of central East Antarctica, *Nature*, *381*, 684–686.

Karl, D. M., D. F. Bird, K. Bjorkman, T. Houlihan, R. Shackelford, and L. Tupas (1999), Microorganisms in the accreted ice of Lake Vostok, Antarctica, *Science*, *286*, 2144–2147.

Killawee, J. A., I. J. Fairchild, J.-L. Tison, L. Janssens, and R. Lorrain (1998), Segregation of solutes and gases in experimental freezing of dilute solutions: Implications for natural glacial systems, *Geochim. Cosmochim. Acta*, *62*, 3637–3655.

Legrand, M. R., C. Lorius, N. I. Barkov, and V. N. Petrov (1988), Vostok (Antarctica) ice core: atmospheric chemistry changes over the last climatic cycle (160,000 years), *Atmos. Environ.*, *22*, 317–331.

Lukin, V., and S. Bulat (2011), Vostok Subglacial Lake: Details of Russian plans/activities for drilling and sampling, in *Antarctic Subglacial Aquatic Environments*, *Geophys. Monogr. Ser.*, doi: 10.1029/2010GM000951, this volume.

Masolov, V. N., V. V. Lukin, A. N. Sheremetyev, and S. V. Popov (2001), Geophysical investigations of the subglacial Lake Vostok in Eastern Antarctica (in Russian), *Dokl. Akad. Nauk*, *379* (5), 680–685. (*Dokl. Earth Sci.*, *379A*(6), 734–738.)

Masolov, V. N., S. V. Popov, V. V. Lukin, A. N. Sheremet'ev, and A. M. Popkov (2006), Russian geophysical studies of Lake Vostok, Central East Antarctica, in *Antarctica—Contributions to Global Earth Sciences*, edited by D. K. Fütterer et al., pp. 135–140, Springer, New York.

Masolov, V. N., V. V. Lukin, S. V. Popov, A. M. Popkov, A. N. Sheremetyev, U. A. Kruglova (2008), Seabed relief of the subglacial lake Vostok, in *Abstract Volume, SCAR/IASC IPY Open Science Conference "Polar Research—Arctic and Antarctic perspectives in the International Polar Year", St. Petersburg, Russia, July 8–11, 2008*, edited by A. Klepikov, S3.1/P08.

Mayer, C., K. Grosfeld, and M. J. Siegert (2003), Salinity impact on water flow and lake ice in Lake Vostok, Antarctica, *Geophys. Res. Lett.*, *30*(14), 1767, doi:10.1029/2003GL017380.

Oswald, G. K. A., and G. D. Robin (1973), Lakes beneath the Antarctic Ice Sheet, *Nature*, *245*, 251–254.

Petit, J. R., et al. (1997), Four climate cycles in Vostok ice core, *Nature*, *387*, 359–360.

Petit, J. R., et al. (1999), Climate and atmospheric history of the past 420,000 years from the Vostok ice core, Antarctica, *Nature*, *399*, 429–436.

Popkov, A. M., G. A. Kudryavtsev, V. A. Shumilov, and N. V. Kondratiev (1998), Methodological features of seismic studies in the region of the Russian station "Vostok" in Lake Vostok Study: Scientific Objectives and Technological Requirements, paper presented at International Workshop, Arct. and Antarct. Res. Inst., St. Petersburg, Russia, 24–26 March.

Popov, S. V., and V. N. Masolov (2007), Forty-seven new subglacial lakes in the 0°–110°E sector of East Antarctica, *J. Glaciol.*, *53*(181), 289–297.

Priscu, J. C., et al. (1999), Geomicrobiology of subglacial ice above Lake Vostok, Antarctica, *Science*, *286*, 2141–2144.

Richter, T. G., et al. (2001), Airborne Gravimetry over the Antarctic Ice Sheet, in *International Symposium on Kinematic Systems in Geodesy, Geomatics and Navigation*, edited by M. E. Cannon and G. Lachapelle, pp. 576–585, Univ. of Calgary, Calgary, Alberta, Canada.

Ridley, J. K., W. Cudlip, and S. W. Laxon (1993), Identification of subglacial lakes using ERS-1 radar altimeter, *J. Glaciol.*, *39*, 625–634.

Robin, G. D. (1955), Ice movement and temperature distribution in glaciers and ice sheets, *J. Glaciol.*, *3*, 589–606.

Robin, G. D., C. W. M. Swithinbank, B. M. E. Smith (1970), Radio echo exploration of the Antarctic ice sheet, in *International Symposium on Antarctic Glaciological Exploration (ISAGE)*, *IAHS Publ.*, *86*, 97–115.

Robin, G. D., D. J. Drewry, and D. T. Meldrum (1977), International studies of ice sheet and bedrock, *Philos. Trans. R. Soc. London, Ser. B*, *279*, 185–196.

Robinson, R. V. (1961), Experiment in visual orientation during flights in the Antarctic, *Sov. Antarct. Exped. Inf. Bull.*, *18*, 28–29.

Ross, N., et al. (2011), Ellsworth Subglacial Lake, West Antarctica: Its history, recent field campaigns, and plans for its exploration, in *Antarctic Subglacial Aquatic Environments*, *Geophys. Monogr. Ser.*, doi: 10.1029/2010GM000936, this volume.

Royston-Bishop, G., M. Tranter, M. J. Siegert, V. Lee, and P. Bates (2004), Is Lake Vostok in chemical and physical steady-state?, *Ann. Glaciol.*, *39*, 490–494.

Royston-Bishop, G., J. C. Priscu, M. Tranter, B. Christner, M. J. Siegert, and V. Lee (2005), Incorporation of particulates into accreted ice above subglacial Lake Vostok, Antarctica, *Ann. Glaciol.*, *40*, 145–150.

Salamatin, A. N. (Ed.) (2000), *Physics of Ice Core Records*, edited by T. Hondoh, pp. 243–282, Hokkaido Univ. Press, Sapporo, Japan.

Salamatin, A. N., V. Y. Lipenkov, N. I. Barkov, J. Jouzel, J. R. Petit, and D. Raynaud (1998), Ice core age dating and paleothermometer calibration based on isotope and temperature profiles from deep boreholes at Vostok Station (East Antarctica), *J. Geophys. Res.*, *103*(D8), 8963–8977, doi:10.1029/97JD02253.

Siegert, M. J. (2005), Lakes beneath the ice sheet: The occurrence, analysis and future exploration of Lake Vostok and other Antarctic subglacial lakes, *Annu. Rev. Earth Planet. Sci.*, *33*, 215–245.

Siegert, M. J., and J. A. Dowdeswell (1996), Spatial variations in heat at the base of the Antarctic Ice Sheet from analysis of the thermal regime above sub-glacial lakes, *J. Glaciol.*, *42*, 501–509.

Siegert, M. J., R. Kwok, C. Mayer, and B. Hubbard (2000), Water exchange between the subglacial Lake Vostok and the overlying ice sheet, *Nature*, *403*, 643–646.

Siegert, M. J., et al. (2001), Physical, chemical and biological processes in Lake Vostok and other Antarctic subglacial lakes, *Nature*, *414*(6864), 603–609.

Siegert, M. J., S. Carter, I. Tabacco, S. Popov, and D. Blankenship (2005), A revised inventory of Antarctic subglacial lakes, *Antarct. Sci.*, *17*(3), 453–460.

Siegert, M. J., A. Le Brocq, and A. Payne (2007), Hydrological connections between Antarctic subglacial lakes and the flow of water beneath the East Antarctic Ice Sheet, in *Glacial Sedimentary Processes and Products*, edited by M. J. Hambrey et al., *Spec. Publ. Int. Assoc. Sedimentol.*, *39*, 3–10.

Skidmore, M. (2011), Microbial communities in Antarctic subglacial aquatic environments, in *Antarctic Subglacial Aquatic Environments*, *Geophys. Monogr. Ser.*, doi: 10.1029/2010 GM000995, this volume.

Souchez, R., R. Lorrain, J. L. Tison, and J. Jouzel (1988), Co-isotopic signature of two mechanisms of basal ice formation in Arctic outlet glaciers, *Ann. Glaciol.*, *10*, 163–166.

Souchez, R., J. R. Petit, J. L. Tison, J. Jouzel, and V. Verbeke (2000), Ice formation in subglacial Lake Vostok, central Antarctica, *Earth Planet. Sci. Lett.*, *181*, 529–538.

Studinger, M., et al. (2003a), Ice cover, landscape setting, and geological framework of Lake Vostok, East Antarctica, *Earth Planet. Sci. Lett.*, *205*(3–4), 195–210.

Studinger, M. (2003b), Geophysical models for the tectonic framework of the Lake Vostok region, East Antarctica, *Earth Planet. Sci. Lett.*, *216*(4), 663–677.

Studinger, M., R. E. Bell, and A. A. Tikku (2004), Estimating the depth and shape of subglacial Lake Vostok's water cavity from aerogravity data, *Geophys. Res. Lett.*, *31*, L12401, doi:10.1029/ 2004GL019801.

Sugden, D. E. (1992), Antarctic ice sheets at risk?, *Nature*, *359*, 775–776.

Tabacco, I. E., C. Bianchi, A. Zirizzotti, E. Zuccheretti, A. Forieri, and A. Della Vedova (2002), Airborne radar survey above Vostok region, east Antarctica: Ice thickness and Lake Vostok geometry, *J. Glaciol.*, *48*, 62–69.

Thoma, M., K. Grosfeld, C. Mayer, and F. Pattyn (2010), Interaction between ice sheet dynamics and subglacial lake circulation: A coupled modelling approach, *Cryosphere*, *4*, 1–12.

Tikku, A. A., et al. (2004), Ice flow field over Lake Vostok, East Antarctica inferred by structure tracking, *Earth Planet. Sci. Lett.*, *227*(3–4), 249–261.

Williams, M. J. M. (2001), Application of a three-dimensional numerical model to Lake Vostok: An Antarctic subglacial lake, *Geophys. Res. Lett.*, *28*, 531–534.

Wright, A., and M. J. Siegert (2011), The identification and physiographical setting of Antarctic subglacial lakes: An update based on recent discoveries, in *Antarctic Subglacial Aquatic Environments*, *Geophys. Monogr. Ser.*, doi: 10.1029/2010GM000933, this volume.

Wüest, A., and E. Carmack (2000), A priori estimates of mixing and circulation in the hard-to-reach water body of Lake Vostok, *Ocean Modell.*, *2*, 29–43.

Zotikov, I. A. (1963), Bottom melting in the central zone of the ice shield of the Antarctic continent and its influence upon the present balance of the ice mass, *Bull. Int. Assoc. Sci. Hydrol.*, *8*(1), 36.

Zotikov, I. A. (2006), *The Antarctic Subglacial Lake Vostok: Glaciology, Biology and Planetology*, 139 pp., Springer-Praxis, Chichester, U. K.

S. Popov, Polar Marine Geosurvey Expedition, 24, Pobeda Str., St. Petersburg, Lomonosov 198412, Russia. (spopov67@yandex.ru)

M. J. Siegert, School of GeoSciences, University of Edinburgh, Edinburgh EH9 3JW, UK. (m.j.siegert@ed.ac.uk)

M. Studinger, Cryospheric Sciences Branch, Goddard Earth Sciences and Technology Center, NASA Goddard Space Flight Center, Code 614.1, 8800 Greenbelt Road, Greenbelt, MD 20771, USA. (michael.studinger@nasa.gov)

Microbial Communities in Antarctic Subglacial Aquatic Environments

Mark Skidmore

Department of Earth Sciences, Montana State University, Bozeman, Montana, USA

Glaciological processes under ice masses, including ice sheets, produce conditions favorable for microbes by forming subglacial aquatic environments (SAE) through basal melting and providing nutrients and energy for microbes from bedrock comminution. The abundance and interconnectivity of water beneath the Antarctic Ice Sheet, largely demonstrated through remote sensing techniques, indicates significant and varied SAE including lakes, saturated sediments and channelized and linked cavity drainage systems. Microbes have been detected in the two Antarctic SAE sampled to date; accreted ice from Vostok Subglacial Lake in East Antarctica and saturated till from beneath ice streams draining the West Antarctic Ice Sheet. Heterotrophic activity has been measured in these samples at temperatures close to freezing in the laboratory, demonstrating that in situ microbial activity in subglacial environments is plausible. Phylogenetic analysis of 16S rRNA gene sequences suggests that organisms with Fe and S oxidizing metabolisms may also be important members of the microbial community in these environments. This is consistent with the geochemistry of the accreted ice and till pore waters that indicates biologically driven sulfide oxidation coupled to carbonate and silicate mineral weathering as a significant solute source. Exploration of Antarctic SAE is in its infancy, and in a number of the unexplored SAE, the lack of connectivity between oxygenated surface waters and the subglacial environment and the extended water flow paths and water-rock residence times may lead to anoxic conditions. Such anoxic conditions would favor anaerobic microbial metabolisms similar to those documented in other deep terrestrial subsurface environments.

1. INTRODUCTION

Research over the past 15 years has revealed abundant water beneath the Antarctic Ice Sheet [e.g., *Kapitsa et al.*, 1996; *Kamb*, 2001; *Engelhardt*, 2004, *Siegert*, 2005; *Llubes et al.*, 2006; *Wingham et al.*, 2006; *Fricker and Scambos*, 2009]. Over a similar time frame, there has been increasing evidence that aqueous subglacial environments are inhabited by active microbial communities that are involved in the biogeochemical cycling of elements including C, Fe, S, N, Si, and P [e.g., *Sharp et al.*, 1999; *Tranter et al.*, 2002; *Skidmore et al.*, 2005; *Gaidos et al.*, 2009; *Mikucki et al.*, 2009]. Sampling of the microbial characteristics of subglacial aquatic environments (SAE) beneath the Antarctic Ice Sheet has been limited to two sites, the accreted ice from Vostok Subglacial Lake in East Antarctica [e.g., *Karl et al.*, 1999; *Priscu et al.*, 1999; *Christner et al.*, 2006; *Bulat et al.*, 2009] and saturated till from beneath the Kamb Ice Stream, West Antarctica [*Lanoil et al.*, 2009]. Despite the limited sampling of the subglacial environment in Antarctica, it has been highlighted as potentially a large planetary reservoir of both microbes and (microbially derived) organic carbon of the same magnitude as that in the surface oceans [*Priscu and Christner*, 2004; *Priscu et al.*, 2008; *Lanoil et al.*, 2009].

Antarctic Subglacial Aquatic Environments
Geophysical Monograph Series 192
Copyright 2011 by the American Geophysical Union
10.1029/2010GM000995

This chapter will (1) briefly review microbially mediated subglacial processes in the better studied Arctic and Alpine systems, (2) provide a synthesis of the microbial ecology and biogeochemistry of the two Antarctic SAE studied to date, and (3) assess unexplored Antarctic SAE and their potential microbial communities and processes.

1.1. Requirements for Life

Liquid water, carbon and energy sources, and certain key nutrients (N for proteins, nucleic acids, P, for nucleic acids, and S for amino acids, and a range of micronutrients, e.g., K, Mg, Ca, Na, Fe, Cu, Mn, Mo, Ni, Se, W, Zn) are requirements for an environment to support microbial life. Phototrophy is not a potential energy source in subglacial systems where light is lacking, thus microbes will have chemoorganotrophic or chemolithotrophic metabolisms. Chemoorganotrophs gain energy from organic compounds, and chemolithotrophs gain energy from inorganic compounds, which is released when such compounds are oxidized either aerobically or anaerobically. Micronutrients such as Fe are required for enzymes involved in cellular respiration, and Fe, Mo, and Ni are essential elements in proteins required for nitrogen fixation. Where ice masses are wet based, there is abundant water, and glacial comminution of bedrock and sediments is a source for a range of elements required for both energy generation and as nutrient supply, including organic carbon. Gases, such as O_2, N_2, and CO_2 are supplied to the subglacial environment as gas bubbles in the ice are released through basal melting. In Antarctic SAE, this will be the only gas source with an atmospheric origin, compared to valley glacier systems where, in certain cases, surface meltwaters reach the glacier bed providing an additional supply of dissolved gases from the atmosphere. Additional gases such as CO_2, CH_4, H_2S, and H_2 could be supplied to the SAE from crustal sources via hydrothermal vent systems associated with subglacial volcanism [*Blankenship et al.*, 1993]. A chemical potential is required by microbes to utilize available free energy through

reduction and oxidation reactions (Table 1). The amount of energy available depends on the redox couple. The hydrology of the SAE is a key control on the rate of supply of carbon and potential redox couples to a microbial community and evacuation of microbial waste products.

1.2. Microbial Processes in SAE Beneath Alpine, Arctic, and Antarctic Valley Glaciers

Studies on Alpine, Arctic, and Antarctic valley glaciers have demonstrated that active and viable microbes can be found in a range of SAE, including outflowing subglacial waters, basal ice, and subglacial sediments (Table 2). Fine-grained sediments at the bed of temperate-based ice masses provide favorable conditions for bacterial activity, since (1) the sediments are water-saturated as liquid water is present from basal melting and from the delivery of surface meltwater to the glacier bed, (2) sediments consist of freshly comminuted chemically reactive debris that is susceptible to colonization by microbes, and (3) contain organic carbon from in-washed organic material from the glacier surface [*Boon et al.*, 2003], from overridden preglacial sediments and soils by the ice mass [*Skidmore et al.*, 2000] or generated in situ by chemolithoautotrophs.

A range of metabolisms have been documented by enrichment culturing of subglacial materials and by inference from aqueous geochemical and isotopic data (Table 2). One can typically only culture $\sim\leq 1\%$ of the microbial population from natural systems [*Amann et al.*, 1995], thus culture-independent analyses provide additional information on the microbial diversity in a given environment. Culture-independent analysis of the microbial communities in these systems, e.g., using 16S rRNA gene sequences, has revealed phylotypes closely related to aerobic heterotrophs, Fe- and S-oxidizing microorganisms, Fe-reducing organisms, and methanogens [*Skidmore et al.*, 2005; *Mikucki and Priscu*, 2007; *Boyd et al.*, 2010]. However, one must be cautious with inferences made purely from phylogenetic data, but it has

Table 1. Potential Energy-Generating Reactions in Subglacial Systems[a]

Microbial respiration	$(CH_2O)_n + O_2 <=> CO_2 + H_2O <=> H^+ + HCO_3^-$
Pyrite oxidation (oxic)	$FeS_2 + 3.5O_2 + H_2O => Fe^{2+} + 2SO_4^{2-} + 2H^+$
Nitrate reduction	$2NO_3^- + (CH_2O)_n => H^+ + HCO_3^- + 2NO_2^-$
Pyrite oxidation (anoxic)	$FeS_2 + 14Fe^{3+} + 8H_2O => 15Fe^{2+} + 2SO_4^{2-} + 16H^+$
Iron reduction	$4FeOOH + 7H^+ + (CH_2O)_n => HCO_3^- + 4Fe^{2+} + 6H_2O$
Sulfate reduction	$0.5SO_4^{2-} + (CH_2O)_n => HCO_3^- + 0.5H_2S$
Methanogenesis	$CO_2 + 4H_2 => CH_4 + 2H_2O$
	$CH_3COOH => CH_4 + CO_2$

[a]The upper two metabolisms require oxygen and thus are present in aerobic systems. The lower five metabolisms do not require oxygen and are found in anaerobic environments at both micro and macroenvironmental scales.

Table 2. Microbial Metabolic Processes That Have Been Documented by Enrichment Culturing of Subglacial Materials and by Inference From Geochemical and Isotopic Data for Alpine, Arctic, and Antarctic Valley Glacier Systems[a]

Metabolic Process	Culture	Geochemistry	Isotope Geochemistry
Aerobic respiration	4, 7, 8, 12, 13	13, 15	
FeS_2 oxidation (oxic)	3	2, 5,10, 13	6
Nitrogen fixation	7		
Nitrate reduction	4, 7	1, 2	14
FeS_2 oxidation (anoxic)		5	6
Iron reduction	7	13	16
Sulfate reduction	4, 11	16	9
Methanogenesis	4, 17	17	

[a]Sources are as follows: 1, *Tranter et al.* [1994]; 2, *Tranter et al.* [1997]; 3, *Sharp et al.* [1999]; 4, *Skidmore et al.* [2000]; 5, *Tranter et al.* [2002]; 6, *Bottrell and Tranter* [2002]; 7, *Foght et al.* [2004]; 8, *Mikucki et al.* [2004]; 9, *Wadham et al.* [2004]; 10, *Skidmore et al.* [2005]; 11, *Kivimaki* [2005]; 12, *Cheng and Foght* [2007]; 13, *Montross* [2007]; 14, *Wynn et al.* [2007]; 15, *Mitchell and Brown* [2008]; 16, *Mikucki et al.* [2009]; and 17, *Boyd et al.* [2010].

utility, when used in conjunction with other methodologies such as enrichment culturing, geochemical, and isotopic data.

1.3. Microbial Processes in Ice-Covered Lakes

A range of aerobic and anaerobic microbial metabolisms have been documented in the water column and benthic zones of permanently ice-covered lakes of the McMurdo Dry Valleys, Antarctica [e.g., *Ward and Priscu*, 1997; *Voytek et al.*, 1999; *Purdy et al.*, 2001; *Karr et al.*, 2005, 2006; *Sattley and Madigan*, 2007]. However, in these lakes, there is significant carbon fixation by photosynthesis in the water column and in benthic mats beneath the ice cover [e.g., *Takacs et al.*, 2001; *Lawson et al.*, 2004], thus this contrasts with subglacial lake systems where photosynthesis is not a viable process for carbon fixation.

The only subglacial lakes that have been investigated for their microbiology and geochemistry outside of Antarctica are those in Iceland [*Ágústsdóttir and Brantley*, 1994; *Gaidos et al.*, 2004, 2009]. Three subglacial lakes exist beneath the Vatnajokull ice cap maintained by volcanic heat [*Bjornsson*, 2002]. The largest (20 km^2) lies within the Grímsvotn caldera, and two smaller (1 km^2) lakes (western and eastern Skafta lakes) occupy the glacial divide separating Grímsvotn and the Badarbunga volcano [*Gaidos et al.*, 2009]. These lakes drain periodically via jökulhlaups (outburst floods), when rising water levels create an outflow channel beneath an ice barrier [*Gaidos et al.*, 2009]. There are two key differences between these subglacial lakes and those in the Antarctic Dry Valleys. First, the ice cover is 200–300 m thick precluding photosynthesis a mechanism for carbon fixation. Second, the hydrothermal vents at the base of the lake provide a continuing supply of reduced species and CO_2 that can be utilized by microbes as energy and carbon sources. Further, the inflow of warm waters at the lake bottom creates a thermal circulation within the lake. Anoxic bottom waters of the lake contained 5×10^5 cells mL^{-1}, and whole-cell fluorescent in situ hybridization (FISH) and polymerase chain reaction (PCR) with domain-specific probes showed these to be essentially all bacteria, with no detectable archaea [*Gaidos et al.*, 2009]. Culture-independent analyses revealed that the microbial assemblage was dominated by a few groups of putative chemotrophic bacteria whose closest cultivated relatives use sulfide, sulfur, or hydrogen as electron donors, and oxygen, sulfate, or CO_2 as electron acceptors [*Gaidos et al.*, 2009]. The phylogenetic analysis revealed an abundant phylotype closely related to *Acetobacterium bakii* (a psychrotolerant homoacetogen) and, when combined with geochemical data, suggests that acetogenesis, where bacteria extract energy from the conversion of H_2 and CO_2 to acetate, may be important in contributing organic carbon to this system [*Gaidos et al.*, 2009].

2. MICROBIOLOGY AND BIOGEOCHEMISTRY OF ANTARCTIC SAE

Water is abundant beneath the Antarctic Ice Sheet; however, there are only two sites where samples have been analyzed for biogeochemical and microbiological parameters. These sites are the accreted ice from Vostok Subglacial Lake (hereinafter referred to as Lake Vostok) in East Antarctica and saturated till from beneath the ice streams draining the West Antarctic Ice Sheet (WAIS).

2.1. Lake Vostok Accreted Ice

The Vostok ice core (78°27′S, 106°52′E) was drilled initially to generate a climate record for interior East Antarctica [e.g., *Petit et al.*, 1999]. However, radar analysis of the ice sheet in the area surrounding the Vostok ice core site indicated prominent flat reflectors at the base of the ice sheet suggesting the presence of a subglacial lake [*Kapitsa et al.,* 1996]. *Jouzel et al.* [1999] confirmed this idea, reporting approximately 200 m of accreted lake ice at the base of the Vostok ice core based on physical and chemical properties of the ice (Figure 1). The upper portion of the lake ice (termed Type I accretion ice) has a greater crystal size, number of sediment inclusions, and dissolved ionic species relative to the glacier ice above it. There is also a marked change in δD and δ^{18}O of the ice at the glacier/lake ice transition [*Jouzel et al.,* 1999]. Cores of the lake ice from 3538 to 3622 m (below the ice surface) were retrieved in the late 1990s. and microbial and biogeochemical measurements have been

made [e.g., *Priscu et al.,* 1999; *Karl et al.,* 1999; *Abyzov et al.,* 2001; *De Angelis et al.,* 2004; *Royston-Bishop et al.,* 2005; *Christner et al.,* 2006; *D'Elia et al.,* 2009]. However, in the past few years, a team of Russian and French scientists restarted the drilling program and have retrieved cores from deeper horizons, 3622–3659 m (see Figure 1) [*Bulat et al.,* 2009; *Gabrielli et al.,* 2009].

2.1.1. Microbial biomass. Cell counts have been performed on melted accretion ice samples using a range of microscopic techniques [*Priscu et al.,* 1999; *Karl et al.,* 1999; *Abyzov et al.,* 2001; *Christner et al.,* 2006; *D'Elia et al.,* 2008] and flow cytometry [*Bulat et al.,* 2009]. Cell

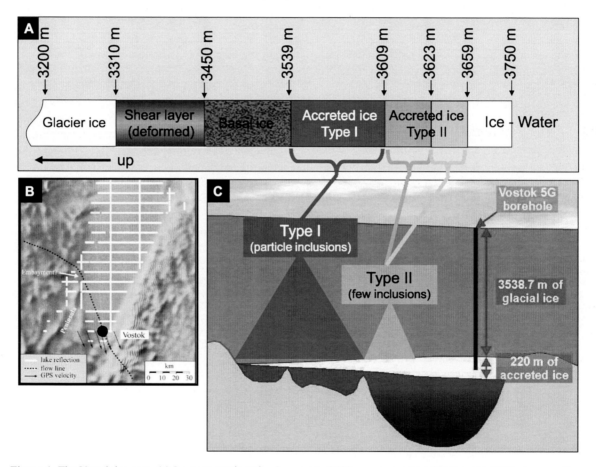

Figure 1. The Vostok ice core. (a) Ice core stratigraphy. Ice core to 3623 m was recovered in the mid-1990s, ice core from 3623 to 3659 m in 2007–2008, below 3659 m, the ice remains uncored. (b) The southern shoreline of Lake Vostok. The ice coring location is marked by a black circle. Solid white lines mark airborne radar echo profiles with lake reflections [*Studinger et al.,* 2003a]. The flow line through the Vostok ice core (black dashed line) has been derived by tracking internal structures in the ice over the lake and the gradient of the ice surface over grounded ice [*Bell et al.,* 2002]. Figure 1b from *Christner et al.* [2006, p. 2487]. Copyright 2006 by the American Society of Limnology and Oceanography, Inc. (c) Schematic cross section along the flow line shown in Figure 1b indicating possible sources for the accreted lake ice. The Type I accretion ice with a high concentration of sediment inclusions is thought to form over the shallow embayment waters, whereas Type II accretion with few sediment inclusions likely reflects formation over open waters in the main body of Lake Vostok. Figures 1a and 1c modified from B. C. Christner (unpublished work).

counts range from a few (2–12) cells mL^{-1} to 10^4 cells mL^{-1} in the more widely studied section of the ice core from 3538 to 3622 m. Only one study exists in the recently collected lower part of the accreted ice, 3623–3659 m, and cell counts are reported as 3–12 mL^{-1} [*Bulat et al.*, 2009]. There is considerable heterogeneity in physical and chemical parameters in ice from the Vostok core by depth [*Souchez et al.*, 2002]; thus, it is not surprising that there would be similar variability in biological parameters (e.g., cell counts). It appears that this variability also exists on a small scale at a given core depth. For example, *D'Elia et al.* [2008] report cell counts of 8 mL^{-1} with viable cells at 7 mL^{-1} for an ice core section at 3582 m. However, for a parallel sample at 3582 m, they report 33 colonies mL^{-1} and 13 unique phylogenetic 16S rRNA gene sequences from the cultured organisms. Typically, one can only recover ~ ≤1 % of viable microbes through culturing [*Amann et al.*, 1995]; thus, this would suggest in the subsample that produced 33 colonies mL^{-1}, the number of viable cells might be much higher, possibly up to 3×10^3 mL^{-1}. In a companion study on fungi from the accretion ice, *D'Elia et al.* [2009] see similarly higher numbers of fungal colonies per milliliter than they report for viable cells in the 3582 m sample. They interpret this to indicate heterogeneity in these samples possibly reflecting a high concentration of microbes adhering to mineral inclusions entrapped in the ice [*D'Elia et al.*, 2009].

2.1.2. Microbial viability and diversity. Karl et al. [1999] and *Christner et al.* [2006] demonstrated that cells from accretion ice at a variety of depths were viable and capable of respiring radiolabeled acetate and glucose at 3°C and 10°C, respectively. Further, viable cells existed at all sample depths of accretion ice analyzed via fluorescent microscopy [*Christner et al.*, 2006; *D'Elia et al.*, 2008]. Aerobic heterotrophs were isolated from ice at 3593 m at 25°C that were also capable of growth down to 4°C [*Christner et al.*, 2001] and from a range of ice depths at temperatures from 4°C to 22°C [*D'Elia et al.*, 2008]. Amplification and sequencing of 16S rRNA genes from extracted DNA and heterotrophic isolates from the accretion ice imply the lake is inhabited by bacteria related to the Proteobacteria (alpha, beta, gamma, and delta subdivisions), Firmicutes (low GC Gram positive), Actinobacteria (high GC Gram positive), and Bacteroidetes [*Priscu et al.*, 1999; *Christner et al.*, 2001, 2006; *Bulat et al.*, 2004] (Figure 2). Likewise, 18 unique bacterial rRNA gene phylotypes, from the Proteobacteria (alpha subdivision), Firmicutes, and Actinobacteria, were determined using 16S–23S rRNA gene intergenic spacer region sequences from heterotrophic isolates cultured from the accretion ice [*D'Elia et al.*, 2008].

D'Elia et al. [2009] have also cultured numerous fungi from the accretion ice, with 270 colonies, 38 of which were

selected for sequence analysis of ribosomal internal transcribed spacers (ITS) region, yielding 28 unique fungal ribosomal DNA sequences. All sequences were obtained directly from cultured isolates as attempts to PCR amplifying fungal ITS DNA directly from ice core melt water were unsuccessful [*D'Elia et al.*, 2009]. Sequences closest to *Rhodotorula mucilaginosa* were the most frequently observed. *Rhodotorula* isolates showed optimal growth at 22°C but also grew well at 15°C and 4°C, indicative of cold tolerance [*D'Elia et al.*, 2009].

The uncultured microbial community, based on partial 16S rRNA gene sequences, amplified from melted accretion ice at various depths, shows evidence for putative chemotrophic metabolisms (Figure 2) [*Bulat et al.*, 2004; *Christner et al.*, 2006; *Lavire et al.*, 2006]. The partial 16S rRNA gene sequences cluster among bacterial species with metabolisms dedicated to iron and sulfur respiration or oxidation (Figure 2) [*Christner et al.*, 2006] and are closely related to H_2 oxidizing chemoautotrophs [*Bulat et al.*, 2004; *Lavire et al.*, 2006]. *Christner et al.* [2006] note that phylogenetic relationships to cultured species especially those that are distant and/or based on partial 16S rRNA gene sequence coverage do not provide robust information on physiology, and therefore, such inferences are equivocal. However, even with such caveats, the data does suggest that in addition to carbon, iron, sulfur, and hydrogen may play a role in the bioenergetics of microorganisms that occur in Lake Vostok [*Bulat et al.*, 2004; *Christner et al.*, 2006]. The results from both culturing and culture-independent methods indicate that there is potentially a reasonably diverse microbial assemblage in the accretion ice and thus by inference in Lake Vostok.

2.1.3. Aqueous geochemistry and particulate materials. Concentrations of solute derived from mineral weathering, e.g., Ca^{2+}, Mg^{2+}, and SO_4^{2-} are significantly higher in Type I accretion ice (3539–3609 m) than in the glacial ice above the accreted ice or the Type II accretion ice below (3610–3622 m), *Christner et al.* [2006] (Figure 1 and Table 3). Similarly, concentrations of rare earth elements, also sourced from weathering of crustal material, are on average an order of magnitude higher in the Type I accretion ice relative to the Type II accretion ice [*Gabrielli et al.*, 2009]. Type I accretion ice contained mineral particulates ranging from 0.1 to 45 µm [*Royston-Bishop et al.*, 2005]. Some of these particulates were aggregated into larger "inclusions" up to a few millimeters in diameter, as observed by *Priscu et al.* [1999] and *Souchez et al.* [2002]. The deepest accretion ice (3622 m) of Type II had the lowest total particle count of all accretion ice samples [*Christner et al.*, 2006]. A substantial decrease in mineral particulates at depths below 3610 m is consistent with the model that the ice accreted over a deep part of the

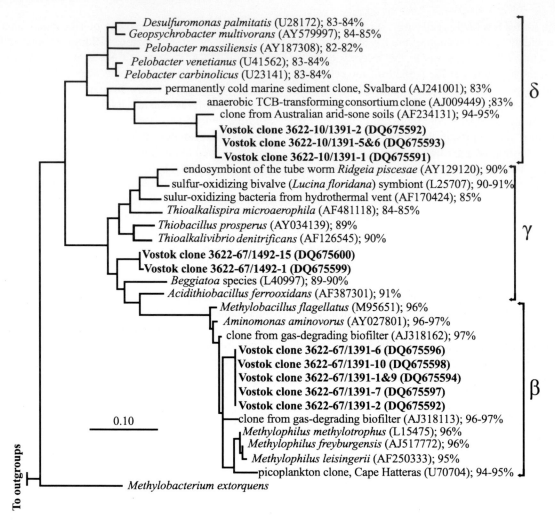

Figure 2. Phylogenetic analysis of β, γ, and δ proteobacterial small subunit rRNA sequences amplified from DNA in the 3622 m Vostok accretion ice sample. The sequences obtained were aligned on the basis of secondary structure [*Ludwig et al.*, 2004] and an 820-nucleotide mask (552–1370, Escherichia coli 16S rRNA gene numbering) of unambiguously aligned positions was constructed. Clones are in bold font and designated by the depth of recovery, the concentration of meltwater added to each initial PCR reaction, the reverse primer used for amplification, and the clone number (i.e., Vostok clone 3622-10/1391-1 is clone no. 1 amplified from sample recovered at 3622 m with the use of 10 mL of meltwater for the template and the small subunit rRNA nucleotide primers 515F and 1391R). GenBank accession numbers are listed in parentheses, followed by percent identity to the nearest Vostok clone. The maximum likelihood tree was generated by fastDNAml [*Olsen et al.*, 1994], and the scale bar indicates 0.1 fixed substitutions per nucleotide position. The 16S rRNA gene sequences of *Aquifex pyrophilus* and *Methanobacterium thermoautotrophicum* were used as outgroups. From *Christner et al.* [2006, p. 2495]. Copyright 2006 by the American Society of Limnology and Oceanography, Inc.

water column east of the bedrock ridge that separates the shallow embayment from the main lake (Figure 1) [*Bell et al.*, 2002]. Concentrations of cells, nonpurgeable organic carbon (NPOC, analogous to dissolved organic carbon (DOC)), and solutes for the uppermost waters of Lake Vostok have been estimated using partition coefficients from Lake Bonney, another permanently ice-covered lake in Antarctica (Table 3) [*Christner et al.*, 2006]. *Christner et al.* [2006]

characterize Lake Vostok as ultraoligotrophic with a total NPOC pool of <250 μmol L^{-1}. Based on this and other calculations, *Christner et al.* [2006] argue that this would only support reproductive growth based on heterotrophy for a proportion of the estimated lake microbial community, and thus, chemolithoautotrophic primary production could be a viable supplement to the microbial food web. This is consistent with aspects of the phylogenetic data that imply that the

Table 3. NPOC, Biomass, and Major Ion Concentrations in Glacial Ice and Accretion Ice From the Vostok 5G Borehole[a]

| Constituent | NPOC μmol L^{-1} | Biomass Cells mL^{-1} | Concentration μmol L^{-1} | | | | | | Total Dissolved Solids mmol L^{-1} |
			Na+	K$^+$	Ca^{2+}	Mg^{2+}	Cl$^-$	SO$_4{}^{2-}$	
Glacial ice (average)	16	120	2.4	0.4	1.1	0.4	2.8	1.8	0.009
Type I accretion ice (average)	65	260	22.0	0.3	6.8	5.8	17.0	9.1	0.061
Type II accretion ice (average)	35	83	0.9	0.1	1.0	0.2	0.9	0.2	0.003
Embayment water[b,c]	160	460	10,000.0	14.0	2,600.0	2,700.0	7,300.0	11,000.0	34.0
Main lake water[c,d]	86	150	430.0	5.9	370.0	69.0	400.0	180.0	1.5
Surface seawater (average)	40–80	0.05–5 $\times$ 10^5	48,000.0	10,000.0	10,000.0	54,000.0	560,000.0	28,000.0	710.0

[a]Modified from the work of *Christner et al.* [2006, p. 2498]. Copyright 2006 by the American Society of Limnology and Oceanography, Inc. Predicted concentrations for lake water were derived as described in the work of *Christner et al.* [2006]. Average seawater concentrations of the various analytes are provided for comparison.

[b]Prediction on the basis of average Type I accretion ice composition.

[c]Partitioning coefficients for NPOC, biomass, Na$^+$, K$^+$, Ca^{2+}, Mg^{2+}, Cl$^-$, and SO$_4{}^{2-}$ are 0.40, 0.56, 0.0021, 0.022, 0.0026, 0.0022, 0.0023, and 0.00083, respectively.

[d]Prediction on the basis of average Type II accretion ice composition.

most closely related organisms to sequences in the gamma subdivision of proteobacteria are Fe and S chemoautotrophs (Figure 2) [*Christner et al.*, 2006].

Clathrates of various compositions of atmospheric gases have been proposed for Lake Vostok based on theoretical calculations using pressure and temperature estimates and a variety of lake water residence times [*McKay et al.*, 2003]. *McKay et al.* [2003] calculations indicate the lake waters may be supersaturated with oxygen and that incorporation of 1% CO_2 into the clathrates would increase their density resulting in their sinkage from the ice-water interface. However, there are numerous unknowns in these calculations, related to both gas production and consumption by biologic processes and also by mineral weathering of sulfides, silicates, and carbonates and the associated chemical oxygen demand and/or dissolved inorganic carbonate production. Direct sampling of the lake waters is required to determine the actual dissolved gas concentrations.

2.2. Basal Sediments (Till) Beneath WAIS Ice Streams (Kamb and Bindschadler)

Forty-three and 68 cm long cores were recovered from the sediments underlying the Kamb Ice Stream (KIS), and Bindschadler Ice Stream (BIS), respectively, during a 2000/2001 expedition by Kamb and colleagues (Figure 3) [*Vogel et al.*, 2005]. The goal of the research was to investigate geophysical aspects of the sediment in relation to motion of the ice streams [*Vogel et al.*, 2005]. The ice at the base of KIS was estimated at ≥20,000 and <100,000 years based on the ice

stratigraphy [*Siegert and Payne*, 2004; *Catania et al.*, 2005; *Lanoil et al.*, 2009]. This location was also most likely ice covered and isolated from the atmosphere for at least the past 400,000 years [*Scherer et al.*, 1998]. The cores were not taken with the goal of microbial analysis; however, since these were the only sediment samples that existed from beneath the WAIS ice streams, they were subject to some basic analyses of the microbial community (KIS only) and pore water geochemistry (KIS and BIS) [*Lanoil et al.*, 2009; *Skidmore et al.*, 2010].

2.2.1. Microbial biomass. Cell counts via fluorescence microscopy were $2.7 \pm 0.1 \times 10^7$ and $1.5 \pm 0.1 \times 10^7$ cells g^{-1} wet sediment in the upper and lower halves of the KIS sediment core, respectively [*Lanoil et al.*, 2009]. It was estimated that the organisms within the KIS sediment core may have undergone a maximum of 4.1 doublings during core storage, at 4°C for 15 months prior to sampling [*Lanoil et al.*, 2009]. Thus, the in situ cell abundance in these sub-ice sheet sediments is likely in the range $2–4 \times 10^5$ cells g^{-1}, similar to values seen in subglacial sediments in Svalbard, $2.1–5.3 \times 10^5$ cells g^{-1}, and less than those observed for New Zealand glaciers, $2.3–7.4 \times 10^6$ cells g^{-1} [*Foght et al.*, 2004; *Kastovska et al.*, 2007; *Lanoil et al.*, 2009].

2.2.2. Microbial viability and diversity. Approximately 0.005% of cells ($1.5 \pm 1.1 \times 10^3$ cells g^{-1}) from the upper half of the sediment core were culturable aerobically at room temperature on R2A plates. Incubation at 4°C increased culturability to ~ 0.2% and 0.1% ($5.6 \pm 0.8 \times 10^4$ and $1.8 \pm 0.9 \times 10^4$ cells g^{-1}) in the upper and lower halves,

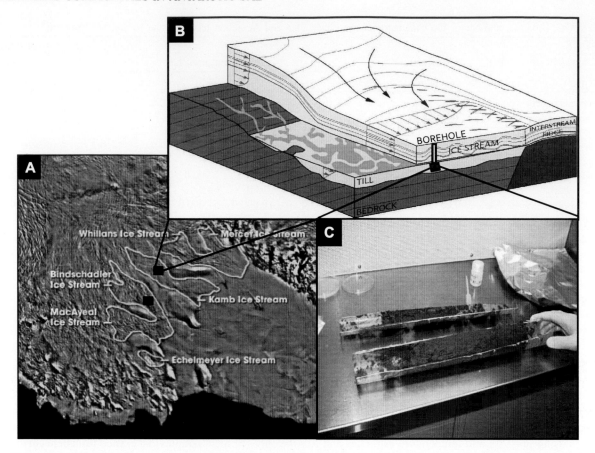

Figure 3. (a) The Siple Coast Ice Streams, West Antarctica, with sampling locations of subglacial sediment cores on Kamb and Bindschadler ice streams. (b) Schematic three-dimensional cross section of the ice stream and the underlying substrate. Flow vectors in the ice and till are also shown. (c) Subglacial sediment core from beneath Kamb Ice Stream. Core is ~26 cm in length. Figure 3a from http://icestories.exploratorium.edu/dispatches/mission-form-siple-dome/. Figure 3b modified from D. D. Blankenship (unpublished work). Figure 3c courtesy of S. Vogel.

respectively [*Lanoil et al.*, 2009]. The higher culturability at 4°C is consistent with adaptation to growth at the low, stable temperatures in the subglacial environment or to adaptation during storage at 4°C. No growth was observed at either temperature under anaerobic conditions on R2A, indicating either a lack of viable fermenters or insufficient electron acceptors in the medium [*Lanoil et al.*, 2009]. Sixty-three isolates from the aerobic 4°C plates all grew well at 9°C; 42% of these also grew at 25°C, while the other 58% were unable to grow at this elevated temperature, further supporting cold tolerance and/or cold adaptation of these isolates [*Lanoil et al.*, 2009]. All isolates fell into three groups based on their

16S rRNA gene sequences (at >98% similarity within each group). The most abundant isolate was closely related to Betaproteobacteria (designated *Comamonas*-like) environmental clone T93 from the KIS sediments and environmental clones and isolates obtained from other subglacial and polar environments (Figure 4) [*Foght et al.*, 2004; *Skidmore et al.*, 2005; *Lanoil et al.*, 2009]. Three sequence clusters, all within the Betaproteobacteria class, were observed in clone libraries of 16S rRNA gene sequences from both the upper and lower sediment samples (Figure 4) [*Lanoil et al.*, 2009]. No observable differences were found between the two clone libraries in either group presence or group representation

Figure 4. (opposite) Neighbor-joining phylogenetic tree showing the relationship of sequences obtained from KIS sediment samples (bold and labeled KIS) to sequences from other subglacial systems (bold) and other sequences from public databases. Characterized isolate names are italicized; uncultivated clone sequences or uncharacterized isolates are in plain text. GenBank accession numbers are included after the names. Bootstrap values (100 replications) generated by the neighbor-joining method are shown above relevant nodes, and those generated by maximum-parsimony analysis are shown below; only bootstrap values above 50 are shown. Scale bar indicates five conserved nucleotide changes per 100 base pairs. The tree is rooted with *E. coli* (J01695). Modified from the work of *Lanoil et al.* [2009].

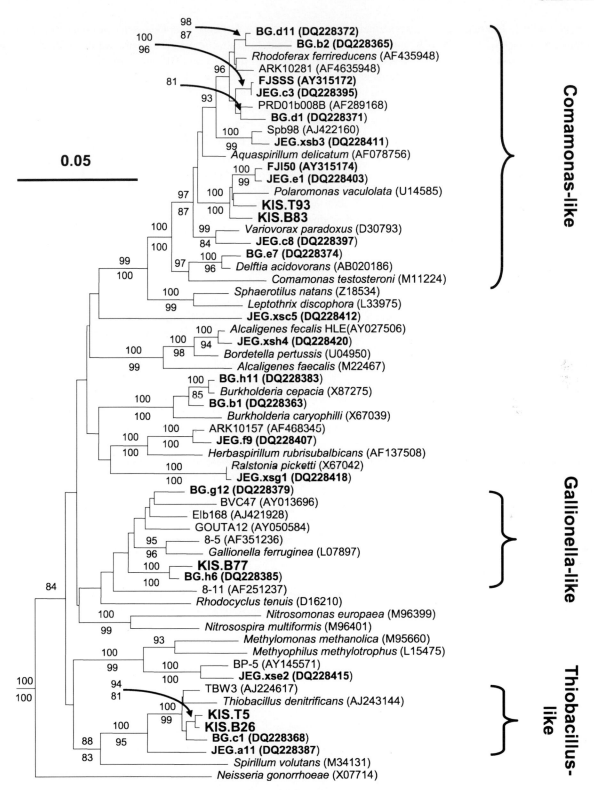

Figure 4

[*Lanoil et al.*, 2009]. Interestingly, these sequences were closely related to groups of organisms (*Comamonas, Gallionella,* and *Thiobacillus*) previously observed as abundant groups in Alpine and Arctic subglacial systems and proglacial soils (FJ, BG, and JEG (John Evans Glacier) sequences in (Figure 4) [*Foght et al.*, 2004; *Skidmore et al.*, 2005; *Nemergut et al.*, 2007; *Lanoil et al.*, 2009]. *Lanoil et al.* [2009] analysis suggests a relatively simple microbial community in this assemblage, likely enriched relative to the in situ environment through incubation at 4°C for 15 months, with both heterotrophs and chemolithoautotrophs. The closest cultured relatives of sequences from the clone libraries are neutrophilic iron oxidizers (i.e., *Gallionella*) and acidophilic iron or sulfur oxidizers (i.e., *Thiobacillus*) (Figure 4) implying that iron and sulfide oxidation may be important in the KIS sediments. Microbial sulfide oxidation is a possible energy source for Lake Vostok [*Christner et al.*, 2006] and has been proposed as a significant process in sulfate generation in subglacial sediments at Bench Glacier, Alaska, where a similar microbial community composition to the KIS sediments has been documented [*Skidmore et al.*, 2005]. If these groups of microorganisms are also prevalent in situ in KIS sediments, such lithotrophic metabolism, combined with autotrophic carbon fixation, could provide the energetic basis of the microbial community and its persistence over the long time period this system has been isolated from direct interaction with the atmosphere [*Lanoil et al.*, 2009]. There is also organic carbon of unknown origin in the sediments (0.5–1.5 wt %) sufficient to support heterotrophic microbial activity, presuming it is in a usable form [*Lanoil et al.*, 2009]. Heterotrophic activity could be carried out by organisms similar to those isolated from the sediments at 4°C on heterotrophic media [*Lanoil et al.*, 2009] and/or inferred present based on the culture-independent analyses.

2.2.3. Aqueous geochemistry and particulate materials. Data for the chemistry of pore waters collected from till beneath the KIS and BIS are compared with those of waters collected from a jökulhlaup near Casey Station, Antarctica [*Goodwin*, 1988] (Table 4). Also included in Table 4 are the compositions of typical glacial runoff [*Brown*, 2002], basal waters from Haut Glacier d'Arolla (HGA), European Alps [*Tranter et al.*, 2002], concentrated late season waters from Robertson Glacier (RG), Canadian Rockies [*Sharp et al.*, 2002], and early season waters from JEG, Canadian High Arctic [*Skidmore and Sharp*, 1999]. The waters from these glaciers have been in longer-term storage at the bed in environments with high rock/water ratios, presumably comparable to the basal tills of the KIS and BIS [*Skidmore et al.*, 2010]. The chemistry of pore waters found in the proglacial zone of Finsterwalderbreen, Svalbard [*Cooper*, 2003] are also included as examples of waters that may be in till pore waters beneath smaller glaciers.

The BIS and KIS pore waters are more concentrated than all but those from FIN and Casey Station by at least an order of magnitude, and additionally have significantly different proportions of ions (Table 4) [*Skidmore et al.*, 2010]. The KIS pore waters are more dilute than the BIS pore waters, but they have similar proportional ion concentrations (Table 4). Cl^- and Na^+ concentrations are ~0.2%–0.4 % and 1.4%–7.4% of seawater, compared with <0.01% for most glacial runoff. The pore waters cannot be a direct dilution of marine water or dissolution of common marine evaporite minerals, since $Na^+ \gg Cl^-$, thus excluding sea salt and $SO_4^{2-} \gg Ca^{2+}$, excluding evaporite gypsum as primary solute sources [*Skidmore et al.*, 2010]. The sub-ice stream pore water type is Na-SO_4-Ca-Mg-HCO_3, compared with the Ca-HCO_3-SO_4-Mg type previously measured in most glacial runoff, including concentrated runoff from HGA and RG [*Skidmore*

Table 4. The Concentration of Major Ions in Typical Glacial Runoff; Concentrated Runoff From Robertson Glacier, Canadian Rockies; John Evans Glacier, Canadian High Arctic, Basal Waters From Haut Glacier d'Arolla (HGA), Switzerland, Waters Draining a SW Sector of the Antarctic Ice Sheet Near Casey Station and Pore Waters in Till in the Proglacial Zone of Finsterwalderbreen (FIN), Svalbard and Beneath the Kamb and Bindschadler Ice Streams (KIS and BIS), West Antarctica[a]

Glacier	pH	Na^+	K^+	Ca^{2+}	Mg^{2+}	Cl^-	SO_4^{2-}	HCO_3^-	$\sum^+$
Typical runoff	7.5	0.1	0.02	0.2	0.1	0.1	0.1	0.4	0.7
Robertson	8.3	0.02	0.01	1.7	0.5	0.01	1.0	2.6	3.4
John Evans	8.3	0.3	0.04	1.3	0.3	0.1	1.5	0.4	3.5
HGA	7.4	0.02	0.02	0.9	0.1	0.01	0.4	0.6	2.0
FIN (Pore)	ND	0.08	0.03	3.1	1.8	0.06	4.1	0.8	9.9
Casey Station	8.4	6.6	0.2	0.5	0.1	2.2	0.3	3.5	8.0
KIS (Pore)	6.5	20.0	0.6	6.4	3.1	1.1	17.0	4.1[b]	39.6
BIS (Pore)	6.5	35.0	0.7	9.0	8.6	2.0	31.0	7.5[b]	70.9

[a]Modified from the work of *Skidmore et al.* [2010]. Published by Wiley-Blackwell. Units are mmol/L (except for pH), values rounded to 1 significant decimal figure and $\sum^+$ denotes the sum of the positive charge in meq/L. Data sources are given in the text. ND, no data;
[b]Derived from charge balance.

et al., 2010]. The pore water type is similar to that reported for high-latitude groundwaters and permafrost, where silicate weathering and sulfide oxidation are argued to be significant solute-generating processes [*Parnachev et al.*, 1999; *Stotler et al.*, 2009].

Subglacial biogeochemical weathering is the most likely process to generate the observed water chemistry for BIS

and KIS pore waters (Table 4) [*Skidmore et al.*, 2010]. In the ice sheet sediments, carbonate and silicate hydrolysis is followed by carbonate dissolution driven by protons from sulfide oxidation and/or the oxidation of organic matter (Table 1) [*Skidmore et al.*, 2010]. Carbonate saturation is rarely reached in melt waters beneath smaller ice masses, since the residence time of the waters is relatively short (0.01–

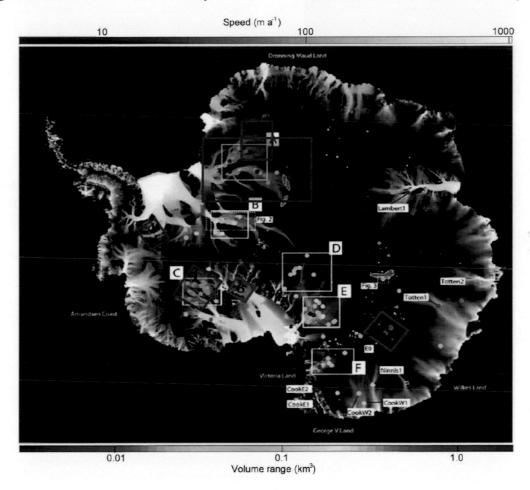

Plate 1. Locations and volume-range estimates for 124 active lakes under the Antarctic ice sheet, shown as points color coded by the volume range. Background grayscale shading shows a combination of satellite-radar-derived surface velocities [*Joughin et al.*, 1999, 2006] and balance velocities [*Bamber et al.*, 2008]. White outlines filled in magenta represent previously published lake locations: outlines are drawn for Vostok subglacial lake [*Studinger et al.*, 2003b], the Recovery Glacier lakes [*Bell et al.*, 2007], and active lakes mapped with radar altimetry in Adventure Trench [*Wingham et al.*, 2006] and with interferometric synthetic aperture radar on Kamb Ice Stream [*Gray et al.*, 2005]. Lakes mapped from airborne radar are represented by circles of diameter equal to the published lake length [*Siegert*, 2005; *Carter et al.*, 2007; *Popov and Masolov*, 2007]. Blue shaded boxes, 1–4. Areas with extensive saturated sediments in the subglacial environment as reported by 1, *Anandakrishnan et al.* [1998]; *Bell et al.* [1998]; *Blankenship et al.* [2001]; *Studinger et al.* [2001]; *Peters et al.* [2007]; 2, *Bamber et al.* [2006]; 3, *Le Brocq et al.* [2008]; 4, *Smith and Murray* [2009]. Red shaded boxes, five to six areas with hydrologically dynamic components involving significant water volume changes (1 km³) over annual time frames. Note the shaded boxes are illustrative and contain the extensive saturated sediments and hydrologically dynamic features; however, they do not imply that the entirety of the shaded area contains a specific feature. Modified from the work of *Smith et al.* [2009]. Reprinted from the *Journal of Glaciology* with permission of the International Glaciological Society.

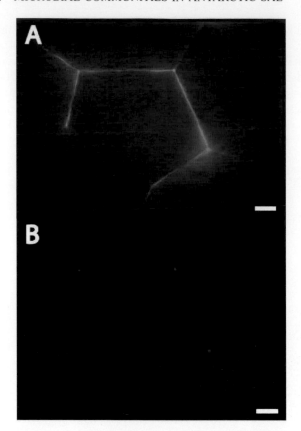

Plate 2. Images of a thin section of ice prepared with MilliQ water, 50 ppm sodium fluorescein, and 10^6 *Sporosarcina* sp. B5 cells mL^{-1}, treated with DAPI (4′,6-diamidino-2-phenylindole, a fluorescent stain that binds strongly to DNA), at $-10°C$. (a) Liquid water veins are highlighted by the presence of high concentrations of fluorescein. (b) DAPI-stained image of the same field of view highlighting higher DAPI concentrations and thus by inference higher *Sporosarcina* sp. B5 cell concentrations at the triple junctions in the liquid vein network. Scale bar = 100 μm on both images. *Sporosarcina* sp. B5 was isolated from basal ice from Taylor Glacier, Antarctica. Image credit: Tim Brox, Skidmore lab group.

10 years). However, this may not be the case beneath ice sheets, where water residence times can be $>10^3$ years [*Tulaczyk et al.*, 2000; *Wadham et al.*, 2010]. The BIS and KIS waters are approaching saturation with respect to carbonates, likely inhibiting further carbonate dissolution, and so allowing protons generated by the oxidation of sulfides and organic matter to be used for silicate weathering [*Skidmore et al.*, 2010]. Research on the microbes present in glacial till, to date, is consistent with the above conceptual model. Microbes appear to catalyze sulfide oxidation in subglacial environments [*Sharp et al.*, 1999; *Skidmore et al.*, 2005; *Mikucki et al.*, 2009], and the closest cultured organisms

relative to the phylotypes from KIS till clone libraries, i.e., *Gallionella* and *Thiobacillus*, have Fe- and sulfide-oxidizing physiologies [*Lanoil et al.*, 2009]. Further, biologically driven sulfide oxidation has been proposed as a significant sulfate source at Bench Glacier, Alaska, an environment with a similar microbial community composition to that from KIS sediments [*Skidmore et al.*, 2005].

Further evidence consistent with sub-ice stream microbial activity is provided by features of the mineral grains from till beneath the neighboring Whillans Ice Stream [*Tulaczyk et al.*, 1998]. Sand-sized grains in the till show little indication of fresh crushing and have predominantly rounded edges. The surfaces of the grains bear evidence of extensive chemical weathering, such as prominent etch pits, which have partially removed the morphological evidence of comminution. These features on aluminosilicate minerals are evidence of microbial activity [*Barker et al.*, 1998]. The microbiological data on the KIS sediments, the proposed geochemical reaction series, and the till morphology all strongly indicate an important role for microbes in driving biogeochemical weathering processes in the KIS till.

3. UNEXPLORED SAE BENEATH THE ANTARCTIC ICE SHEET

Subglacial hydrology has a first-order impact on the oxygen content, and therefore oxygen and redox potential (Eh), of the subglacial environment [*Tranter et al.*, 2005]. The adage, where there is water there is life, holds in subglacial environments as well as other extreme environments at the Earth's surface [*Priscu et al.*, 1998; *Rothschild and Mancinelli*, 2001]. Microbes are normally associated with subglacial debris [*Sharp et al.*, 1999; *Rothschild and Mancinelli*, 2001], since the debris contains potential sources of energy, carbon, and nutrients. Currently, direct observations of subglacial hydrological conditions have only been made in a few locations beneath the Antarctic Ice Sheet. *Gow et al.* [1968] report evidence for a thin film of liquid water beneath ~2000 m of ice at the base of the Byrd ice core, West Antarctica. Similarly, *Boereboom et al.* [2007] note subglacial water entering the base of the 2774 m deep EPICA ice core borehole in Dronning Maud Land, East Antarctica. Drilling by the CalTech research group on the WAIS revealed saturated sediments beneath the Whillans, Kamb, and Bindschadler ice streams (former ice streams B, C, and D) and subglacial water pockets beneath Kamb Ice Stream C [e.g., *Engelhardt et al.*, 1990; *Vogel et al.*, 2003; *Vogel*, 2004].

There is a range of unexplored SAE beneath the Antarctic Ice Sheet: (1) subglacial lake waters, (2) subglacial lake sediments, (3) saturated sediments in non-lake settings, (4) systems connecting subglacial lakes and saturated sediments: hydrologic, with flowing water, conduit, or linked cavity system and solid phase, basal ice, and the liquid water therein as films on particles and vein networks. Freeze-on in upstream locations and melting in downstream locations results in transport of microbes and sediments from potentially contrasting environments. The distribution of some of these SAE is documented in Plate 1. Here follows a brief description of what is known about the distribution of these SAE and potential microbial metabolisms that may exist in these environments.

3.1. Subglacial Lakes and Subglacial Lake Sediments

Accreted lake ice from the base of the Vostok Core, East Antarctica confirms the presence of Lake Vostok and provides samples for inference about lake conditions [e.g., *Petit et al.*, 1999; *Priscu et al.*, 1999; *Karl et al.*, 1999]. The remainder of our knowledge of subglacial hydrology in Antarctica is via remote sensing techniques. These analyses have revealed a large number of subglacial lakes with 386 as the current best estimate, [*Wright and Siegert*, this volume]. However, the authors note there is inherent bias in their inventory, due to the available data coverage. There is potential for sediment accumulation at the base of all the subglacial lakes discovered, to date, and it has been estimated that there may be 300–400 m of unconsolidated sediments in the deepest lake, Lake Vostok [*Filina et al.*, 2008].

Subglacial Antarctic lakes can be characterized into three main types following *Wright and Siegert* [this volume]. Type 1 lakes are those in the ice sheet interior. These can be (1) lakes in areas of low relief, thus with little depth potential for the size of their surface area; (2) lakes that occupy topographic depressions with the elongate, rectilinear morphology characteristic of tectonically controlled features [*Tabacco et al.*, 2006]; and (3) lakes situated on the flanks of subglacial mountain ranges that are characteristically small (<10 km long) features constrained within steep local topography. Type 2 lakes are associated with the onset of enhanced ice flow features. These lakes can be both small (<10 km long) [*Dowdeswell and Siegert*, 2002] or larger, up to (1000 km^2) [*Bell et al.*, 2007]. Type 3 lakes are beneath the trunks of ice streams. These are generally of smaller size and are more transient in nature, with large proportions of their water volume filling and draining on an annual or semiannual basis [*Smith et al.*, 2009].

Given the wide range of lake size and type, it is difficult to characterize a singular model for potential aqueous and benthic microbial communities. However, where there is basal melting of meteoric ice, there is supply of gases to the subglacial environment of atmospheric composition (at the time of ice formation), thus including ~21% oxygen. The process of lake

ice accretion to the base of the ice sheet results in the rejection of gases into the lake waters. Thus, where basal melting and refreezing occur in the same lake system, e.g., at Lake Vostok, there is a process concentrating oxygen into the lake waters. This supply of atmospheric oxygen and DOC would facilitate aerobic heterotrophic microbial processes in the upper parts of the water column and the rate of oxygen consumption down the water column would be determined by both chemical and biological oxygen demand. Based on inferences from the accretion ice from Lake Vostok, this lake would be considered ultraoligotrophic, i.e., highly nutrient limited [*Christner et al.*, 2006]; however, the authors note that inputs of DOC and other nutrients that are supplied from lake marginal environments by subglacial processes are not accounted for in this analysis. *Priscu et al.* [2008] suggest that chemolithoautotrophs may also be an important source of new organic carbon in subglacial lake systems. The same would also hold true in other subglacial environments. The new carbon produced by this process can then provide a substrate to support heterotrophic activity. Clathrates of various compositions of atmospheric gases have been proposed for Lake Vostok based on theoretical calculations using pressure and temperature estimates [*McKay et al.*, 2003], but their existence and thus possible more widespread distribution, i.e., in other subglacial lakes remains unknown.

One also has to consider the possibility that some of the lakes could be highly stratified with concentrated brines overlain by a freshwater cap, as are some of the permanently ice-covered lakes in the McMurdo Dry Valleys (see review by *Green and Lyons* [2009]). Brines can be as high as $3 \times$ seawater in the base of these McMurdo Dry Valley lakes such as at Lake Vida [*Doran et al.*, 2003]. If this is the case, then microbes in waters at these lower levels would need to be halotolerant or halophilic [*Mondino et al.*, 2009]. Where sediments do accumulate in these lakes, one can envisage that there may be an oxygen gradient in the sediments if oxygen is still present in the water column at the sediment-water interface. Oxygen concentrations would likely decrease with sediment depth as a function of both chemical and biological oxygen demand as in subaerial freshwater lakes and in marine sediments [*Wetzel*, 2001; *Fry et al.*, 2008]. As oxygen is depleted, organisms with metabolisms that do not require oxygen (Table 1) would be favored as in freshwater and marine sediments.

3.2. Subglacial Sediments Not Associated With Subglacial Lakes

Significant subglacial sediment thicknesses have been inferred for certain regions of the Antarctic Ice Sheet that are not necessarily linked to the presence of subglacial lakes (Plate 1). The sediment accumulations are often dependent

Table 5. Metabolisms Measured and Inferred in Deep Terrestrial and Marine Subsurface Lithoautotrophic Microbial Ecosystems and a Subglacial Volcanic Lake[a]

Environment	Depth (km)	Maximum T (°C)	Maximum Cell Count mL^{-1}	Metabolisms	Methods	Ref
Marine (mudrocks)	1.6	100	1.5×10^6	Methanogenesis, ANME	16S rRNA geochemistry	1
Terrestrial (shale)	2.7	76	1×10^5	FeRB, SRB, fermentation	Culture	2
Terrestrial (granite)	5.3	75	ND	Fermentation, FeRB (possible)	Culture	3
Terrestrial (quartzite)	1.4	40	5×10^5	SRB, acetogenesis, methanogenesis	16S rRNA geochemistry	4
Terrestrial (subglacial lake)	0.4	4	5×10^5	SRB, FeRB, fermentation, acetogenesis, S, and H_2 oxidation	16S rRNA geochemistry, FISH, culture	5

[a]References are as follows: 1, *Roussel et al.* [2008]; 2, *Onstott et al.* [1998]; 3, *Szewzyk et al.* [1994]; 4, *Lin et al.* [2006]; 5, *Gaidos et al.* [2009]. Metabolism abbreviations are as follows: FeRB, iron reduction; SRB, sulfate reduction; ANME, anaerobic methane oxidation. Method abbreviations are as follows: 16S rRNA, metabolism inferred based on similarity in 16S rRNA gene sequence with known organisms with that metabolism or group in the case of ANME; Geochemistry, a metabolism inferred based on aqueous geochemical measurements and, in some cases, stable isotope geochemistry; FISH, fluorescent in situ hybridization with group specific probes targeting the 16S rRNA gene.

on the underlying bedrock geometry. Sediment thicknesses of 400–600 m have been reported beneath an ice stream draining the WAIS [*Anandakrishnan et al.*, 1998] with thicknesses of 1–2.4 km in the ice stream onset region [*Bell et al.*, 1998], and up to 5 km have been inferred in the deepest basins beneath the WAIS [*Blankenship et al.*, 2001; *Studinger et al.*, 2001]. Similarly in East Antarctica, sediment thicknesses of up to 3 km are reported in subglacial troughs in the onset area of an enhanced flow unit, Slessor Glacier, Coats Land [*Bamber et al.*, 2006]. Further, *Le Brocq et al.* [2008] note the potential for a large subglacial basin more than 1500 m below present-day sea level, beneath the Recovery Glacier and its catchment in East Antarctica that may contain substantial thicknesses of saturated sediment. The discoveries of these large volumes of subglacial sediment have occurred in regions where there is available data. However, as noted by *Bamber et al.* [2006], there is a lack of appropriate geophysical data for the interior of the East Antarctic Ice Sheet (EAIS) and limited studies addressing the relationship between ice dynamics and geological boundary conditions in these regions. *Smith and Murray* [2009] note a patchwork of basal conditions beneath the Rutford Ice Stream and the mobile nature of these saturated sediments. In the deep troughs, the bed consists of spatially extensive areas of soft sediment, deforming to a depth of at least a few meters, whereas on the bed high point in the center of the ice stream, regions of basal sliding and bed deformation co-exist adjacent to one another [*Smith and Murray*, 2009].

The significant sediment thicknesses described above would also likely be depleted of oxygen in the upper portion of the sediment package, thus favoring anaerobic metabo-

lisms in the likely greater anoxic lower portion (Table 1). Here one could envisage similar anaerobic metabolisms as documented in other deep terrestrial and marine sedimentary environments (Table 5). Whether the low rate of water circulation and thus the relatively high isolation time for organisms in these deep sediments may lead to ecosystems with extremely low diversity microbial communities, as has been documented in a terrestrial system [*Chivian et al.*, 2008] is unknown, but is certainly a possibility.

Temperatures at depth in the sedimentary package would be above 0°C due to the Earth's background geothermal heat flux. In certain sub-ice locations, where there is evidence of sub-ice sheet volcanism [*Blankenship et al.*, 1993; *Behrendt et al.*, 1998], the heat flux may be elevated and may also be accompanied by hydrothermal fluids. These fluids are often rich in reduced metals and gases such as H_2, CO_2, and H_2S that are potential chemical energy sources that could be introduced into either the subglacial sediments or lake waters depending on connectivity. If the hydrothermal fluid input is in a subglacial lake setting, then the basal waters of Icelandic subglacial lakes may serve as a reasonable analog. Acetoclastic organisms are implied as an abundant member of the microbial community in one of the Icelandic subglacial lakes (Table 5) [*Gaidos et al.*, 2009]. This is interesting, as few other Earth surface systems have niche environments, where acetoclastic organisms can outcompete methanogens. These volcanic gases could support a microbial ecosystem driven purely by crustal sources of H_2 and CO_2 that are used as energy and carbon sources, which is independent of photosynthetic carbon [*Pedersen*, 2000].

3.3. Subglacial Lakes and Their Hydrological Connections

One hundred and twenty-four active subglacial lakes were detected from laser altimetry data between 2003 and 2008 and estimates of volume changes for each lake made [*Smith et al.*, 2009]. Lakes sometimes appear to transfer water from one to another, but also often exchange water with distributed sources undetectable from the laser altimetry data, suggesting that the lakes may provide water to, or withdraw water from, the hydrologic systems that lubricate glacier flow [*Smith et al.*, 2009]. Previous studies of subglacial lakes [*Gray et al.*, 2005; *Wingham et al.*, 2006; *Fricker and Scambos*, 2009] have concluded that subglacial lakes are hydrologically linked, especially when the surface displacements for adjacent lakes suggested draining upstream and filling downstream. The water volumes and timescales for these lake drainage events can be considerable. *Wingham et al.* [2006] document 1.8 km^3 of water travelling a distance of 290 km over 16 months beneath the EAIS. Similarly, *Fricker and Scambos* [2009] document 0.1 to 2 km^3 draining over a range of timescales from several months to years beneath the WAIS. Smaller water volumes and shorter timeframes for these events have also been noted. For example, *Gray et al.* [2005] measured 0.01 and 0.02 km^3 for the BIS and KIS, respectively, moving in under a month. However, it remains unknown what the morphology of the drainage system is between these lakes, whether it is via single or multiple arterial channels, or distributed channels or cavities, or some combination thereof. *Peters et al.* [2009] model for the Adventure Trench outburst notes a single conduit is viable for water drainage; however, *Carter et al.* [2009] note a distributed system is also consistent with the 3 month delay between water release at the source lake and water arrival at the destination lake. Observations of intermittent flat bright bed reflections in radar data acquired along the flow path are consistent with the presence of a broad shallow water system [*Carter et al.*, 2009]. Geomorphological evidence of some of these components, e.g., large channelized systems, exists in current ice-free areas in the Labyrinth in the Wright Valley, Antarctica [*Lewis et al.*, 2006]. A key hydrological difference exists between the Antarctic Ice Sheet and the Greenland Ice Sheet, the other large ice sheet on Earth. At present, there is no evidence for significant surface meltwater input to the subglacial hydrological system around the margins of the Antarctic Ice Sheet as for the Greenland Ice Sheet [*Box and Ski*, 2007; *Das et al.*, 2008].

Movement of large volumes of water subglacially, as documented above, has the ability to significantly change the geochemistry and microbial ecology of the upstream source environments, transitional environments in the flow path, and the downstream environment, often another lake or depending on proximity to the ice margin, the ocean. *Statham et al.* [2008] note that the release of subglacially derived aqueous iron, likely generated through biogeochemical weathering processes, into the coastal oceans around Antarctica and Greenland, may be an important nutrient in these waters. The greatest impact of this source of glacial meltwater iron is anticipated to be in Antarctic high nutrient, low chlorophyll waters, where phytoplankton productivity is typically limited by availability of iron [*Statham et al.*, 2008]. The estimated total dissolved meltwater iron input for Antarctic waters is about 10% of that suggested to come from sea ice melting, but glacial inputs continue throughout the austral summer ablation period after sea ice melt is complete [*Statham et al.*, 2008]. Further, it is likely that in many cases, significant sediment movement is associated with these large subglacial hydrologic events. Subglacial sediments contain significant microbial biomass as documented in all subglacial systems investigated, to date [e.g., *Sharp et al.*, 1999; *Foght et al.*, 2004; *Skidmore et al.*, 2005] and thus, these hydrologic events have the potential to move large quantities of biomass beneath the ice sheet.

3.4. Basal Ice

Basal ice is ice that forms by freeze-on at the glacier base (see recent review by *Hubbard et al.* [2009]) and can be up to ~89 m thick in ice sheets [*Simoes et al.*, 2002]. It often has a significantly higher sediment content than the overlying meteoric ice. Basal ices are a component of the subglacial aquatic system, since they form by freeze-on in certain places but may melt in downstream locations when the basal thermal regime changes. Laboratory experiments demonstrate the liquid vein network in polycrystalline ice prepared at −10°C and show it as a viable microbial habitat (Plate 2). Thus, the liquid vein network in basal ices, close to the freezing point (>−10°C) [*Siegert*, 2005], are an additional SAE.

Antarctic ice cores show a range of debris-rich basal ice thicknesses, with 4–5 m in the Byrd ice core [*Gow et al.*, 1968], 2 m at Siple Dome [*Gow and Meese*, 2007], 10 m at the Kamb Ice Stream [*Vogel et al.*, 2003], and up to 89 m in the Vostok core [*Simoes et al.*, 2002]. Debris in the basal ice of the Byrd core included bands of silt, sand, and pebbles, with scattered larger fragments, all interspersed with ice [*Gow et al.*, 1968]. Most of the particles identified as pebbles in the Byrd core actually consisted of sedimentary aggregates composed of clay, silt, and sand that disintegrated upon melting. Similarly, it was found that many of the sand-sized particles entrained in the basal ice at Siple Dome were also composed of frozen aggregates of silt and clay [*Gow and*

Meese, 2007]. The glacial flour in the 89 m of basal ice in the Vostok core is also all silt and clay sized particles, largely in the 1–10 μm range.

The basal ice of the world glaciers and ice sheets has higher concentrations of sediment, solute, nutrients, and cells relative to the overlying glacier ice [e.g., *Sharp et al.*, 1999; *Skidmore et al.*, 2000; *Souchez et al.*, 2006; *Miteva et al.*, 2009], and biogeochemical weathering of the sediments provides significant chemical and energy sources for microbes. The fine-grained sediments also provide significant surface area for trapping unfrozen water at subfreezing temperatures on mineral surfaces [*Price*, 2007] also enhancing these environments as microbial habitat. Anomalous gas concentrations in the basal ice zones of terrestrial ice masses at temperatures of −11°C or warmer are argued to be the product of in situ microbial activity under both aerobic and anaerobic conditions [*Campen et al.*, 2003; *Tung et al.*, 2005; *Miteva et al.*, 2007; *Rohde et al.*, 2008]. These arguments are plausible, as certain organisms have been documented growing aerobically at −10°C and −12°C in liquid cultures in laboratory studies [*Bakermans et al.*, 2003; *Breezee et al.*, 2004]. However, no study has demonstrated or quantified microbial respiration or autotrophy in situ in these basal ices. Basal ice is certainly a viable SAE; however, organisms that would be active in these environment systems would need to be adapted to the high solute content of the liquid veins or films and the slightly lower environmental temperatures of the basal ice than other Antarctic SAE.

4. SUMMARY

There are some common findings in the two Antarctic SAE explored to date, Lake Vostok accretion ice and BIS and KIS subglacial sediments and pore waters:

1. Aerobic heterotrophs have been cultured and growth and activity of these organisms demonstrated at 4°C in the laboratory.

2. 16S rRNA gene sequences imply iron and sulfur oxidizers may be important, raising the potential for autotrophy

3. The aqueous geochemistry is consistent with carbonate and silicate dissolution driven by protons derived from sulfide oxidation and carbonic acid

Significant research on anaerobic metabolisms has not been undertaken for these environmental samples; thus, it remains largely unknown the role such organisms may have in these systems. Fungi have been isolated from the Lake Vostok accretion ice and thus may be a component of the microbial food web in this ecosystem, but their presence has not been analyzed in the WAIS ice stream tills, the only other Antarctic SAE sampled to date.

There are a range of unexplored SAE in Antarctica; including the ~384 lakes documented to date and their sediments, saturated sediments in non-lake settings, hydrologically dynamic components, including channelized drainage and linked cavity systems and debris-rich basal ice. Given the volume and depth of sediments in Antarctic subglacial basins, suboxic and anoxic conditions are likely. Therefore, a range of anaerobic metabolisms, iron reduction, sulfate reduction, acetogenesis, methanogenesis, fermentation, and anaerobic methane oxidation, may exist and should be investigated in future research. The role of viruses in both the explored and unexplored Antarctic SAE remains unknown, but is likely important based on our knowledge of other microbially dominated ecosystems. Given there is only data for microbial and biogeochemical processes in SAE from two sites for 10 million km^2 of the Antarctic Ice Sheet, this indicates the relative infancy of our knowledge on this topic and provides significant opportunities for advancement of our scientific understanding of Antarctic SAE over the coming decades.

Acknowledgments. Research related to this topic in my laboratory has been supported by the National Science Foundation, grants EAR 0525567 and OPP 0636770. T. Brox provided Plate 2 and B. Christner and D. Blankenship provided components for Figures 1 and 3 from unpublished work. S. Vogel provided the photograph in Figure 3c. An anonymous review improved the clarity of the manuscript.

REFERENCES

Abyzov, S. S., I. N. Mitskevich, M. N. Poglazova, N. I. Barkov, V. Y. Lipenkov, N. E. Bobin, B. B. Koudryashov, V. M. Pashkevich, and M. V. Ivanov (2001), Microflora in the basal strata at Antarctic ice core above the Vostok Lake, in *Space Life Sciences: Living Organisms, Biological Processes and the Limits of Life*, edited by Y. Mogami et al., *Adv. Space Res., 28*, 701–706.

Ágústsdóttir, A. M., and S. L. Brantley (1994), Volatile Fluxes Integrated over 4 Decades at Grimsvotn Volcano, Iceland, *J. Geophys. Res., 99*, 9505–9522.

Amann, R. I., W. Ludwig, and K.-H. Schleifer (1995), Phylogenetic identification and in situ detection of individual microbial cells without cultivation, *Microbiol. Rev., 59*, 143–169.

Anandakrishnan, S., D. D. Blankenship, R. B. Alley, and P. L. Stoffa (1998), Influence of subglacial geology on the position of a West Antarctic ice stream from seismic observations, *Nature, 394*, 62–65.

Bakermans, C., A. I. Tsapin, V. Souza-Egipsy, D. A. Gilichinsky, and K. H. Nealson (2003), Reproduction and metabolism at-10 degrees C of bacteria isolated from Siberian permafrost, *Environ. Microbiol., 5*, 321–326.

Bamber, J. L., F. Ferraccioli, I. Joughin, T. Shepherd, D. M. Rippin, M. J. Siegert, and D. G. Vaughan (2006), East Antarctic ice stream tributary underlain by major sedimentary basin, *Geology*, *34*, 33–36.

Bamber, J. L., J. L. Gomez-Dans, and J. A. Griggs (2008), A new 1 km digital elevation model of the Antarctic derived from combined satellite radar and laser data - Part 1: Data and methods, *Cryosphere Discuss.*, *2*, 811–841.

Barker, W. W., S. A. Welch, S. Chu, and J. F. Banfield (1998), Experimental observations of the effects of bacteria on aluminosilicate weathering, *Am. Mineral.*, *83*, 1551–1563.

Behrendt, J. C., C. A. Finn, D. Blankenship, and R. E. Bell (1998), Aeromagnetic evidence for a volcanic caldera(?) complex beneath the divide of the West Antarctic Ice Sheet, *Geophys. Res. Lett.*, *25*, 4385–4388.

Bell, R. E., D. D. Blankenship, C. A. Finn, D. L. Morse, T. A. Scambos, J. M. Brozena, and S. M. Hodge (1998), Influence of subglacial geology on the onset of a West Antarctic ice stream from aerogeophysical observations, *Nature*, *394*, 58–62.

Bell, R. E., M. Studinger, A. A. Tikku, G. K. C. Clarke, M. M. Gutner, and C. Meertens (2002), Origin and fate of Lake Vostok water frozen to the base of the East Antarctic ice sheet, *Nature*, *416*, 307–310.

Bell, R. E., M. Studinger, C. A. Shuman, M. A. Fahnestock, and I. Joughin (2007), Large subglacial lakes in East Antarctica at the onset of fast-flowing ice streams, *Nature*, *445*, 904–907.

Bjornsson, H. (2002), Subglacial lakes and jokulhlaups in Iceland, *Global Planet. Change*, *35*, 255–271.

Blankenship, D. D., R. E. Bell, S. M. Hodge, J. M. Brozena, J. C. Behrendt, and C. A. Finn (1993), Active volcanism beneath the West Antarctic Ice-sheet and implications for Ice-sheet stability, *Nature*, *361*, 526–529.

Blankenship, D. D., D. L. Morse, C. A. Finn, R. E. Bell, M. E. Peters, S. D. Kempf, S. M. Hodge, M. Studinger, J. C. Behrendt, and J. M. Brozena (2001), Geological controls on the initiation of rapid basal motion for West Antarctic ice streams: A geophysical perspective including new airborne radar sounding and laser altimetry results, in *The West Antarctic Ice Sheet: Behavior and Environment*, *Antarct. Res. Ser.*, vol. 77, edited by R. B. Alley and R. A. Bindschadler, pp. 105–121, AGU, Washington, D. C.

Boereboom, T., D. Samyn, S. Kipfstuhl, F. Wilhelms, and J. L. Tison (2007), Gas properties of EPICA Dronning Maud Land (EDML) basal refrozen water, *Geophys. Res. Abstr.*, *9*, EGU2007-A-00897

Boon, S., M. Sharp, and P. Nienow (2003), Impact of an extreme melt event on the runoff and hydrology of a high Arctic glacier, *Hydrol. Processes*, *17*, 1051–1072.

Bottrell, S. H., and M. Tranter (2002), Sulphide oxidation under partially anoxic conditions at the bed of the Haut Glacier d'Arolla, Switzerland, *Hydrol. Processes*, *16*, 2363–2368.

Box, J. E., and K. Ski (2007), Remote sounding of Greenland supraglacial melt lakes: Implications for subglacial hydraulics, *J. Glaciol.*, *53*, 257–265.

Boyd, E. S., M. Skidmore, C. Bakermans, A. Mitchell, and J. W. Peters (2010), Methanogenesis in subglacial sediments, *Environ. Microbiol. Rep.*, *2*(5), 685–692.

Breezee, J., N. Cady, and J. T. Staley (2004), Subfreezing growth of the sea ice bacterium "Psychromonas ingrahamii", *Microb. Ecol.*, *47*, 300–304.

Brown, G. H. (2002), Glacier meltwater hydrochemistry, *Appl. Geochem.*, *17*, 855–883.

Bulat, S., et al. (2004), DNA signature of thermophilic bacteria from the aged accretion ice of Lake Vostok, Antarctica: Implications for searching for life in extreme icy environments, *Int. J. Astrobiol.*, *3*, 1–12.

Bulat, S. A., I. A. Alekhina, V. Y. Lipenkov, V. V. Lukin, D. Marie, and J. R. Petit (2009), Cell concentrations of microorganisms in glacial and lake ice of the Vostok ice core, East Antarctica, *Microbiology*, *78*, 808–810.

Campen, R. K., T. Sowers, and R. B. Alley (2003), Evidence of microbial consortia metabolizing within a low-latitude mountain glacier, *Geology*, *31*, 231–234.

Carter, S. P., D. D. Blankenship, M. E. Peters, D. A. Young, J. W. Holt, and D. L. Morse (2007), Radar-based subglacial lake classification in Antarctica, *Geochem. Geophys. Geosyst.*, *8*, Q03016, doi:10.1029/2006GC001408.

Carter, S. P., D. D. Blankenship, D. A. Young, M. E. Peters, J. W. Holt, and M. J. Siegert (2009), Dynamic distributed drainage implied by the flow evolution of the 1996–1998 Adventure Trench subglacial lake discharge, *Earth Planet. Sci. Lett.*, *283*, 24–37.

Catania, G. A., H. Conway, C. F. Raymond, and T. A. Scambos (2005), Surface morphology and internal layer stratigraphy in the downstream end of Kamb Ice Stream, West Antarctica, *J. Glaciol.*, *51*, 423–431.

Cheng, S. M., and J. M. Foght (2007), Cultivation-independent and -dependent characterization of Bacteria resident beneath John Evans Glacier, *FEMS Microbiol. Ecol.*, *59*, 318–330.

Chivian, D., et al. (2008), Environmental genomics reveals a single-species ecosystem deep within earth, *Science*, *322*, 275–278.

Christner, B. C., E. Mosley-Thompson, L. G. Thompson, and J. N. Reeve (2001), Isolation of bacteria and 16S rDNAs from Lake Vostok accretion ice, *Environ. Microbiol.*, *3*, 570–577.

Christner, B. C., G. Royston-Bishop, C. M. Foreman, B. R. Arnold, M. Tranter, K. A. Welch, W. B. Lyons, A. I. Tsapin, M. Studinger, and J. C. Priscu (2006), Limnological conditions in subglacial Lake Vostok, Antarctica, *Limnol. Oceanogr.*, *51*(6), 2485–2501.

Cooper, R. J. (2003), Chemical denudation in the proglacial zone of Finsterwalderbreen, Svalbard. Ph.D. thesis, Univ. of Bristol, Bristol, U. K.

Das, S. B., I. Joughin, M. D. Behn, I. M. Howat, M. A. King, D. Lizarralde, and M. P. Bhatia (2008), Fracture propagation to the base of the Greenland Ice Sheet during supraglacial lake drainage, *Science*, *320*, 778–781.

De Angelis, M., J. R. Petit, J. Savarino, R. Souchez, and M. H. Thiemens (2004), Contributions of an ancient evaporitic-type reservoir to subglacial Lake Vostok chemistry, *Earth Planet. Sci. Lett.*, *222*, 751–765.

D'Elia, T., R. Veerapaneni, and S. O. Rogers (2008), Isolation of microbes from Lake Vostok accretion ice, *Appl. Environ. Microbiol.*, *74*, 4962–4965.

D'Elia, T., R. Veerapaneni, V. Theraisnathan, and S. O. Rogers (2009), Isolation of fungi from Lake Vostok accretion ice, *Mycologia*, *101*, 751–763.

Doran, P. T., C. H. Fritsen, C. P. McKay, J. C. Priscu, and E. E. Adams (2003), Formation and character of an ancient 19-m ice cover and underlying trapped brine in an "ice-sealed" east Antarctic lake, *Proc. Natl. Acad. Sci. U. S. A.*, *100*, 26–31.

Dowdeswell, J. A., and M. J. Siegert (2002), The physiography of modern Antarctic subglacial lakes, *Global Planet. Change*, *35*, 221–236.

Engelhardt, H. (2004), Thermal regime and dynamics of the West Antarctic ice sheet, *Ann. Glaciol.*, *39*, 85–92.

Engelhardt, H., N. Humphrey, B. Kamb, and M. Fahnestock (1990), Physical conditions at the base of a fast moving Antarctic ice stream, *Science*, *248*, 57–59.

Filina, I. Y., D. D. Blankenship, M. Thoma, V. V. Lukin, V. N. Masolov, and M. K. Sen (2008), New 3D bathymetry and sediment distribution in Lake Vostok: Implication for pre-glacial origin and numerical modeling of the internal processes within the lake, *Earth Planet. Sci. Lett.*, *276*, 106–114.

Foght, J., J. Aislabie, S. Turner, C. E. Brown, J. Ryburn, D. J. Saul, and W. Lawson (2004), Culturable bacteria in subglacial sediments and ice from two Southern Hemisphere glaciers, *Microb. Ecol.*, *47*, 329–340.

Fricker, H. A., and T. Scambos (2009), Connected subglacial lake activity on lower Mercer and Whillans ice streams, West Antarctica, 2003–2008, *J. Glaciol.*, *55*, 303–315.

Fry, J. C., R. J. Parkes, B. A. Cragg, A. J. Weightman, and G. Webster (2008), Prokaryotic biodiversity and activity in the deep subseafloor biosphere, *FEMS Microbiol. Ecol.*, *66*, 181–196.

Gabrielli, P., F. Planchon, C. Barbante, C. F. Boutron, J. R. Petit, S. Bulat, S. Hong, G. Cozzi, and P. Cescon (2009), Ultra-low rare earth element content in accreted ice from sub-glacial Lake Vostok, Antarctica, *Geochim. Cosmochim. Acta*, *73*, 5959–5974.

Gaidos, E., B. Lanoil, T. Thorsteinsson, A. Graham, M. Skidmore, S. K. Han, T. Rust, and B. Popp (2004), A viable microbial community in a subglacial volcanic crater lake, Iceland, *Astrobiology*, *4*, 327–344.

Gaidos, E., et al. (2009), An oligarchic microbial assemblage in the anoxic bottom waters of a volcanic subglacial lake, *ISME J.*, *3*, 486–497.

Goodwin, I. D. (1988), The nature and origin of a jokulhlaup near Casey Station, Antarctica, *J. Glaciol.*, *34*, 95–101.

Gow, A. J., and D. Meese (2007), Physical properties, crystalline textures and c-axis fabrics of the Siple Dome (Antarctica) ice core, *J. Glaciol.*, *53*, 573–584.

Gow, A. J., H. T. Ueda, and D. E. Garfield (1968), Antarctic ice sheet: Preliminary results of first core hole to bedrock, *Science*, *161*, 1011–1013.

Gray, L., I. Joughin, S. Tulaczyk, V. B. Spikes, R. Bindschadler, and K. Jezek (2005), Evidence for subglacial water transport in the West Antarctic Ice Sheet through three-dimensional satellite radar interferometry, *Geophys. Res. Lett.*, *32*, L03501, doi:10.1029/2004GL021387.

Green, W. J., and W. B. Lyons (2009), The Saline Lakes of the McMurdo Dry Valleys, Antarctica, *Aquat. Geochem.*, *15*, 321–348.

Hubbard, B., S. Cook, and H. Coulson (2009), Basal ice facies: A review and unifying approach, *Quat. Sci. Rev.*, *28*, 1956–1969.

Joughin, I., L. Gray, R. Bindschadler, S. Price, D. Morse, C. Hulbe, K. Mattar, and C. Werner (1999), Tributaries of West Antarctic ice streams revealed by RADARSAT interferometry, *Science*, *286*, 283–286.

Joughin, I., J. L. Bamber, T. Scambos, S. Tulaczyk, M. Fahnestock, and D. R. MacAyeal (2006), Integrating satellite observations with modelling: Basal shear stress of the Filcher-Ronne ice streams, Antarctica, *Philos. Trans. R. Soc. A*, *364*, 1795–1814.

Jouzel, J., J. R. Petit, R. Souchez, N. I. Barkov, V. Y. Lipenkov, D. Raynaud, M. Stievenard, N. I. Vassiliev, V. Verbeke, and F. Vimeux (1999), More than 200 meters of lake ice above subglacial Lake Vostok, Antarctica, *Science*, *286*, 2138–2141.

Kamb, B. (2001), Basal zone of the West Antarctic ice streams and its role in lubrication of their rapid motion, in *The West Antarctic Ice Sheet: Behavior and Environment, Antarct. Res. Ser.*, vol. 77, edited by R. B. Alley and R. A. Bindschadler, pp. 157–200, AGU, Washington, D. C.

Kapitsa, A. P., J. K. Ridley, G. d. Q. Robin, M. J. Siegert, and I. A. Zitikov (1996), A large deep freshwater lake beneath the ice of central East Antarctica, *Nature*, *381*, 684–686.

Karl, D. M., D. F. Bird, K. Bjorkman, T. Houlihan, R. Shackelford, and L. Tupas (1999), Microorganisms in the accreted ice of Lake Vostok, Antarctica, *Science*, *286*, 2144–2147.

Karr, E. A., W. M. Sattley, M. R. Rice, D. O. Jung, M. T. Madigan, and L. A. Achenbach (2005), Diversity and distribution of sulfate-reducing bacteria in permanently frozen Lake Fryxell, McMurdo Dry Valleys, Antarctica, *Appl. Environ. Microbiol.*, *71*, 6353–6359.

Karr, E. A., J. M. Ng, S. M. Belchik, W. M. Sattley, M. T. Madigan, and L. A. Achenbach (2006), Biodiversity of methanogenic and other Archaea in the permanently frozen Lake Fryxell, Antarctica, *Appl. Environ. Microbiol.*, *72*, 1663–1666.

Kastovska, K., M. Stibal, M. Sabacka, B. Cerna, H. Santruckova, and J. Elster (2007), Microbial community structure and ecology of subglacial sediments in two polythermal Svalbard glaciers characterized by epifluorescence microscopy and PLFA, *Polar Biol.*, *30*, 277–287.

Kivimaki, A.-L. (2005), Presence and activity of microbial populations in glaciers and their impact on rock weathering at glacial beds, Ph.D. thesis, Univ. of Bristol, Bristol, U. K.

Lanoil, B., M. Skidmore, J. C. Priscu, S. Han, W. Foo, S. W. Vogel, S. Tulaczyk, and H. Engelhardt (2009), Bacteria beneath the West Antarctic Ice Sheet, *Environ. Microbiol.*, *11*, 609–615.

Lavire, C., P. Normand, I. Alekhina, S. Bulat, D. Prieur, J. L. Birrien, P. Fournier, C. Hanni, and J. R. Petit (2006), Presence of Hydrogenophilus thermoluteolus DNA in accretion ice in the subglacial Lake Vostok, Antarctica, assessed using rrs, cbb and hox, *Environ. Microbiol.*, *8*, 2106–2114.

Lawson, J., P. T. Doran, F. Kenig, D. J. Des Marais, and J. C. Priscu (2004), Stable carbon and nitrogen isotopic composition of benthic and pelagic organic matter in lakes of the McMurdo Dry Valleys, Antarctica, *Aquat. Geochem.*, *10*, 269–301.

Le Brocq, A. M., A. Hubbard, M. J. Bentley, and J. L. Bamber (2008), Subglacial topography inferred from ice surface terrain analysis reveals a large un-surveyed basin below sea level in East Antarctica, *Geophys. Res. Lett.*, *35*, L16503, doi:10.1029/2008GL034728.

Lewis, A. R., D. R. Marchant, D. E. Kowalewski, S. L. Baldwin, and L. E. Webb (2006), The age and origin of the Labyrinth, western Dry Valleys, Antarctica: Evidence for extensive middle Miocene subglacial floods and freshwater discharge to the Southern Ocean, *Geology*, *34*, 513–516.

Lin, L. H., et al. (2006), Long-term sustainability of a high-energy, low-diversity crustal biome, *Science*, *314*, 479–482.

Llubes, M., C. Lanseau, and F. Remy (2006), Relations between basal condition, subglacial hydrological networks and geothermal flux in Antarctica, *Earth Planet. Sci. Lett.*, *241*, 655–662.

Ludwig, W., et al. (2004), ARB: A software environment for sequence data, *Nucleic Acids Res.*, *32*, 1363–1371.

McKay, C. P., K. P. Hand, P. T. Doran, D. T. Andersen, and J. C. Priscu (2003), Clathrate formation and the fate of noble and biologically useful gases in Lake Vostok, Antarctica, *Geophys. Res. Lett.*, *30*(13), 1702, doi:10.1029/2003GL017490.

Mikucki, J. A., and J. C. Priscu (2007), Bacterial diversity associated with blood falls, a subglacial outflow from the Taylor Glacier, Antarctica, *Appl. Environ. Microbiol.*, *73*, 4029–4039.

Mikucki, J. A., C. M. Foreman, B. Sattler, W. B. Lyons, and J. C. Priscu (2004), Geomicrobiology of Blood Falls: An iron-rich saline discharge at the terminus of the Taylor Glacier, Antarctica, *Aquat. Geochem.*, *10*, 199–220.

Mikucki, J. A., A. Pearson, D. T. Johnston, A. V. Turchyn, J. Farquhar, D. P. Schrag, A. D. Anbar, J. C. Priscu, and P. A. Lee (2009), A contemporary microbially maintained subglacial ferrous "ocean", *Science*, *324*, 397–400.

Mitchell, A. C., and G. H. Brown (2008), Modeling geochemical and biogeochemical reactions in subglacial environments, *Arct. Antarct. Alp. Res.*, *40*, 531–547.

Miteva, V., T. Sowers, and J. Brenchley (2007), Production of N2O by ammonia oxidizing bacteria at subfreezing temperatures as a model for assessing the N2O anomalies in the Vostok ice core, *Geomicrobiol. J.*, *24*, 451–459.

Miteva, V., C. Teacher, T. Sowers, and J. Brenchley (2009), Comparison of the microbial diversity at different depths of the GISP2 Greenland ice core in relationship to deposition climates, *Environ. Microbiol.*, *11*, 640–656.

Mondino, L. J., M. Asao, and M. T. Madigan (2009), Cold-active halophilic bacteria from the ice-sealed Lake Vida, Antarctica, *Arch. Microbiol.*, *191*, 785–790.

Montross, S. N. (2007), Geochemical evidence for microbially mediated subglacial mineral weathering, M.Sc. thesis, Mont. State Univ., Bozeman.

Nemergut, D. R., S. P. Anderson, C. C. Cleveland, A. P. Martin, A. E. Miller, A. Seimon, and S. K. Schmidt (2007), Microbial community succession in an unvegetated, recently deglaciated soil, *Microb. Ecol.*, *53*, 110–122.

Olsen, G. J., H. Matsuda, R. Hagstrom, and R. Overbeek (1994), FastDNAml: A tool for construction of phylogenetic trees of DNA sequences using maximum likelihood, *Comput. Appl. Biosci.*, *10*, 41–48.

Onstott, T. C., et al. (1998), Observations pertaining to the origin and ecology of microorganisms recovered from the deep subsurface of Taylorsville Basin, Virginia, *Geomicrobiol. J.*, *15*, 353–385.

Parnachev, V. P., D. Banks, A. Y. Berezovsky, and D. Garbe-Schonberg (1999), Hydrochemical evolution of Na-SO4-Cl groundwaters in a cold, semi-arid region of southern Siberia, *Hydrogeol. J.*, *7*, 546–560.

Pedersen, K. (2000), Exploration of deep intraterrestrial microbial life: Current perspectives, *FEMS Microbiol. Lett.*, *185*, 9–16.

Peters, L. E., S. Anandakrishnan, R. B. Alley, and A. M. Smith (2007), Extensive storage of basal meltwater in the onset region of a major West Antarctic ice stream, *Geology*, *35*, 251–254.

Peters, N. J., I. C. Willis, and N. S. Arnold (2009), Numerical analysis of rapid water transfer beneath Antarctica, *J. Glaciol.*, *55*, 640–650.

Petit, J. R., et al. (1999), Climate and atmospheric history of the past 420,000 years from the Vostok ice core, Antarctica, *Nature*, *399*, 429–436.

Popov, S. V., and V. N. Masolov (2007), Forty-seven new subglacial lakes in the 0-110 degrees E sector of East Antarctica, *J. Glaciol.*, *53*, 289–297.

Price, P. B. (2007), Microbial life in glacial ice and implications for a cold origin of life, *FEMS Microbiol. Ecol.*, *59*, 217–231.

Priscu, J. C., and B. C. Christner (2004), Earth's icy biosphere, in *Microbial Biodiversity and Bioprospecting*, edited by A. T. Bull, pp. 130–145, Am. Soc. for Microbiol. Press, Washington, D. C.

Priscu, J. C., C. H. Fritsen, E. E. Adams, S. J. Giovannoni, H. W. Paerl, C. P. McKay, P. T. Doran, D. A. Gordon, B. D. Lanoil, and J. L. Pinckney (1998), Perennial Antarctic lake ice: An oasis for life in a polar desert, *Science*, *280*, 2095–2098.

Priscu, J. C., et al. (1999), Geomicrobiology of subglacial ice above Lake Vostok, Antarctica, *Science*, *286*, 2141–2144.

Priscu, J. C., S. Tulaczyk, M. Studinger, M. C. Kennicutt, II, B. Christner, and C. M. Foreman (2008), Antarctic subglacial water: Origin, evolution, and ecology, in *Polar Lakes and Rivers*, edited by W. F. Vincent and J. Laybourn-Parry, pp. 119–135, Oxford Univ. Press, Oxford, U. K.

Purdy, K. J., I. Hawes, C. L. Bryant, A. E. Fallick, and D. B. Nedwell (2001), Estimates of sulphate reduction rates in Lake Vanda, Antarctica support the proposed recent history of the lake, *Antarct. Sci.*, *13*, 393–399.

Rohde, R. A., P. B. Price, R. C. Bay, and N. E. Bramall (2008), In situ microbial metabolism as a cause of gas anomalies in ice, *Proc. Natl. Acad. Sci. U. S. A.*, *105*, 8667–8672.

Rothschild, L. J., and R. L. Mancinelli (2001), Life in extreme environments, *Nature*, *409*, 1092–1101.

Roussel, E. G., M. A. Cambon-Bonavita, J. Querellou, B. A. Cragg, G. Webster, D. Prieur, and R. J. Parkes (2008), Extending the sub-sea-floor biosphere, *Science*, *320*, 1046.

Royston-Bishop, G., J. C. Priscu, M. Tranter, B. Christner, M. J. Siegert, and V. Lee (2005), Incorporation of particulates into accreted ice above subglacial Vostok Lake, Antarctica, *Ann. Glaciol.*, *40*, 145–150.

Sattley, W. M., and M. T. Madigan (2007), Cold-active acetogenic bacteria from surficial sediments of perennially ice-covered Lake Fryxell, Antarctica, *FEMS Microbiol. Lett.*, *272*, 48–54.

Scherer, R. P., A. Aldahan, S. Tulaczyk, G. Possnert, H. Engelhardt, and B. Kamb (1998), Pleistocene collapse of the West Antarctic ice sheet, *Science*, *281*, 82–85.

Sharp, M., J. Parkes, B. Cragg, I. J. Fairchild, H. Lamb, and M. Tranter (1999), Widespread bacterial populations at glacier beds and their relationship to rock weathering and carbon cycling, *Geology*, *27*, 107–110.

Sharp, M., R. A. Creaser, and M. Skidmore (2002), Strontium isotope composition of runoff from a glaciated carbonate terrain, *Geochim. Cosmochim. Acta*, *66*, 595–614.

Siegert, M. J. (2005), Lakes beneath the ice sheet: The occurrence, analysis, and future exploration of Lake Vostok and other Antarctic subglacial lakes, *Annu. Rev. Earth Planet. Sci.*, *33*, 215–245.

Siegert, M. J., and A. J. Payne (2004), Past rates of accumulation in central West Antarctica, *Geophys. Res. Lett.*, *31*, L12403, doi:10.1029/2004GL020290.

Simoes, J. C., J. R. Petit, R. Souchez, V. Y. Lipenkov, M. De Angelis, L. B. Liu, J. Jouzel, and P. Duval (2002), Evidence of glacial flour in the deepest 89 m of the Vostok ice core, *Ann. Glaciol.*, *35*, 340–346.

Skidmore, M. L., and M. J. Sharp (1999), Drainage behaviour of a high Arctic polythermal glacier, *Ann. Glaciol.*, *28*, 209–215.

Skidmore, M. L., J. M. Foght, and M. J. Sharp (2000), Microbial life beneath a high Arctic glacier, *Appl. Environ. Microbiol.*, *66*, 3214–3220.

Skidmore, M., S. P. Anderson, M. Sharp, J. Foght, and B. D. Lanoil (2005), Comparison of microbial community compositions of two subglacial environments reveals a possible role for microbes in chemical weathering processes, *Appl. Environ. Microbiol.*, *71*, 6986–6997.

Skidmore, M., M. Tranter, S. Tulaczyk, and B. Lanoil (2010), Hydrochemistry of ice stream beds—evaporitic or microbial effects?, *Hydrol. Processes*, *24*, 517–523.

Smith, A. M., and T. Murray (2009), Bedform topography and basal conditions beneath a fast-flowing West Antarctic ice stream, *Quat. Sci. Rev.*, *28*, 584–596.

Smith, B. E., H. A. Fricker, I. R. Joughin, and S. Tulaczyk (2009), An inventory of active subglacial lakes in Antarctica detected by ICESat (2003–2008), *J. Glaciol.*, *55*, 573–595.

Souchez, R., J. R. Petit, J. Jouzel, J. Simões, M. de Angelis, N. Barkov, M. Stiévenard, F. Vimeux, S. Sleewaegen, and R. Lorrain (2002), Highly deformed basal ice in the Vostok core, Antarctica, *Geophys. Res. Lett.*, *29*(7), 1136, doi:10.1029/2001GL014192.

Souchez, R., J. Jouzel, A. Landais, J. Chappellaz, R. Lorrain, and J.-L. Tison (2006), Gas isotopes in ice reveal a vegetated central Greenland during ice sheet invasion, *Geophys. Res. Lett.*, *33*, L24503, doi:10.1029/2006GL028424.

Statham, P. J., M. Skidmore, and M. Tranter (2008), Inputs of glacially derived dissolved and colloidal iron to the coastal ocean and implications for primary productivity, *Global Biogeochem. Cycles*, *22*, GB3013, doi:10.1029/2007GB003106.

Stotler, R. L., S. K. Frape, T. Ruskeeniemi, L. Ahonen, T. C. Onstott, and M. Y. Hobbs (2009), Hydrogeochemistry of ground-waters in and below the base of thick permafrost at Lupin, Nunavut, Canada, *J. Hydrol.*, *373*, 80–95.

Studinger, M., R. E. Bell, D. D. Blankenship, C. A. Finn, R. A. Arko, D. L. Morse, and I. Joughin (2001), Subglacial sediments: A regional geological template for ice flow in West Antarctica, *Geophys. Res. Lett.*, *28*, 3493–3496.

Studinger, M., et al. (2003a), Ice cover, landscape setting, and geological framework of Lake Vostok, East Antarctica, *Earth Planet. Sci. Lett.*, *205*, 195–210.

Studinger, M., G. D. Karner, R. E. Bell, V. Levin, C. A. Raymond, and A. A. Tikku (2003b), Geophysical models for the tectonic framework of the Lake Vostok region, East Antarctica, *Earth Planet. Sci. Lett.*, *216*, 663–677.

Szewzyk, U., R. Szewzyk, and T. A. Stenstrom (1994), Thermo-philic, anaerobic-bacteria isolated from a deep borehole in granite in Sweden, *Proc. Natl. Acad. Sci. U. S. A.*, *91*, 1810–1813.

Tabacco, I. E., P. Cianfarra, A. Forieri, F. Salvini, and A. Zirizotti (2006), Physiography and tectonic setting of the subglacial lake district between Vostok and Belgica subglacial highlands (Antarctica), *Geophys. J. Int.*, *165*, 1029–1040.

Takacs, C. D., J. C. Priscu, and D. M. McKnight (2001), Bacterial dissolved organic carbon demand in McMurdo Dry Valley lakes, Antarctica, *Limnol. Oceanogr.*, *46*(5), 1189–1194.

Tranter, M., G. H. Brown, A. Hodson, A. M. Gurnell, and M. J. Sharp (1994), Variation in the nitrate concentration of glacial runoff in alpine and sub-polar environments, *IAHS Publ.*, *223*, 299–311.

Tranter, M., M. J. Sharp, G. H. Brown, I. C. Willis, B. P. Hubbard, M. K. Nielsen, C. C. Smart, S. Gordon, M. Tulley, and H. R. Lamb (1997), Variability in the chemical composition of in situ subglacial meltwaters, *Hydrol. Processes*, *11*, 59–77.

Tranter, M., M. J. Sharp, H. R. Lamb, G. H. Brown, B. P. Hubbard, and I. C. Willis (2002), Geochemical weathering at the bed of Haut Glacier d'Arolla, Switzerland—a new model, *Hydrol. Processes*, *16*, 959–993.

Tranter, M., M. Skidmore, and J. Wadham (2005), Hydrological controls on microbial communities in subglacial environments, *Hydrol. Processes*, *19*, 995–998.

Tulaczyk, S., B. Kamb, R. P. Scherer, and H. F. Engelhardt (1998), Sedimentary processes at the base of a West Antarctic ice stream: Constraints from textural and compositional properties of sub-glacial debris, *J. Sediment. Res.*, *68*, 487–496.

Tulaczyk, S., B. Kamb, and H. Engelhardt (2000), Basal mechanics of Ice Stream B, West Antarctica 2. Undrained plastic bed model, *J. Geophys. Res.*, *105*, 483–494.

Tung, H. C., N. E. Bramall, and P. B. Price (2005), Microbial origin of excess methane in glacial ice and implications for life on Mars, *Proc. Natl. Acad. Sci. U. S. A.*, *102*, 18,292–18,296.

Vogel, S. W. (2004), The basal regime of the West Antarctic Ice Sheet. Interaction of subglacial geology with ice dynamics, Ph.D. thesis, Univ. of Calif. at Santa Cruz, Santa Cruz.

Vogel, S. W., S. Tulaczyk, and I. R. Joughin (2003), Distribution of basal melting and freezing beneath tributaries of ice stream C: Implication for the Holocene decay of the West Antarctic ice sheet, *Ann. Glaciol.*, *36*, 273–282.

Vogel, S. W., S. Tulaczyk, B. Kamb, H. Engelhardt, F. D. Carsey, A. E. Behar, A. L. Lane, and I. Joughin (2005), Subglacial conditions during and after stoppage of an Antarctic ice stream: Is reactivation imminent?, *Geophys. Res. Lett.*, *32*, L14502, doi:10.1029/2005GL022563.

Voytek, M. A., J. C. Priscu, and B. B. Ward (1999), The distribution and relative abundance of ammonia-oxidizing bacteria in lakes of the McMurdo Dry Valley, Antarctica, *Hydrobiologia*, *401*, 113–130.

Wadham, J. L., S. Bottrell, M. Tranter, and R. Raiswell (2004), Stable isotope evidence for microbial sulphate reduction at the bed of a polythermal high Arctic glacier, *Earth Planet. Sci. Lett.*, *219*, 341–355.

Wadham, J. L., M. Tranter, M. Skidmore, A. J. Hodson, J. Priscu, W. B. Lyons, M. J. Sharp, P. Wynn, and M. Jackson (2010), Biogeochemical weathering under ice: Size matters, *Global Biogeochem. Cycles*, *24*, GB3025, doi:10.1029/2009GB003688.

Ward, B. B., and J. C. Priscu (1997), Detection and characterization of denitrifying bacteria from a permanently ice-covered Antarctic lake, *Hydrobiologia*, *347*, 57–68.

Wetzel, R. G. (2001), *Limnology: Lake and River Ecosystems*, 3rd ed., Academic Press, San Diego, Calif.

Wingham, D. J., M. J. Siegert, A. Shepherd, and A. S. Muir (2006), Rapid discharge connects Antarctic subglacial lakes, *Nature*, *440*, 1033–1036.

Wright, A., and M. J. Siegert (2011), The identification and physiographical setting of Antarctic subglacial lakes: An update based on recent discoveries, in *Antarctic Subglacial Aquatic Environments, Geophys. Monogr. Ser.*, doi: 10.1029/2010GM000933, this volume.

Wynn, P. M., A. J. Hodson, T. H. E. Heaton, and S. R. Chenery (2007), Nitrate production beneath a High Arctic Glacier, Svalbard, *Chem. Geol.*, *244*, 88–102.

M. Skidmore, Department of Earth Sciences, Montana State University, Bozeman, MT 59717, USA. (skidmore@montana.edu)

Subglacial Lake Sediments and Sedimentary Processes: Potential Archives of Ice Sheet Evolution, Past Environmental Change, and the Presence of Life

M. J. Bentley,[1] P. Christoffersen,[2] D. A. Hodgson,[3] A. M. Smith,[3] S. Tulaczyk,[4] and A. M. Le Brocq[1,5]

The development of funded programs to drill into deep continental Antarctic subglacial lakes means that there is an imminent prospect of retrieving the first sediments from isolated and unexplored aquatic environments beneath Earth's largest ice sheet. Here we demonstrate how these sediments are potential archives of information on past ice sheet configurations, paleoenvironmental change, and the presence of life in the deep, dark, and cold setting of subglacial lakes. We review what is known about the physical characteristics of subglacial lake sediments, indirectly from remote geophysical surveys of modern and former subglacial lakes and directly from sampling of sediments in former subglacial lakes emerging at the margin of the Antarctic ice sheet and terrestrial settings elsewhere. We show that mean sedimentation rates in subglacial lakes can vary from near zero to several millimeters per year and that the thickness of sedimentary sequences in subglacial lakes ranges from a few centimeters to several hundred meters. The most important control on sedimentation rate and sediment thickness is likely to be the amount of sediment-laden water delivered by regional basal water systems. Episodic lake discharge will also likely have a significant impact on some sedimentary sequences. On the basis of our review and previously documented limnological and glaciological processes, we propose a conceptual model of subglacial lake sedimentation that we apply to lakes Vostok, Ellsworth, and Whillans. We discuss the implications of the likely sediment physical characteristics for designing coring technologies and for the choice of analytical procedures to examine sediment samples from subglacial lakes.

[1]Department of Geography, University of Durham, Durham, UK.

[2]Department of Geography, University of Cambridge, Cambridge, UK.

[3]British Antarctic Survey, Cambridge, UK.

[4]Department of Earth and Planetary Sciences, University of California, Santa Cruz, California, USA.

[5]School of Geography, University of Exeter, Exeter, UK.

Antarctic Subglacial Aquatic Environments
Geophysical Monograph Series 192
Copyright 2011 by the American Geophysical Union
10.1029/2010GM000940

1. INTRODUCTION

It has long been recognized that sedimentary sequences preserved in Antarctic subglacial lakes will likely record information on past environmental change. For example, the early inference, from seismic evidence, of thick sediments beneath Vostok Subglacial Lake (hereinafter referred to as Lake Vostok) was used to discuss the potential for retrieving long paleoenvironmental records potentially dating back to at least the Late Cenozoic [*Siegert et al.*, 2001].

There are three primary interests in retrieving and analyzing such sediments [*Siegert et al.*, 2006]. First, the sediments may record changes over time (cyclic or otherwise) in conditions at the base of the ice sheet. For example, it is possible that the composition of subglacial lake sediments may vary in response to the succession of glacial-interglacial cycles, with changes in both the volume of sediments, and in sedimentary composition. Second, some lakes may preserve nonglacial sediments deposited during intervals of ice sheet retreat or collapse. For example, the presence of marine sediment units in sediment cores from subglacial lakes beneath the West Antarctic Ice Sheet (WAIS) might be reasonably interpreted as evidence of former WAIS collapse events. A major attraction of using subglacial sediments to infer paleoenvironmental or ice sheet change is that many lakes are thought likely to contain sediments substantially older than the maximum length of the record provided by ice cores (currently ~780 ka [e.g., *Augustin et al.*, 2004]). In fact, because much of the sediment is derived from the ice sheet base, the sedimentary record "should pick up where the ice cores leave off" [*Doran et al.*, 2004]. For example, it has been speculated that the sedimentary record in Lake Vostok (Figure 1) may extend back to, and perhaps even predate, the initiation of East Antarctic Ice Sheet (EAIS) growth at >30 Ma [*Siegert et al.*, 2001]. A third justification for the study of subglacial sediments is that direct detection of organic geochemical and geochemical fingerprints in the sediments may provide evidence of life surviving in subglacial environments, particularly at the relatively nutrient-rich sediment-water interface, and the sediments may also record changes in these microbial communities through time [*Siegert et al.*, 2006].

Understanding of sediments and sedimentary processes in deep subglacial aquatic environments is so far based either on remote geophysical measurements of lake sediments, or on lake sediments with inferred subglacial origin. Although the exact nature of sedimentary sequences in subglacial lakes will not be fully established until sediment samples from modern lakes are available, there is a growing body of evidence emerging from the geophysical surveys of modern subglacial lakes and paleoanalogs and from sampling of lakes previously overridden by glaciers or ice sheets. Here we bring together the results from these studies, along with a knowledge of limnological processes with the aim to develop a conceptual model of sedimentation in subglacial lakes.

At present, there are three funded programs to drill into subglacial lakes. The lakes targeted for direct exploration are Lake Vostok (Russian-led) under the EAIS and Subglacial Lake Whillans (hereinafter referred to as Lake Whillans) (U.S.-led WISSARD project, www.wissard.org) and Ellsworth Subglacial Lake (hereinafter referred to as Lake Ellsworth) (U.K.-led Ellsworth Consortium, www.geos.ed.ac.uk/ellsworth) under the WAIS (Figure 1). The latter two of these projects have specifically identified subglacial sediment retrieval as a priority objective, and so we focus on the likely sediment characteristics in these lakes. The Lake Vos-

tok program currently has no published plans to penetrate the sedimentary sequence on the lake floor. However, the existing geophysical surveys of the deep sediment deposits, along with important questions about the long-term environmental history of the EAIS make Lake Vostok a likely long-term target for future coring, so we include it in the discussion of sedimentary processes in this review.

2. SUBGLACIAL LAKE SEDIMENTS AS ARCHIVES OF PALEOENVIRONMENTAL CHANGE

Aerogeophysical surveys have been carried out in the Antarctic since the International Geophysical Year in 1957–1958, and it was surveys from the 1960s that led to the discovery of subglacial lakes beneath the ice sheet [*Oswald and Robin*, 1973]. Subsequent studies of these lakes were intermittent, but interest in subglacial Antarctic lakes increased after microorganisms were detected in the accreted ice layer over Lake Vostok [*Jouzel et al.*, 1999; *Karl et al.*, 1999; *Siegert*, 2000]. A second stimulus came from satellite imaging of surface elevation change showing lake interconnections and rapid hydrological discharge [*Gray et al.*, 2005; *Wingham et al.*, 2006; *Fricker et al.*, 2007]. Recent research has focused largely on (1) the nature and sustainability of living organisms in the dark and atmospherically decoupled subglacial aquatic environment [*Christner et al.*, 2006], (2) bathymetry and circulation of water masses [*Mayer et al.*, 2003; *Studinger et al.*, 2004; *Thoma et al.*, 2007, 2008b], and (3) ice accretion and the physical interaction of lakes with the overlying ice sheet [*Bell et al.*, 2002; *Tikku et al.*, 2004; *Pattyn et al.*, 2004; *Pattyn*, 2008; *Thoma et al.*, 2010]. Sedimentary sequences in subglacial lakes are central to all three of these foci for subglacial lake research. First, the lake floor is a key site for future biological investigations because studies of oligotrophic (nutrient-poor) subaerial lakes show that sedimentary biomass and activity rates are generally much higher than those in the overlying water column. Second, geochemical analyses of lake sediments may yield evidence of past changes in water composition and circulation. Third, the physical properties and lateral distribution of lake sediments can be used to identify hydrologic and physical interactions with the overlying ice sheet.

2.1. Subglacial Lake Sediments Inferred From Antarctic Geophysical Surveys

Airborne surveys have been carried out with the specific intent to elucidate the setting and physiography of subglacial lakes [*Bell et al.*, 2002; *Studinger et al.*, 2003, 2004; *Tikku et al.*, 2004; *Carter et al.*, 2007; *Vaughan et al.*, 2007; *Filina et al.*, 2008]. These surveys have identified ice flow directions over subglacial lakes, constrained the rates of melting and freezing of the ice-water interface, and produced estimates of lake water residence time. However, the potential of subglacial lakes to act as proxy for climatic and other environmental processes relies on sedimentary processes and the preservation of sediment accumulations on the lake floor. Whereas excellent new constraints on subglacial lake physiography have been derived from recent surveys, the detailed extent and thickness of sedimentary sequences remain largely unknown. This conspicuous lack of information is related to the need for ground-based seismic investigations, which are difficult and logistically challenging undertakings in remote polar regions, but nonetheless central to the elucidation of subglacial sediment characteristics.

Seismic soundings have, however, been carried out over Lake Vostok for several decades, and these reveal the possible presence of several hundreds of meters of unconsolidated sediments below the main body of water [*Masolov et al.*, 2006; *Filina et al.*, 2007]. *Filina et al.* [2008] integrated these soundings in a gravity model that includes the presence of unconsolidated lake sediment with assumed density of 1850 kg/m^3. The model was developed from airborne geophysical data collected in 2000/2001. The same data set was used in previous gravity models, but unconsolidated lake sediment was not included in the earlier studies, which focused either on the tectonic setting of the lake [*Studinger et al.*, 2003] or simply the bathymetry of the lake itself [*Studinger et al.*, 2004]. The integration of unconsolidated lake sediment in a gravity model shows that Lake Vostok may contain 2600 km^3 of such sediment, including sequences in excess of 300 m thick [*Filina et al.*, 2008]. The modeled distribution of unconsolidated sediment and the associated lake bathymetry are shown in Plate 1. Although gravity modeling offers a novel means to examine the possible presence and thickness of a sedimentary body, it does not inform us of the physical character of sedimentary units or sequences within the body.

The physical character and lateral distribution of subglacial sediments can be inferred from seismic experiments. Seismic reflections have been successfully used to identify soft till with a porosity of 40% beneath Antarctic ice streams [*Blankenship et al.*, 1986, 1987]. Till cores sampled from boreholes subsequently confirmed this porosity [*Engelhardt et al.*, 1990], and the close match between seismic estimate and direct observations showed that seismic facies can be successfully converted to physical properties. Seismic profiling has more recently been used in Antarctica to determine the lateral distribution of sediment beneath ice streams [*Smith*, 1997; *Anandakrishnan et al.*, 1998; *Peters et al.*, 2008], observe the formation of subglacial sedimentary landforms [*Smith et al.*, 2007; *King et al.*, 2009; *Smith and Murray*, 2009], and examine subglacial hydrologic systems [*Peters et al.*, 2007; *Smith et al.*, 2007; *Winberry et al.*, 2009]. *Peters et al.* [2008] used

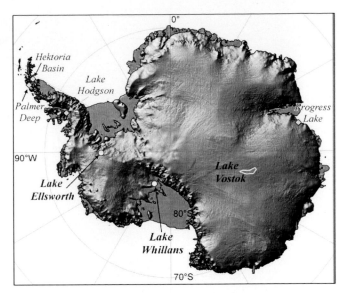

Figure 1. Location map. Background is a hillshade of a modified version of the digital elevation model of *Bamber et al.* [2009]. Circles and lake outline mark contemporary subglacial lakes discussed in text. Diamonds mark former subglacial lakes. Coastline is from MOA (T. Haran et al., MODIS mosaic of Antarctica (MOA) image map, National Snow and Ice Data Center, Boulder, Colorado, 2005, available at http://nsidc.org/data/nsidc-0280.html).

seismic investigations to determine the extent and depth of a subglacial lake near the South Pole. Figure 2 shows a seismic profile across Lake Ellsworth. This 10 km long and 3 km wide lake is situated in a topographic depression and below more than 3 km of ice in central West Antarctica [*Siegert et al.*, 2004; *Vaughan et al.*, 2007]. These seismic profiles reveal a water column up to 156 m [*Woodward et al.*, 2010] and the presence of underlying unconsolidated lake sediments [*Smith et al.*, 2008]. The seismic character of sediment in Lake Ellsworth is not yet fully established, but it includes reflectors from geological boundaries at and below the lake floor and along shorelines. The seismic profiles from Lake Ellsworth differ from profiles of subaerial lakes because the lack of direct sampling means that the acoustic properties cannot be directly calibrated and interpreted.

2.2. Subglacial Paleolakes as Analogs of the Modern Environment

The widespread presence of subglacial lakes below the Antarctic ice sheet suggests that geological remains of former subglacial lakes (henceforth referred to as subglacial paleolakes) should exist in previously glaciated regions. Identification and investigations of subglacial paleolakes are important because their exposed state offers a means to determine the extent and composition of sedimentary sequences

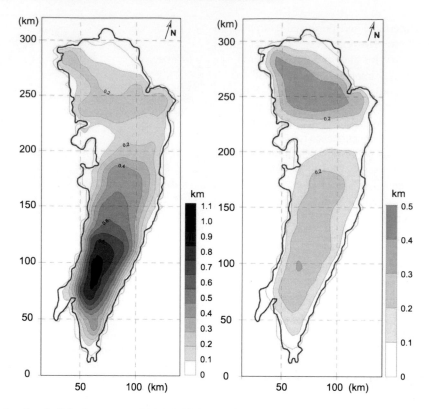

Plate 1. Maps showing (left) bathymetry and (right) thickness of unconsolidated sediment from gravity modeling of Lake Vostok. Bathymetry has been calculated with respect to the ice sheet base as a datum, and so it shows water column thickness. Red line is the lake coastline from radar data. Modified from *Filina et al.* [2008], reprinted with permission from Elsevier.

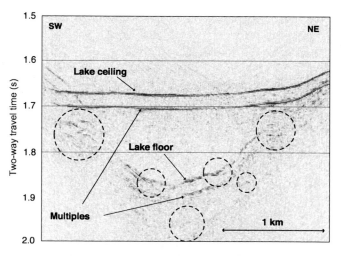

Figure 2. Seismic profile across Lake Ellsworth showing water depth up to approximately 150 m. Strong reflectors mark lake ceiling and lake floor and their ghosts (labeled "multiples"). Dashed circles highlight internal seismic indicators of sedimentary features on and below the lake floor.

formed in subglacial aquatic environments. Geophysical surveys and sediment samples from these paleoanalogs provide a means to guide the interpretation of seismic facies from geophysical surveys over modern subglacial lakes because the absence of ice cover improves the resolution and quality of seismic reflection data. Accurate interpretations of seismic facies are crucial because they inform the choice of drill sites for direct exploration. Moreover, the analysis of subglacial paleolakes offers a cost-effective opportunity to test hypotheses relating to subglacial biological activity and the colonization, succession, and extinction of microorganisms in subglacial environments. They may also serve as testing grounds for the technology needed for direct exploration of deep continental subglacial lakes.

The presence of low energy subglacial lacustrine environments beneath former ice sheets has been inferred from geological studies in southern Norway [*Gjessing*, 1960], eastern Ireland [*McCabe and Ó Cofaigh*, 1994], central Europe [*van Rensbergen et al.*, 1999], and southwest Canada [*Munro-Stasiuk*, 2003; *Russell et al.*, 2003]. These observations yield important insights into sedimentary processes associated with subglacial meltwater. Geological evidence

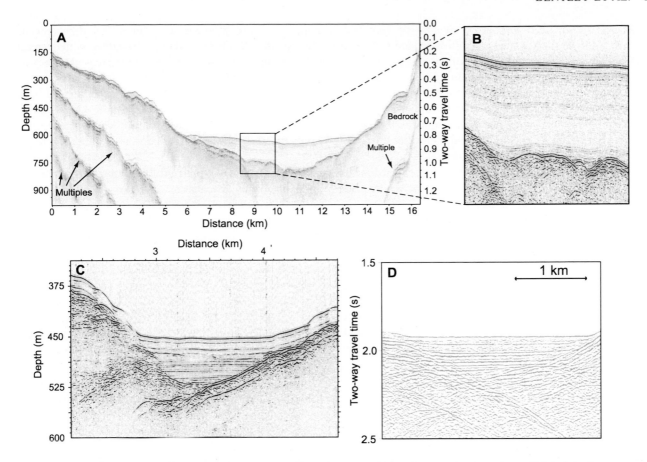

Figure 3. Seismic profiles showing (a–c) subglacial lake sediment in the Great Slave Lake, Canada, and (d) Palmer Deep, Antarctica. Figures 3a and 3b show transect across 620 m deep lake trough where 150 m of subglacial lake sediments forms a seismically stratified sedimentary body. Figures 3c and 3d show seismically stratified and deltaic subglacial sediments. Figures 3a–3c are modified from *Christoffersen et al.* [2008]. Figure 3d is modified from *Rebesco et al.* [1998], reprinted with permission from Elsevier.

in these studies was, however, found along the relatively warm margins of former ice sheets or in Alpine regions, and these settings are unlikely to offer a true analog of Antarctic subglacial lakes (see section 3).

Better analogs of Antarctic subglacial aquatic environments should ideally be found beneath the central parts of former ice sheets. Previous studies suggested that subglacial lakes may have existed in Hudson Bay, Canada [e.g., *Shoemaker*, 1991], but subsequent work suggested that evidence of subglacial floods at this location was caused by abrupt drainage of the proglacial lakes Agassiz and Ojibway 8500 years ago and is not associated with subglacial lakes [*Clarke*, 2003; *Lajeunesse and St-Onge*, 2008]. *Christoffersen et al.* [2008] proposed that an ideal analog of modern subglacial Antarctic lakes may be found in the deep troughs located in the east arm of the Great Slave Lake, Canada. The deepest of these troughs is Christie Bay, where the lake floor is situated 620 m

below the lake's surface and 480 m below sea level. The location of this 100 km long trough (and the comparably deep and adjacent McCloud Bay) is tectonically linked to brittle splay faults from the Great Slave Lake shear zone. This shear zone is a striking magnetic anomaly on the Canadian shield, and it is the ancient root of a continental transform fault that cut through the entire crust approximately 1.9 billion years ago [*Hoffman*, 1987]. The geology and tectonic setting of Christie and McCloud Bay is similar to the gravity-modeled setting of Lake Vostok [*Studinger et al.*, 2003].

The geological setting of the Great Slave Lake and the formation of the >4 km thick Keewatin Ice Dome over the Slave region during the Last Glacial Maximum [*Dyke et al.*, 2002; *Peltier*, 2002] make Christie and McCloud Bays ideal candidates for finding Pleistocene environmental analogs of deeply buried and tectonically associated subglacial lakes on the central East Antarctic continent [*Studinger et al.*, 2003;

Bell et al., 2006]. Seismic surveys of Christie Bay have identified a 150 m thick sequence of fine-grained sediments below a 20 m thick drape consisting of varves of late glacial and postglacial sediments (Figures 3a and 3b). The lower sequence was interpreted to represent material accumulated from sediment-laden turbidity currents (underflows) in a subglacial aquatic environment where ambient lake water was warmer and lighter than subglacial meltwater [*Christoffersen et al.*, 2008]. The sedimentary sequence imaged in the Great Slave Lake is similar to seismically inferred sediments in the >1400 m deep Palmer Deep basin on the western side of the Antarctic Peninsula (Figures 3c and 3d). Here *Rebesco et al.* [1998] estimated that a 200 m thick sequence of silt and clay was deposited in a subglacial cavity. This interpretation is supported by geomorphologic constraints, which further indicate that the lake was up to 800 m deep and contained about 30 km^3 of water [*Domack et al.*, 2006]. Another similar sequence of sediments is found in the >800 m deep Hektoria basin, exposed after the collapse of the Larsen B ice shelf in 2002: sediment samples from this site may be available in 2010 (E. Domack, personal communication, 2009). *Christoffersen et al.* [2008] used the seismic profiles of the Great Slave Lake sedimentary sequences to interpret a change in dominant depositional mechanism from underflow to suspension settling due to the transition from subglacial to proglacial setting. This transition may be more complex: underflows may, for example, remain the dominant depositional mechanism if dissolved or suspended sediment loads make inflowing water denser than the ambient lake water. The underflows in proglacial lakes should, however, produce different sedimentary sequences compared to underflows in subglacial lakes because seasonal variations in the production of surface meltwater and lake-ice cover are likely to produce rhythmically laminated structures such as varves [e.g., *Lewis et al.*, 2002]. Varves are unlikely to exist in sedimentary sequences in deep subglacial lakes unaffected by surface processes. A transition from dominant deposition by underflows to suspension settling within subglacial lakes may, on the other hand, occur if sediment load of meltwater is small and if the ice cover is less than a critical limit. As pointed out by *Wuest and Carmack* [2000], there is a threshold of ice thickness that fundamentally defines the circulation regime of subglacial lakes. In deep subglacial lakes where ice cover is greater than ~3 km, the water column should be convectively unstable such that warming increases buoyancy and causes cold water at the ice-water interface to sink. Sedimentation is thus likely to be underflow-dominated. In cases where the ice cover is less than this critical limit, the water column can become stratified because the temperature-buoyancy relationship reverses so that geothermally heated bottom water becomes denser than cold meltwater at the lake

ceiling. In this case, the water column should become stratified, and the stratification should strengthen with decreasing water column height owing to the increased effect from geothermal heat [*Thoma et al.*, 2008b; *Williams*, 2001]. Sedimentation within stratified lakes may be influenced by suspension settling, although this would require a small sediment load of the subglacial meltwater. Lake Vostok has >3 km of ice overburden, but Lake Whillans has ~1 km. The implication of *Wuest and Carmack*'s [2000] model is that lakes beneath thinner ice, such as Whillans, may experience different sedimentation mechanisms (overflows). Lake Ellsworth is unique in that the critical thermodynamic boundary (where the thermal expansivity of water changes sign) actually lies within the lake [*Woodward et al.*, 2010]. Potential consequences of this for water circulation and sedimentation are uncertain, but the processes may be different to both deeper and shallower subglacial lakes.

3. DIRECT INFORMATION FROM SURFACE EXPOSURES AND SEDIMENT CORES ACQUIRED FROM SUBGLACIAL PALEOLAKES

A great deal is known about the physical properties of subglacial sediments from the reworked and discharged deposits found in glacial forelands and in glaciomarine sediments offshore of most of the world's major ice sheets. In contrast, very little is known about sediments accumulating in deep continental subglacial lakes. This dearth of information is related to the fact that subglacial lakes have not yet been directly explored, and there have been only a few opportunities for direct studies of subglacial lake sediments exposed in previously glaciated terrain or around the margin of an existing ice sheet. To our knowledge, there are only two environments in Antarctica where direct sampling and analysis of former subglacial lakes have been achieved (Palmer Deep has been interpreted as a former subglacial lake, but coring has not penetrated into the subglacial sediment units [*Domack et al.*, 2006]). The first of these is in shallow subglacial lakes emerging from under the margins of the EAIS near Larsemann Hills (e.g., Progress Lake [*Hodgson et al.*, 2006]), and the second is around the margins of the Antarctic Peninsula Ice Sheet on Alexander Island (e.g., Hodgson Lake [*Hodgson et al.*, 2009b, 2009a]). These lakes have emerged from under the retreating margins of the ice sheet since the Last Glacial Maximum and retain lake sediment records deposited when they were overridden by the maximum ice sheet.

In Progress Lake, the most marked feature of the subglacial sediment was the very low volume of deposition, with 4 cm deposited during the onset of glaciation after Marine Isotope Stage 5e, <1 cm deposited under subglacial conditions, and 6 cm deposited during Termination 1 before the lake once again was exposed by retreating ice in the late Holocene and accumulated >16 cm of sediment. On the basis of the <1 cm of accumulated subglacial sediments and the near absence of fossil and biogeochemical markers of biological activity, the lake appears to have experienced a near-hiatus in biological activity and sediment deposition for a period of approximately 90,000 years, with the exception of a few submillimeter thick sand layers. Although an erosional unconformity cannot be ruled out, it is unlikely because the Marine Isotope Stage 5e deposit consists of soft decayed compressed cyanobacterial mats, which would have been easily removed on ice contact and thus suggests that the stratigraphy is intact.

In contrast, in Hodgson Lake, which today remains sealed beneath thick perennial lake ice (Plate 2), a more active overriding ice sheet delivered sediment at a faster rate with 3.8 m of sediment accumulating during the last approximately 93 ka. This has enabled detailed physical, chemical, and biological analysis to be carried out on samples of subglacial lake sediments [*Hodgson et al.*, 2009b, 2009a]. These analyses revealed (Figure 4) that the sediment is composed of fine-grained sediments together with sands, gravels, and small clasts whose changing relative abundances were interpreted as tracking changes in the subglacial depositional environment linked principally to changing glacier dynamics and mass

transport and indirectly to climate change. There was no evidence from sediment cores of overriding glaciers being in contact with the bed, reworking the stratigraphy or eroding this sediment, suggesting that the lake existed through the last glaciation in a subglacial cavity. To confirm this, however, requires geophysical data or more cores from the site to determine the size and potential temporal variability of the subglacial cavity by looking for unconformities in the sediment stratigraphy.

As there have been no direct observations from subglacial lakes to date, we cannot yet directly compare the sedimentary characteristics of Progress Lake and Hodgson Lake with observations from any modern deep continental subglacial lake. However, the demise of ice sheets in the Northern Hemisphere has, as pointed out above, exposed sedimentary environments inferred as being derived from subglacial paleolakes. For example, the preglacial McGregor and Tee Pee Valleys in south-central Alberta, Canada, contain a complex sequence of deposits that record the regional glacial history from initial Laurentide Ice Sheet advance through to deglaciation. It has been suggested from these deposits that large volumes of water accumulated as subglacial lakes in the preglacial valleys, that acted as a large-scale interconnected cavity system that both stored and transported water during occupation by the Laurentide Ice Sheet [*Munro-Stasiuk*, 2003]. Unlike the sedimentary sequence in the 620 m deep

Plate 2. Hodgson Lake (72°00.55′S, 068°27.71′W), center, has emerged from >300 m of overlying glacial ice during the last few thousand years. The modern geomorphology and limnology are described by *Hodgson et al.* [2009b]. The paleolimnology was studied through analysis of a 3.8 m long sediment core extracted at a depth of 93.4 m below the ice surface. The core was dated using a combination of radiocarbon, optically stimulated luminescence and RPI dating incorporated into a chronological model.

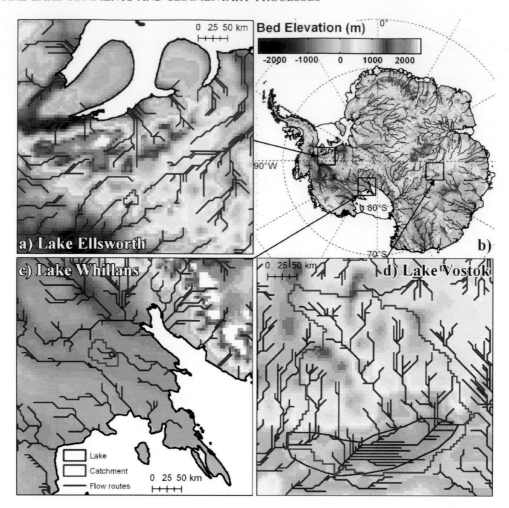

Plate 3. Estimated catchment sizes for lakes Vostok, Ellsworth, and Whillans. The maps use BEDMAP topography [*Lythe and Vaughan*, 2001] and the ice sheet digital elevation model from *Bamber et al.* [2009]. Coastline is from MOA (T. Haran et al., MODIS mosaic of Antarctica (MOA) image map, National Snow and Ice Data Center, Boulder, Colorado, 2005, available at http://nsidc.org/data/nsidc-0280.html). Catchments are calculated using the approach of *Wright et al.* [2008]. (a) Lake Ellsworth, (b) Location map, (c) Lake Whillans, and (d) Lake Vostok. See text for discussion.

east arm of the Great Slave Lake [*Christoffersen et al.*, 2008], the inferred subglacial lake sediments in south-central Alberta are now exposed at the surface and can therefore be sampled directly. Analyses of these sediments showed that they consist of massive silty diamictons composed of approximately 55% silt, approximately 30% sand, and approximately 15% clay, which in various stratigraphic units include blocks of clayey diamicton; small soft sediment clasts; diamicton and silt couplets with large, rounded, soft-sediment clasts silt/clay couplets; and rippled silt and sand beds. These document a subaqueous environment subject to a range of sedimentary processes including gravity flow, water transport, and suspension settling. During the early stages of lake existence, numerous sets of shear planes in

the sedimentary beds indicate that the base of the ice often contacted the bed. During the later stages of lake existence, the lake environment was relatively stable and accumulated fine-grained sediments. Gradual drainage of the lake resulted in lowering of the ice onto the lake beds resulting in subglacial till deposition. Drainage was not a single continuous event, but was characterized by multiple phases of near total drainage (till deposition), followed by water accumulation (lake sedimentation). Water accumulation events became successively less significant, reflected by thinning of lake beds and thickening of till beds higher in the stratigraphic sequence [*Munro-Stasiuk*, 2003]. Since subglacial lake sedimentation appears to be restricted to the subglacial valleys, it has been suggested that these acted as major routes of

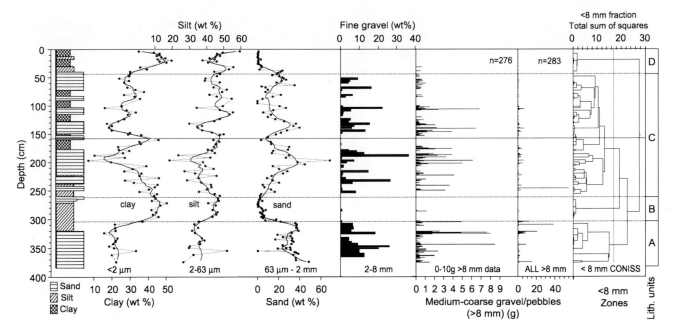

Figure 4. Sedimentological data from the 3.8 m long Hodgson Lake sediment core. The sediments in zones A to C, which date from the last glacial, consist of fine-grained sediments together with sands, gravels, and small clasts. In zone D, there is a transition to finer grained sediments characteristic of lower-energy delivery, which is interpreted as the beginning of the Holocene interglacial sediment regime [*Hodgson et al.*, 2009a].

water flow beneath the southern Laurentide Ice Sheet. However, this interpretation of the Alberta Lakes has been challenged by *Evans et al.* [2006], who suggest that the sedimentary sequences could be the product of proglacial rather than subglacial lakes. The work by *Evans et al.* [2006] highlights one of the major difficulties with inferring subglacial lake sediments, namely, that we do not have well-studied modern analogs from which to determine characteristic facies models. Without these, there is always the risk of circular reasoning: inferring subglacial sediment characteristics from sedimentary sequences assumed to be subglacial. The retrieval of sediment cores from active subglacial lakes will begin to address this problem.

McCabe and Ó Cofaigh [1994] also inferred a former subglacial lake as part of the subglacial drainage network identified in Enniskerry, eastern Ireland. Many of their sediment exposures were associated with meltwater efflux (entry) points into the lake and are characterized by complex sequences of coarse unsorted material up to and including boulder-sized deposits, overlain by finer (sand to clay-sized) material with sedimentary structures recording deposition from waning flows. They noted the lack of Gilbert-type deltas and suggested that this therefore may be a former subglacial, rather than proglacial, lake site. The implication of the *McCabe and Ó Cofaigh* [1994] study is that the margins of subglacial lakes will accumulate a complex record,

containing significant coarse material. There is a relatively small volume of fine-grained (silt-clay) sediment in the Enniskerry example: this has been attributed to a continuously active subglacial throughflow, which is also supported by the presence of unidirectional paleoflow indicators in the center of the basin [*McCabe and Ó Cofaigh*, 1994]. Instead, there are numerous layers of interfingered winnowed diamict. Like the Lake McGregor example in Canada, the Enniskerry sediment has been periodically deformed by the drag of the ice roof, especially in exposures where there is evidence for substantial accumulations of sediment at efflux points, and which may have shoaled upward to reach the basal ice. In places, these sequences pass upward into basal till.

The subglacial lake environments from beneath former ice sheets are useful in the sense that they have identified important sedimentary processes and products. However, for the reasons outlined above, it is not immediately clear that they can be demonstrably classed as subglacial lake sediments or that they offer a true analog of deep continental subglacial lakes in the Antarctic interior. For instance, climate reconstructions show that the southern margin of the Laurentide Ice Sheet was warm and wet [*Bromwich et al.*, 2004, 2005], so it is likely that surface meltwater and precipitation falling as rain at rates up to several millimeters per day during summer months [*Bromwich et al.*, 2005] fed subglacial water systems including lakes constrained by valley topography in ice marginal settings. This environment resembles modern conditions in Greenland, where surface meltwater penetrates to the bed during warm summer months [*Zwally et al.*, 2002; *Das et al.*, 2008; *Shepherd et al.*, 2009] and possibly the environment on the Antarctic Peninsula where subglacial lakes may have existed in Palmer Deep and in Hektoria Basin. In stark contrast, subglacial lakes in East and West Antarctica are isolated from the atmosphere and fed exclusively by subglacial water forming from relatively slow rates of basal melting (a few millimeters per year or less).

From these three examples, it can be seen that deposition in former inferred subglacial environments ranges from what appears to be a highly restricted deposition seen in the passive environment of Progress Lake (sedimentation rate close to zero), to several meters of undisturbed subglacial sedimentation observed in Hodgson Lake at an average linear rate of 0.04 mm yr^{-1}, through to the highly dynamic subglacial environments in the Palmer Deep, Great Slave Lake, McGregor Lake, and Enniskerry, producing as much as 100 m thick subglacial lake sedimentary sequences at estimated linear rates >1 mm yr^{-1}. These respective rates of sediment accumulation will influence the composition and resolution of paleoenvironmental records potentially stored in subglacial lake sediments.

4. SEDIMENTARY PROCESSES IN SUBGLACIAL LAKES

Since there are no direct measurements of processes in Antarctic subglacial lakes, we have to use a combination of indirect geophysical information and models, the known composition of accreted ice above Lake Vostok and theoretical understanding of water column stability and general circulation. It is, in addition, important to draw on analogies from a comprehensive literature on glacial and lacustrine processes. Here we use such information to speculate on the most likely processes and to develop a conceptual model of subglacial lake sedimentation.

4.1. Water Circulation

The geophysical surveys over Lake Vostok have significantly improved the boundary conditions of numerical models used to simulate the circulation of water in subglacial lakes. The stability of the water column and the circulation of water masses are geologically important because vertical mixing and current velocity control the partitioning of sedimentary particles in suspension and those deposited on the lake floor. Early studies suggested that water in Lake Vostok would be convectively unstable because the temperature of maximum density of subglacial lake water is lower than the

freezing point when the overlying ice sheet is thicker than 3 km [*Wuest and Carmack*, 2000]. Meltwater forming at the lake ceiling may sink because it is colder and therefore denser than the ambient lake water. Application of a modified ocean general circulation model confirmed this potential for vertical mixing, but showed that a stable circulation pattern can be maintained only by temperature-induced density contrasts resulting from variations of the freezing point along the inclined lake ceiling [*Mayer et al.*, 2003]. Improved boundary conditions and model physics have since showed that water circulation in Lake Vostok is likely more stable than previously thought [*Thoma et al.*, 2007, 2008b]. Whereas the magnitudes of geothermal heat and conductive heat loss into the overlying ice sheet affect the rates and spatial distribution of melting and freezing, they exert little effect on the turbulent kinetic energy of water in circulation. Salinity, however, influences the freezing point and stabilizes the water column because it reduces the density contrast between cold fresh meltwater and less dense ambient lake water. Geochemical analyses of accreted ice over Lake Vostok indicate that salinity is unlikely to be higher than a few per mille [*Souchez et al.*, 2000] and probably between 1/20th and 1/470th of seawater values [*Christner et al.*, 2006], but even small variations in salinity can modulate circulation and potentially cause a change in the style of sedimentation. The dampening effect of salinity on circulation is evident in the numerical experiments of *Thoma et al.* [2008b], where turbulent kinetic energy of water with 2‰ salinity is only half of that of freshwater circulation.

4.2. Sediment Pathways

There are five likely sediment sources for subglacial lakes. The first two are (1) material originally deposited on the ice sheet surface and transported downward via surface accumulation and flow to reach the lake ceiling and eventually "rain-out" [*Siegert*, 2000] and (2) subglacially eroded material frozen into basal layers of the ice sheet and melted out at the ice-water interface. Both of these pathways deliver sediment at the lake ceiling and only where the base of the ice sheet is melting (often indicated by a very sharp basal reflector on radar profiles). (3) A third source is transport of sediment in subglacial water systems. An active hydrological network beneath ice sheets has long been predicted by theory [e.g., *Shreve*, 1972; *Nye*, 1973] and shown under and at the margins of former ice sheets by geomorphologists [e.g., *Sugden et al.*, 1991; *Lowe and Anderson*, 2003; *Sawagaki and Hirakawa*, 1997]. An active subglacial hydrological system beneath the Antarctic ice sheet has now been demonstrated by satellite measurements of transient surface elevation changes [*Gray et al.*, 2005; *Wingham et al.*, 2006; *Fricker et* al., 2007]. On the basis of analogy with studies of glacier hydrology and sedimentation beneath and at the margins of contemporary glaciers [Ashley et al., 1985], this latter pathway may be highly effective in flushing large volumes of sediment into subglacial lakes through influx points and subsequent dispersal through the lake. Indeed, it is possible that it may be volumetrically dominant over the subglacial melt out of basal debris and rain-out of englacial and supraglacial debris. (4) Advection of subglacial sediment into lakes may occur where the ice sheet is underlain by easily deformable sediment, such as below ice streams. This will not be present for lakes surrounded by bedrock. (5) The role of chemical sedimentation within subglacial lakes is difficult to assess, and although it is likely to be volumetrically minor, it may create distinctive authigenic minerals. The potential for chemical sedimentation will depend strongly on the lake waters, on biogeochemical weathering processes [Skidmore et al., 2010], and whether the lake is a closed hydrological system and relatively rapid concentration and saturation can occur.

Estimating the relative contribution of these sediment pathways can be attempted with a knowledge of ice composition and flow, basal topography, and basal thermal regime. For the supraglacial material (mostly dust), the sediment flux into the lake will depend primarily on the flux onto the ice surface (and thus dust concentration in the ice) and melt rates at the ice sheet base. This source is likely to be volumetrically very small, as shown by an order of magnitude estimate from the dust content of ice cores. For example, the dust content of the EPICA ice core mostly lies in the range 10 to 1000 ng dust/g of ice [Lambert et al., 2008]. Thoma et al. [2008a] estimate a melt rate at the ceiling of Lake Vostok of 2.65 ± 0.10 cm yr^{-1}. With an ice density of 0.9 g cm^{-3}, this suggests a dust flux of 20 to 2000 ng cm^{-2} yr^{-1}. If all the material is clay ($\rho \approx 2$ g cm^{-3}), this is equivalent to a maximum sedimentation rate of ~0.01 μm yr^{-1}, or 10 cm Ma^{-1}. This rate would likely be slightly higher in West Antarctica, where dust concentrations in ice tend to be higher. The supraglacial flux may contain extraterrestrial material: micrometeorites have been observed in Lake Vostok accretion ice [Doran et al., 2004]. This means that the sediments are likely to contain meteorites, micrometeorites, and cosmic dust, and so subglacial lake sediments may offer an opportunity to measure long-term flux of extraterrestrial material [Doran et al., 2004].

For the subglacially eroded material, the flux into the lake will depend primarily on the melt rate at the lake ceiling and the concentration of basal debris in the ice sheet. The latter is dependent on the basal thermal regime: the presence of subglacial water promotes erosion, regelation, and freeze-on of the sediment-bearing basal water film, but recent observations under glaciers show that the water is not necessarily a

requirement [Cuffey et al., 2000]. In broad terms, it might be speculated that the larger the area of basal melting directly upstream of a lake, the greater the likely delivery of subglacially eroded material into the lake. Deposition of the subglacially eroded material will probably be concentrated close to the up-flow margin of the lake because progressive melting as the ice flows across the lake will exhaust the basal ice layer. The lack of any coarse sediment material in that part of the Vostok ice core at the base of the ice sheet (immediately above accretion ice) shows that any subglacial debris layer in Vostok has been completely melted out by the time it flows to this point [Christner et al., 2006].

Thoma et al. [2008a] estimate that only 36% of the Lake Vostok upper surface area is in contact with meteoric ice (i.e., 64% of ceiling is accretion ice). This means that there will be a correspondingly small area of the lake affected by direct rain-out sedimentation. In the case of lakes Ellsworth and Whillans, no accretion ice layer has yet been detected, although [Woodward et al., 2010] discuss one model that suggests there may be an area of accretion ice at the downstream end of Lake Ellsworth. In these lakes, the area of melting/meteoric ice may be up to 100%, in which case rain-out might affect most if not all of the lake area.

Estimating the volume of sediment from subglacial hydrological transport is more difficult. It will be dependent on the rate of erosion of bedrock and any preexisting sediments in the hydrological catchment of the lake (itself a function of the basal thermal regime, flow velocity, and the erosive power of water flow), as well as the velocity and thus stream power [Knighton, 1999] of the subglacial drainage network. The subglacial network seems to operate episodically [Wingham et al., 2006; Smith et al., 2009], and as yet we do not have sufficient time series to understand whether the drainage events of individual lakes are periodic. This, together with a lack of knowledge on flow conduit dimensions, means that providing firm estimates of "background" and peak flow is challenging.

Subaerial stream catchments show a positive (log-log) relationship between catchment size and sediment discharge (mass per unit time) [Hay, 1998]. If such a relationship exists for subglacial catchments, then estimating the size of catchment might be instructive for estimating relative sediment discharge. We have used the approach of Wright et al. [2008] to estimate the catchment sizes of Lake Vostok, Lake Ellsworth, and Lake Whillans (Plate 3). Using the BEDMAP topography [Lythe and Vaughan, 2001] and the ice sheet surface digital elevation model of Bamber et al. [2009], it is possible to calculate the basal drainage network and to partition this into catchments for each lake. From this we estimate the hydrological catchment sizes: Vostok, 70,925 km^2; Whillans, 1500 km^2; and Ellsworth, 450 km^2. Although this is only a crude estimate, it nonetheless illustrates that there are more than 2 order-of-magnitude differences in catchment size that may translate into differences in sediment discharge from meltwater. In detail, the sediment flux estimates at a particular point in each lake will be affected by the number of discharge points, lake area, and bathymetry. There will also be other controls on hydrologically delivered sediment to lakes, including how and where sediment is stored in catchments, its composition, and whether peak flows from upstream lake discharge would be enough to flush sediment storage areas.

4.3. Analogous Processes in Subaerial Lakes

There is a rich literature on limnology and the processes operating in lakes and coastal environments, including lakes and fjords in contact with an ice margin (proglacial) and those fed by meltwater (glacier-fed). Comprehensive reviews are provided in a number of volumes [e.g., *Ashley et al.*, 1985; *Drewry*, 1986; *Benn and Evans*, 1998; *Hambrey*, 1994]. From these we can reasonably infer the likely sedimentary processes occurring in subglacial lakes.

One of the dominant controls on the style of sedimentation delivered via inflow into lakes or fjords is the level at which sediment is introduced into the water column. There are three possibilities, namely, near the lake surface (overflows), at midlevels (interflows), or at the lake bed (underflows). The level of inflow is controlled by the relative density of the incoming sediment-laden water and the lake water/seawater itself, so that, for example, meltwater with a relatively small sediment load arriving in seawater will spread as an overflow, while a meltwater plume with high sediment load will spread through a fresh lake as an underflow [*Ashley et al.*, 1985; *Drewry*, 1986]. *Lamoureux and Gilbert* [2004] have quantified the variation in density of fresh waters with suspended sediment load and showed that the suspended sediment load dominates over temperature- and dissolved load-driven density variations. In subglacial lakes, salinities are thought to be relatively low [*Souchez et al.*, 2000; *Christner et al.*, 2006; *Vaughan et al.*, 2007]. Direct measurement of the waters in formerly subglacial Lake Hodgson shows that conductivity is 0.02 mS cm^{-1}, with anions and cations just higher than concentrations recorded in continental rain [*Hodgson et al.*, 2009b]. For these reasons, the dominant sedimentation mechanism for subglacial lakes will be underflows (also called hyperpycnal flows) along the lake floor. Overflows will occur only when clean, sediment-poor waters flow into the lake and thus are not likely to be sedimentologically important. Limited evidence for this also comes from the relative lack of sediment in the accreted ice zone immediately above

Lake Vostok [*Christner et al.*, 2006]: if overflows did exist in the lake, then coarser sediment inclusions at this high level might be expected.

Observations of underflows in glacier-fed and proglacial lakes have demonstrated that underflows can travel more than 6 km in Lillooet Lake, British Columbia [*Ashley et al.*, 1985], possibly up to 60 km in Lake Wakatipu, New Zealand [*Pickrill and Irwin*, 1982] and shown to transport fine sand over 15 km in Lake Geneva [*Houbolt and Jonker*, 1968]. Underflows in glacial settings including subglacial lakes are inferred from sediment core samples from Lake Zurich [*Lister*, 1984] and seismic profiles of Lake Le Bourget in the European Alps [*van Rensbergen et al.*, 1999]. Descending underflows may often behave as turbidity currents where particles are suspended by turbulent suspension. Turbidity currents deposit a well-known sequence ("Bouma cycle") of sediment with a basal erosional unconformity overlain by normally graded fine gravels, sand, and silt, with a clay drape on top [*Bouma*, 1962]. However, the full Bouma sequence is unlikely to be recorded by lacustrine turbidites, and we might expect to see only the graded top part of the sequence. Measured underflow velocities are mostly a few centimeters to tens of centimeters per second [*Ashley et al.*, 1985; *Crookshanks and Gilbert*, 2008], but at least two studies have reported sporadic underflow velocities in excess of 1 m s^{-1} (e.g., 1 km from mouth of Rhine in Lake Constance [*Lambert*, 1982] and quasi-continuous flows up to 0.6 m s^{-1} [*Gilbert and Crookshanks*, 2009]. Such strong flows are capable of sediment scouring and are the likely cause of subaqueous channels (some with levees) in glacier-fed lakes [*Ashley et al.*, 1985]. Underflows can persist for several days, and measurements in several lakes have demonstrated that underflows are also directed by lake floor topography [*Ashley et al.*, 1985] such that dominant sediment transport paths can develop along the lake bottom, and high points can be starved of sediment. Underflows are also deflected by the Coriolis effect [*Smith*, 1978], causing sedimentation to be concentrated along the right hand side of the lake (relative to dominant throughflow) in Northern Hemisphere lakes. Thus, we might expect sedimentation focused along the left hand side of Antarctic subglacial lakes. Underflows can also be pulsed, even in the presence of a constant input source and thus lead to complex stratigraphic deposits [*Best et al.*, 2005]. Observations from former subglacial lakes are consistent with a dominant underflow sedimentation mechanism (see section 3).

4.4. Sediment-Landform Associations at the Grounding Line

Depending on the delivery rate of sediment, stability of the influx point, and stability of lake level, various landforms can develop at the influx point of proglacial and glacier-fed lakes.

In some respects, these can act as analogs for what may be deposited at the grounding line (margin) of subglacial lakes. One critical difference, however, is that subglacial lacustrine landforms cannot grade to the water level to form a subaerial landform such as a delta; such vertical aggradation will be prevented when sedimentation eventually reaches the lake ceiling, in which case deposition will be forced to move laterally or sediment will prograde. This shoaling of sedimentation has been described for the Enniskerry former subglacial lake [*McCabe and Ó Cofaigh*, 1994].

Possible sediment-landform associations that might occur along the grounding line at the margin of subglacial lakes include morainal banks and grounding-line fans. These are a continuum of landforms deposited at the grounding line [*Benn and Evans*, 1998]. Grounding-line fans are isolated landforms, developed at meltwater efflux points. They are analogous to the subaqueous fans described by [*Houbolt and Jonker*, 1968], who suggested that they are formed where there is a stable efflux point, and thus, persistent underflows will eventually develop fans with a system of distributary channels, levees, and lobes. *McCabe and Ó Cofaigh* [1994] described exposures of subaqueous fans in the Enniskerry Lake. Morainal banks are elongate masses of sediment deposited at the grounding line, sometimes formed by amalgamation of grounding-line fans. They can be very small where sedimentation is low; for example, *Evans* [1990] reported subaqueous moraines <1 m high in northern Ellesmere Island, Arctic Canada, and yet which mark the former Last Glacial Maximum position of the ice sheet.

The Salpausselkä moraines in Finland constitute a 600 km long set of narrow, sinuous landforms deposited at the eastern margin of the Fennoscandian ice sheet when it was grounded in the proglacial Baltic Ice Lake during deglaciation [*Hambrey*, 1994]. This landform belt has been interpreted as a coalesced series of grounding-line landforms ranging from large individual deltas with braided tops built up to water level (conduit-focused sedimentation); lower, narrower coalescing grounding-line fans of finer material (distributed drainage system); and small laterally overlapping subaqueous fans (unstable subglacial conduits) [*Fyfe*, 1990]. We can do no more than speculate at this point, but it is possible that linear landform belts similar to Salpausselkä may exist along the margins of some Antarctic subglacial lakes. Clearly any change in size of subglacial lakes will also lead to changes in the position of such belts of grounding-line landforms, and so if these could ever be imaged with geophysical methods or directly sampled, then the position of such landforms could yield insights into the long-term history of the lake.

The margins of some subglacial lakes are analogous to the grounding line of an ice sheet, such as along the Siple Coast, where ice streams flow into the Ross Ice Shelf. Here grounding zone wedges (sometimes called grounding-line wedges [*Benn and Evans*, 1998] or "till deltas" [*Alley et al.*, 1987; *Larter and Vanneste*, 1995]) develop from the delivery of (deforming) sediment at the base of the ice stream [*Anandakrishnan et al.*, 2007; *Alley et al.*, 2007], which is redeposited by gravity flows down the front of the wedge. Overriding of the grounding zone wedge will lead to erosion of the top surface by the active deforming layer and progradation of "foresets" of freshly deposited or reworked sediment on the front face of the wedge [*Alley et al.*, 1989]. Retreat will preserve this sequence, while oscillation of the ice margin will be recorded by interbedding of subglacial (deformation till) and subaqueous (debris flow) deposits [*Benn and Evans*, 1998]. Grounding zone wedges are characteristic of areas where meltwater discharge is small; as meltwater influence increases, a grounding-line fan or moraine bank will form [*Benn and Evans*, 1998]. This analogy is most appropriate for those subglacial lakes beneath ice streams where there is rapid ice flow and delivery of abundant deforming sediment at the grounding line or lake margin (e.g., Lake Whillans).

The gravity flows at the margins of lakes, and the deposits they produce could be spatially and temporally variable. For example, in subaerial settings, *Middleton and Hampton* [1976] developed a fourfold classification of gravity flows, each producing a distinctive stratigraphic sequence. These can include both normal and reverse-graded deposits and a wide range of grain sizes and sorting characteristics. Processes operating on steep fans have also been detailed by [*Nemec*, 1990]. The field criteria for recognizing deposits of these flows are beyond the scope of this review but have been detailed in a wide range of publications (see references in *Nemec* [1990] and *Eyles et al.* [1983]).

4.5. Distal Sedimentation

There will be a decrease in flow competence away from the influx point, such that sediment grain sizes will progressively decrease with distance. Lake floor sedimentation will be dominated by the progressive settling out of distal sediment from underflows, turbidity, or slump-derived currents [*Ashley et al.*, 1985]. These distal sediments will show a transition from cross-laminated sand and silt deposits with occasional gravel layers through to planar laminated silt and clay deposits. The laminated deposits will develop as silt-clay couplets. These are normally graded, relating to waning flow after each underflow current. In many subaerial lakes, these couplets derive from the annual cycle of sedimentation (varves), but in subglacial lakes, annual variations are highly unlikely, and so individual couplets will be related to individual underflows from episodic hydrological input or gravitational failures. A key difference to varved lake sediment is that the thickness of the silt and clay components will be in proportion to one another (in varves, the silt layer thickness can vary, but the clay layer thickness is uniform across the lake). The total thickness of these couplets will decrease away from the source region and will likely be in the range of a few centimeters (proximal) to millimeters (distal). Successive couplets will be separated by a sharp contact, representing an interval of nondeposition between flows. The deposition of the finest grain sizes will depend on turbulent flow or circulation within the lake. In the case of Lake Vostok, one of the few studies of subglacial lake sediment relates to this finest sediment. The accreted ice at the base of the Lake Vostok ice core contains fine sediment particles up to ~23 μm [Royston-Bishop et al., 2005]. Calculations of Stokes settling velocities, corrected for particle shape, have shown that 98% of these particle sizes could be maintained in the water column by vertical current velocities of 0.0003 m s^{-1}, and the coarsest 2% by velocities of up to 0.0013 m s^{-1}. There are variations in the grain size within the accretion ice, and these imply perturbations in vertical velocity [Royston-Bishop et al., 2005], perhaps related to individual hydrological "flushing" events. The suspension of particles in the <20 μm range has led some to suggest that the water column in Lake Vostok will be turbid/cloudy and analogous to the water found in montane glacier-fed lakes [Wuest and Carmack, 2000]. However, more recent work has shown that primary particles can flocculate into large aggregations in lakes [Hodder and Gilbert, 2007], with the implication that settling velocities may be greater than suggested from particle size measurements of individual grains. The presence of flocs has been used to explain the observation that the water column appears to clear over the winter in some deep arctic lakes [Lamoureux and Gilbert, 2004]. The mechanism for incorporation of fine sediment into basal ice above subglacial lakes may be analogous to the sediment incorporation into sea ice by "suspension freezing" [Nürnberg et al., 1994], which requires formation of either frazil or anchor ice. In smaller lakes with persistent throughflows, there may be less finer grained sedimentation because of the continuous influence of coarser underflows and gravity flows, as suggested for subglacial paleolake Enniskerry [McCabe and Ó Cofaigh, 1994].

4.6. Preglacial Sediments

Many Antarctic subglacial lakes will contain a lower sequence of preglacial sediments. In some cases, the basins are well below sea level, even when glacially unloaded (e.g., Lake Ellsworth), and so preglacial sediments will be marine. In Lake Vostok, a rift valley origin for the basin has been suggested [Siegert et al., 2001; Studinger et al., 2003], and so preglacial sediments may contain rift sequences. The total sediment thickness in Lake Vostok has been suggested to be up to 300 m [Filina et al., 2008], 500 m [Wuest and Carmack, 2000] or as much as 2 km [Studinger et al., 2003]. In the case of Vostok, it has been suggested that the preglacial sediments may date to before EAIS initiation, >30 Ma [Siegert et al., 2001], and thus could provide a mid-Cenozoic record for the interior of the continent. However, a major unknown aspect is the degree to which preglacial sediments survived early glacial inundation during ice sheet growth [Doran et al., 2004]. Shearing of sediment in inferred subglacial paleolakes shows that the early stages of lake formation may be accompanied by substantial erosion and deformation [Munro-Stasiuk, 2003].

In the case of West Antarctic subglacial lakes (e.g., Ellsworth), there is the potential for preservation of a record of ice sheet collapse, in the form of nonglacial/marine sediment in presently ice-covered locations. For example, Scherer [1993] and Scherer et al. [1998] used the [10]Be and microfossil content of subglacial sediment from beneath Whillans Ice Stream to argue for Pleistocene collapse of the ice sheet. This has been disputed [Burckle, 1993], partly because of the difficulty of dating deforming subglacial sediment, but the future retrieval of continuous sedimentary sequences from subglacial lakes raises the prospect of being able to date ice sheet collapse events more easily [Siegert et al., 2006]. However, as with ice sheet initiation over Lake Vostok, the effect of ice sheet retreat and regrowth on the sedimentary sequences in West Antarctic subglacial lakes is not yet known. As well as preglacial sediments, subglacial lakes may contain glacial sediments derived from proglacial or direct subglacial sedimentation prior to (re)formation of a subglacial lake.

4.7. A Conceptual Model for Sedimentation in Subglacial Lakes

With an understanding of the likely processes operating in and around subglacial lakes, we can develop a conceptual model of the nature and spatial distribution of sediments (and lake-floor landforms) in subglacial lakes. In Figure 5, we illustrate this model, modified for the likely processes operating in each of the three lakes considered: Vostok, Whillans, and Ellsworth. We present this model here as a hypothesis that can be tested during direct exploration of subglacial lakes and in some cases by further geophysical data acquisition. These models are drawn on the basis of a single influx point for hydrological flushing events. If there are multiple influx points, perhaps even on different sides of the lake, then there will be complex interactions between the proximal-distal sedimentary sequences for each site, leading to more complex sediment distribution with the potential for interfingering deposits. Differences in geometry, setting, and stability will promote different sedimentation conditions in different subglacial lakes. Lakes Vostok and Ellsworth are both located beneath thick ice (>3 km) in the deep interior of the Antarctic ice sheet, but whereas the former has formed on an ancient craton reminiscent of the Canadian Shield, the latter is located in a more mountainous subglacial setting on the edge of the Ellsworth-Whitmore block. Whillans Lake is located beneath a fast flowing ice stream and below relatively thin ice (~1 km).

In Lake Vostok (Figure 5a), the upstream side of the lake is likely to be affected by rain-out of supraglacial, englacial, and subglacial material, but this rain-out stops when melting of the ice roof ceases and water starts to accrete onto it.

Dropstones and coarser rain-out sediment should be found concentrated along the upstream shoreline because this material will have to come from basal ice layers, which are finite in thickness and quickly lost by basal melting. For instance, a 10 m thick basal ice layer would be lost in roughly 500 years and within a few kilometers of the shoreline, assuming that the modeled average rate of melting is correct (2 cm yr^{-1} [*Thoma et al.*, 2008a]). At the grounding line, meltwater efflux points and subglacial rain-out may be building a grounding-line fan, composed of subaqueous outwash, and mass flow diamictons. If this is prograding, then it may lead to gravitational instability and thus result in episodic underflows. If grounding-line deposition is occurring along much of the upstream margin, then a Salpausselkä-type feature may be present. If so, then past fluctuations of the lake margin will

(a) Vostok

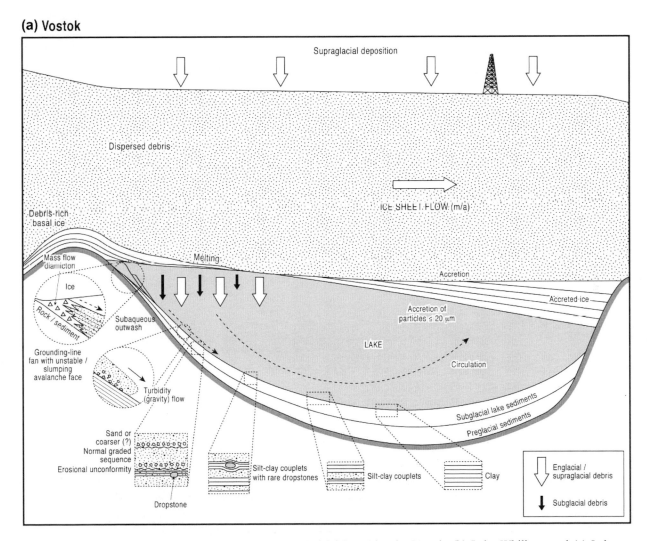

Figure 5. Conceptual model of sedimentation in subglacial lakes: (a) Lake Vostok, (b) Lake Whillans, and (c) Lake Ellsworth. See text for detailed discussion of the basis for these models.

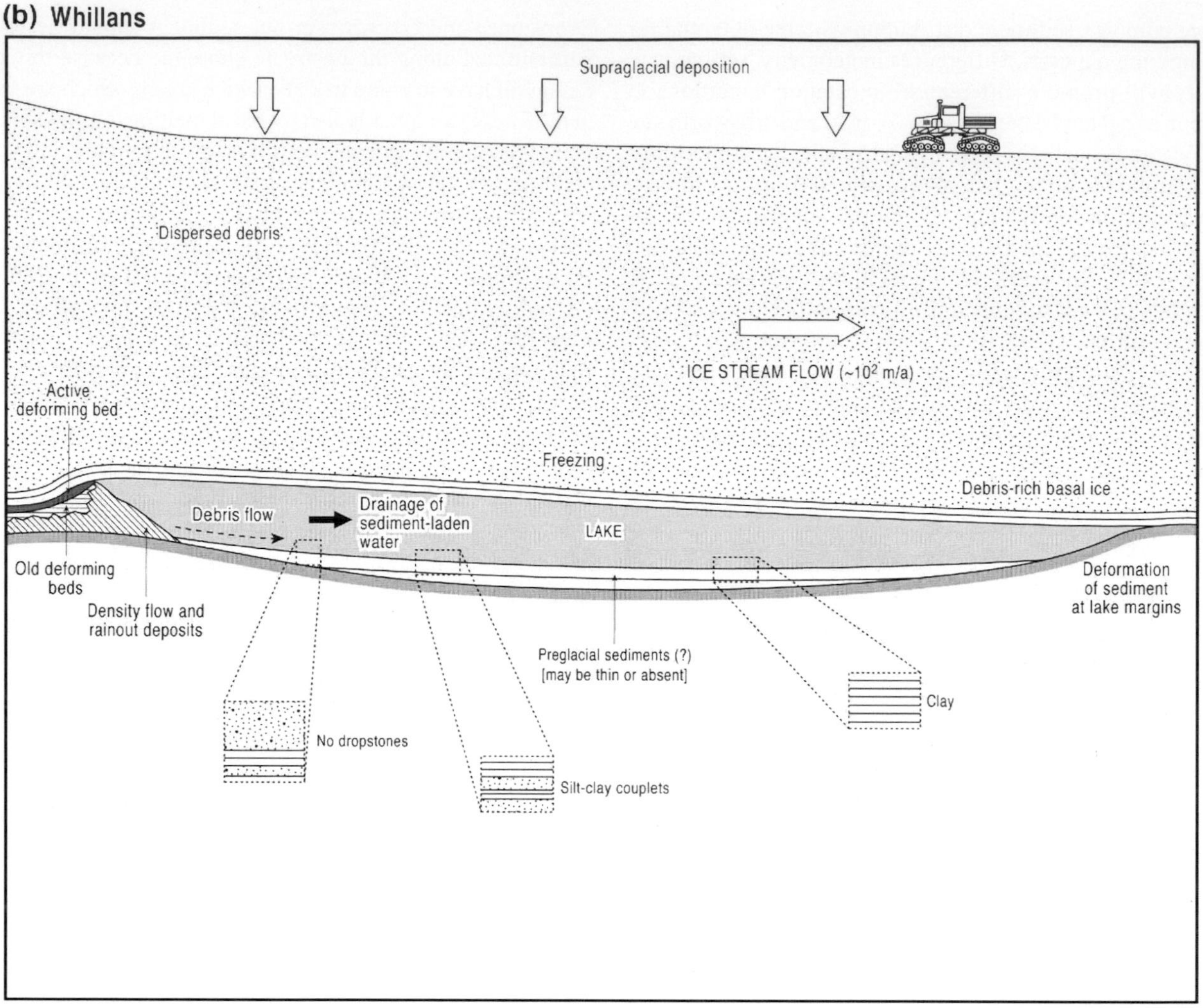

(b) Whillans

Figure 5. (continued)

have created multiple landforms in this zone. The laterally continuous distribution of unconsolidated sediment reported by *Filina et al.* [2008] shows widespread and even deposition of sediments. These sediments may be brought to the lake floor in underflows affecting large areas of the lake bed and with a small component of rain-out from the ice roof. The finest material is kept in suspension and incorporated into the accreted ice of the lake ceiling [*Souchez et al.*, 2000; *Royston-Bishop et al.*, 2005].

In Lake Whillans (Figure 5b), the transport of deforming sediment into the upstream part of the lake may be creating a grounding zone wedge, similar to those along the Siple Coast. Fluctuations in lake position (and sediment supply) may have created a complex architecture within the wedge. Prior estimates of subglacial till fluxes beneath Whillans Ice Stream fall mostly within the range of 10 to 100 m³ yr⁻¹ per

meter width of ice stream [e.g., *Tulaczyk et al.*, 2001]. Taking the width of Lake Whillans as ~5 km, the anticipated till input rate into the basin is then 0.05 to 0.5 km³ yr⁻¹. The other major mode of sediment input into Lake Whillans should be influx of sediments with water entering during the filling phase of this active subglacial lake. Available ICESat elevation data suggest that subglacial water enters Lake Whillans at the rate of ~0.1 km³ yr⁻¹ [*Fricker and Scambos*, 2009]. Sediment concentrations in glacial streams vary significantly, and we choose the broad range of 0.01 to 1% volume fraction to arrive at an estimate of 0.00001 to 0.001 km³ yr⁻¹ as the possible average glaciofluvial input into Lake Whillans. Previous modeling studies indicate that the lower part of Whillans Ice Stream experiences basal freezing of several millimeters per year [*Joughin and Padman*, 2003; *Joughin et al.*, 2004]. Basal freezing should

(c) Ellsworth

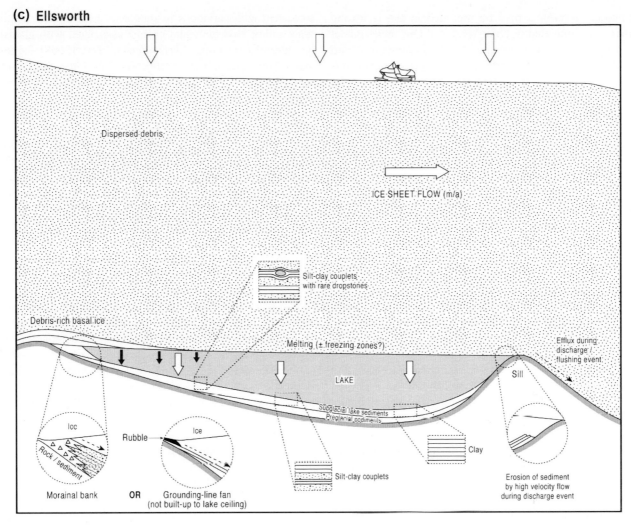

Figure 5. (continued)

be even stronger locally over Lake Whillans, where basal shear heating goes to nil. In the prior modeling work, basal shear heating in the area contributed heat equivalent to 2 to 4 mm yr^{-1} of basal freezing. Hence, the ice ceiling over Lake Whillans may be experiencing freezing at a rate of about 1 cm yr^{-1}, even when one neglects any additional ice accretion owing to supercooled subglacial water entering the lake from upstream. This is a fundamental difference between Lake Whillans and lakes Vostok and Ellsworth, which are located beneath 3 to 4 km thick ice cover, which acts as an effective thermal insulator. The conductive heat loss from the deep lakes into the overlying ice is smaller than any reasonable input of geothermal heat, so that the lake basins are generally in a positive heat balance resulting in sediment input from melting of the debris-laden basal ice. For Lake Whillans, basal freezing should be predominant, and we do not expect

basal melting to provide an appreciable input of sediments. One implication of this is that coarser debris (dropstones) deposited directly from basal debris layers will be rare or absent. This leaves influx of deforming subglacial till from upstream and subglacial glaciofluvial fluxes as the two main mechanisms of sediment delivery to the lake. Both processes are paced by temporal transients beneath Whillans Ice Stream. The subglacial till deformation rates will vary on the same time scales as ice stream velocity, which is changing on tidal scales [e.g., *Bindschadler et al.*, 2003] and experiences long-term deceleration [*Joughin et al.*, 2002]. Inflation and deflation of Lake Whillans, which are interpreted as changes in subglacial water influx/drainage rates, are constrained only since 2003, and over this period, the lake experienced two cycles, which had similar periods to within about 1 year [*Fricker and Scambos*, 2009]. These deformational and hydrological transients may be recorded in the lacustrine sediments accumulated in Lake Whillans. Our facies model envisions a grounding-zone-type till wedge on the upstream side of the lake that is shedding poorly sorted subaqueous debris flows, which may expand across the lake basin. The subglacial water input will likely enter the lake in one or more discrete points where subglacial channel(s) intersects the basin. Sand and gravel are likely to accumulate in fans proximal to the channel outlet(s), with silts and clays traveling to more distal parts of the basin. The (quasi) periodic inflation/deflation of Lake Whillans will produce a zone of deformational modification of lake sediments at least in areas proximal to the basin boundaries where the ice ceiling has grounded on the sediments during deflation.

In Lake Ellsworth (Figure 5c), the narrow nature, and likely connection to a drainage network upstream [*Ross et al.*, this volume], may mean that hydrological inputs play a strong role [*Vaughan et al.*, 2007]. If the upstream lake(s) episodically drains into the lake, then they may create a grounding-line fan or morainal bank. Drainage of Lake Ellsworth itself through the efflux point may cause erosion of sediments at the downstream end. Rain-out sedimentation of supraglacial material will occur throughout the lake, and subglacial material including dropstones will be deposited close to the upstream end. The proportion of fine sediment will be dependent on the strength of the hydrological throughflow and whether it is maintained continuously. A strong, continuous throughflow will lead to coarser sediments in the distal parts of the lake and to unidirectional sedimentary indicators of paleocurrents. Recent geophysical surveys of Lake Ellsworth have demonstrated a maximum water column thickness of 156 m [*Woodward et al.*, 2010]. Thermodynamic modeling of the water cavity suggests that Lake Ellsworth may have both zones of melting and freezing, although this result is dependent on the assumption of the lake being a closed hydrological system [*Woodward et al.*, 2010]. This assumption may not be valid, given the lake's position in a cascade of lakes [*Ross et al.*, this volume].

5. SEDIMENTARY ARCHIVES OF LAKE FORMATION AND HYDROLOGICAL DISCHARGE

The discussion of subglacial lake sedimentation has highlighted the crucial role of hydrological inputs and outputs. A key question for each of the sequences that could be sampled is in what way they may have been affected by lake drainage events. For example, the seismic data sets from the Great Slave Lake, Palmer Deep, and Hektoria Basin contain laterally continuous internal reflectors, which show substantial changes in the size or character of subglacial lake sediment. The cause of these transitions is not yet known, but they are consistent with variations in subglacial hydrological conditions and/or environmental change associated with glacial-interglacial transitions. *Christoffersen et al.* [2008] suggest that densely spaced bottom reflectors in the Great Slave Lake sediment represent a high-energy environment in the early phase of subglacial lake formation, and that stronger, but less densely spaced reflectors, higher in the sequence, are associated with coarse-grained material deposited during the Last Glacial Maximum, i.e., when the inflow of water and sediments from subglacial conduits was particularly high. The latter could be related to enlargement of the subglacial catchment feeding the lake and an increased extent and magnitude of basal melting in response to ice sheet growth during marine-isotope stage 2, about 18 to 22 ka BP. *Domack et al.* [2006] report a similar sequence of events from seismic surveys of Palmer Deep, which also contains evidence of ice sheet grounding during the most extreme phase of sea-level lowering in Marine Isotope Stage 2. *Christoffersen et al.* [2008] found no evidence of ice contact after subglacial lake formation, indicating direct transition from subglacial to proglacial lake. Whereas the Palmer Deep subglacial lake formed by entrapment of water, originally part of an ice-shelf cavity, the Great Slave subglacial lake appears to have formed subglacially, from ice-growth-induced basal melting in a central part of the Laurentide Ice Sheet. The different evolution indicates that formation and decay of subglacial lakes depend on whether they are centrally positioned within the ice sheet or located near ice sheet margins. Results from numerical modeling and the common occurrence of lakes in the deep interior of Antarctica [*Dowdeswell and Siegert*, 2003; *Siegert et al.*, 2005] indicate that the interior lakes filled when ice viscosity and surface slopes were low, i.e., when the ice sheet was in full expansion [*Pattyn*, 2008]. Subglacial lakes in ice marginal settings may evolve from ice-shelf cavities, as proposed by *Alley et al.* [2006].

The model experiments by *Pattyn* [2008] illustrate the crucial role of ice surface gradients in controlling subglacial lake stability. The main control on stability is the air-ice surface slope, which prescribes the hydrological gradient and determines whether subglacial water will pool or flow out of topographic depressions. The higher surface slopes near ice sheet margins may cause subglacial lakes in this setting to fill and drain frequently, as observed in several satellite remote sensing studies [*Gray et al.*, 2005; *Wingham et al.*, 2006; *Fricker et al.*, 2007]. Theory suggests that interior lakes may also drain episodically [*Evatt et al.*, 2006], and numerical modeling predicts that lakes in central regions, e.g., Lake Ellsworth and Vostok, might drain easily. Numerical experiments suggest that smaller lakes drain slower than larger ones and not completely [*Pattyn*, 2008]. Thus, the deepest parts of lakes could remain water-filled if drainage events are not complete, and this may protect distal sedimentary records from erosion by ice contact with the bed, while yielding complex sequence of erosion and deposition around lake margins where the water is shallower. This inherent instability is exacerbated by ice viscosity changes in warm interglacial periods including the Holocene and the present day [*Pattyn*, 2008]. It is clear that subglacial lake sediment may therefore contain evidence of drainage events as well as environmental change during glacial-interglacial periods.

5.1. Potential Sedimentary Archives of Hydrological Discharge in Lakes Vostok, Ellsworth, and Whillans

The sensitivity experiments by *Pattyn* [2008] suggest that large lakes, shallow ice, small surface slopes, and fast ice flow increase lake water retention. An implication is that small lakes under thick ice with slow flow, such as Lake Ellsworth, may be particularly sensitive to hydrological outflow. This connectivity to a regional water system could yield geological evidence in the form of compaction of previously unconsolidated lake sediment from periods when the ice sheet and the lake floor are in contact. However, the acoustic characteristics of the sediment in Lake Ellsworth suggest that it is unconsolidated [*Smith et al.*, 2008], similar to deep ocean sediment, and thus unlikely to have been subject to overriding ice making contact with the bed. However, it might be feasible that there has been partial drainage [*Pattyn*, 2008] of Lake Ellsworth on occasion, but that the lake floor remained free of grounded ice.

The large size of Lake Vostok makes it less sensitive to changes in surface slope, but modeling suggests that drainage of large lakes occurs faster than drainage of shallower lakes. This indicates that truly immense subglacial floods may occur. A subglacial flood from Lake Vostok should occur less frequently than drainage events in smaller lakes as restoration of the lake volume will take longer. The available topographic constraints suggest that the subglacial catchment feeding meltwater to Lake Vostok is 71,000 km^2 and assumed melting of 1 mm yr^{-1} yield annual inflow on the order of 0.07 km^3. The water volume is estimated to 5000 km^3 [*Filina et al.*, 2008; *Studinger et al.*, 2004], so filling the lake with the assumed annual inflow would take about 70,000 years, if melting and freezing of the lake ceiling are in balance, or 38,000 years if modeled net ice loss of 0.06 km^3 yr^{-1} is taken into account [*Studinger et al.*, 2004; *Thoma et al.*, 2008a]. The similar values of catchment inflow and local melt are coincidental, but indicate potentially even contributions.

An important question remains: What proportion of the lake volume would be discharged in the case of a subgla-

cial drainage event in Lake Vostok? *Evatt et al.* [2006] use *Nye*'s [1976] theory for subglacial floods through tunnels and veins to show that Antarctic subglacial lakes may flood catastrophically, despite their clustered location near ice divides where the ice sheet surface becomes flat. Their subglacial outburst flood model suggest that 600 km^3 of water could drain from Lake Vostok in about 4 years, with peak discharge of 10^5 m^3 s^{-1}. Such a flood would be several orders of magnitude larger than the observed discharge between subglacial lakes in the Adventure subglacial trough [*Wingham et al.*, 2006]. The recurrence of such an event is most likely limited by the lake refilling time, as discussed above.

Seismic records of subglacial lake sediment may contain evidence of these floods, e.g., by presence of detectable seismic reflections from sedimentary flood deposits or strong reflectors from consolidated sediment strengthened from ice sheet contact after lake drainage. In this case, the accumulated thickness of sediment between these reflectors is an indicator of sedimentation rate and possibly lake refilling time. It is important to consider the possibility of erosion if periodic drainage causes ice to be in direct contact with the lake floor or a portion of the lake floor. To minimize the influence of erosion, core samples should be collected in the deepest part of subglacial lakes or where the water column is thickest. Ice sheet contact will, on the other hand, yield valuable information about past subglacial hydrology, e.g., flood recurrence intervals and magnitudes. Geological evidence of past hydrology may thus be present in the littoral zone where ice-bed contact is more likely to occur and create unconformities but with unknown amounts of erosion during events. Interpreting such a record from core data alone would be challenging and would probably require some detailed geophysical surveys of sedimentary architecture. The disturbance of sediments in a ring around most or all of a lake would serve as a source of gravity processes initiated by the disturbance, and so the deep-water record, distal from influx points, may contain a continuous record of the frequency and magnitude (amount of lowering of ice ceiling) of the draining events. Grounding-line landforms may also record changes in lake position. Large-scale subglacial flood landforms eroded into bedrock (e.g., the Labyrinth, Transantarctic Mountains) [*Sugden et al.*, 1991] demonstrate how powerful the erosive power of flood events might be.

The setting of subglacial lakes beneath fast flowing ice streams is distinctly different to the setting of deep continental lakes. The large glaciers along the Siple Coast including Kamb and Whillans ice streams are particularly dynamic features in the Antarctic ice sheet. Kamb Ice Stream stopped 150 years ago, and Whillans Ice Stream is in the process of shutting down too [*Bougamont et al.*, 2003a; *Joughin et al.*, 2005]. The cause of these major discharge fluctuations has been a matter of debate. *Anandakrishnan and Alley* [1997] proposed that the stoppage of Kamb Ice Stream was caused by diversion of subglacial meltwater to a neighboring ice stream (Whillans). *Tulaczyk et al.* [2000] proposed that non-linear bed properties induce ice stream instability. The nature of this instability was examined theoretically in subsequent studies [*Bougamont et al.*, 2003a, 2003b; *Christoffersen and Tulaczyk*, 2003], but the processes controlling the flow switches that characterize the Siple Coast ice streams are still not fully established. The duration of active and stagnant phases is not known, and this causes uncertainty in predictions of ice sheet mass balance [*Joughin and Tulaczyk*, 2002]. The planned drilling to Lake Whillans is likely to answer important questions related to the transient state of ice stream flow. For instance, direct investigations will determine whether sub-ice stream lakes are transient or semipermanent features, identify the factors that control their formation and elucidate the impact they have on ice stream dynamics.

6. TECHNOLOGIES FOR EXPLORATION AND ANALYSIS OF SUBGLACIAL LAKE SEDIMENTS

Sediment coring at several kilometers water depth is a near-daily occurrence in the world's oceans, but the technologies used will have to be substantially adapted for use in subglacial lakes. Their deployment needs to be preceded by drilling an access hole through the overlying (several kilometers) glacial ice, which must be kept open for the duration of the drilling operation. Two main types of marine sediment retrieval are commonly used: (1) coring by propulsion of a sediment corer barrel into the sediment by gravity, percussion hammering, or vibrating; and (2) drilling using a rotating drill such as that used by ANDRILL at Cape Roberts. Corers currently under development for subglacial lakes include percussion corers that are lowered to the lake bed and then use a self-propelled hammer weight to achieve penetration. To achieve a sufficient depth of penetration will require a design compromise between corer diameter (a narrow corer penetrates further into [older] sediment), corer weight (greater weight gives greater penetration but requires a stronger cable to retrieve), and a consideration of the amount of sediment needed from each depth interval for robust sediment analysis (corer-diameter-related). However, in all cases, if coarse clasts are present, then penetration will be compromised. Retrieval and retention of the sediment core in the (lined) core barrel are typically achieved through the use of core catchers at the lower end and a closing valve at the upper end. However, these are not fail-safe where very flocculent sediments, sandy sediments, or clathrates are pres-

ent. The latter is a particular problem as the core is susceptible to degassing as it loses pressure during transit to the surface. In order to detect life at very low concentrations and to prevent contamination of subglacial lake environments, these corers also need to be designed to allow for sterilization and cleaning.

The use of a conventional rotating drill, such as the wireline system with a drill derrick that was deployed at Cape Roberts (Drilling) Project between 1997 and 1999 and the ANDRILL project in 2008, has not yet been seriously considered in subglacial lake settings, largely because of logistic difficulty and cost. However, this type of system has proved capable of recovering more than 1000 m long drill cores and offers the potential, where deep subglacial sediments are present, to reconstruct the full glaciation history of continental Antarctica and its impact on the world's oceans currents and the atmosphere. As such sediments have been inferred from seismic evidence at Lake Vostok [*Siegert et al.*, 2001], wireline drilling of subglacial sedimentary sequences must remain a top future priority for Antarctic geologists as it will allow the Cenozoic history of the interior of the continental ice sheet to be reconstructed in situ rather than indirectly from reworked and redeposited sediments deposited at continental margins.

There is a wide range of analytical methods that can be applied to sediments retrieved from subglacial lakes once the essential chronological analyses have been carried out. These will be applied to address the three areas of interest, namely, past ice sheet configurations, paleoenvironmental change, and the presence of life (Table 1).

6.1. Establishing a Chronology in Subglacial Lake Sediments

Establishing a geochronology for subglacial sedimentary deposits will be one of the major technological challenges to be addressed in interpreting the paleoenvironmental information contained in sediment cores. A number of the conventional methods used in lake sediment analyses such as radiocarbon dating will not be applicable in subglacial environments as the provenance of the carbon will be unknown and likely dominated by the signal from old CO_2 in meteoric ice. Similarly, exposure dating methods such as optically stimulated luminescence dating of quartz grains and analysis of cosmogenic isotopes (e.g., ^{10}Be and ^{3}He) will, in most cases, be compromised because of the unknown provenance of the material being dated. However, they do have an important application in identifying when the sediments were last exposed to light or cosmogenic isotope bombardment and hence may have the potential to constrain the age(s) of ice sheet inception.

Table 1. Examples of Analytical Methods for Subglacial Lake Sediments

Measurement	Result	Potential Outcome
Detecting past ice sheet configurations and paleoenvironmental change		
Relative paleomagnetic intensity dating	identifying magnetic excursions and polarity reversals	provides a chronology
Exposure dating methods such as optically stimulated luminescence and analysis of cosmogenic isotopes such as ^{10}Be, ^{3}He	identifying when the sediments were last exposed to light and measuring rates of cosmogenic isotope bombardment	constraints on the age(s) of ice sheet inception
Radiometric U-Th dating of sediment if carbonates present	identifying quaternary age sedimentary deposits	dating quaternary periods of (West Antarctic) ice sheet loss
Mineralogy (XRD and XRF) and grain size	mineralogical classification of sediment and identification of extraterrestrial material	sediment flux/source origin and information on age if grain size changes follow glacial cycles
Scanning electron microscopy and micromorphology	morphology of sediment	distinguish aeolian (dust) vs. subglacially transported grains
Radiometric Nd-Sr	provenance of sediment	identification of sediment delivery pathways
Carbonate isotopes	environmental conditions in lake	requires presence of carbonate precipitate in closed system (difficult), but might find ikaite (cf. calcite or aragonite)
Bedrock sample	local geology	provides information on the geological setting of the lake
Sedimentology and microfossil analysis	identification of subglacial and marine sediment units	determines when the West Antarctic Ice Sheet last collapsed
Oxygen isotope analysis of microfossils and minerals	oxygen isotope stratigraphy	comparison with marine oxygen isotope stack to provide age constraints
Detecting life		
Microscopy (LM and SEM)	enumeration and description of morphological remains	list of morphologically recognizable life forms
Nucleic acid staining	determination of total cell numbers in lake sediments	estimate of biological production
Lipid analysis	total cell number determinations	estimate of biological production
Molecular biological analyses	determination of phylogenetic groups, metabolic activity, and biogeochemical pathways	identification of biological assemblages, functional groups, and biological interactions
Sediment geochemistry	detection of changes in sediment geochemistry attributed to biological activity	identification of biological interactions with bedrock and sediment

Methods based on the magnetic properties of the sediments offer the potential to establish a sediment chronology. These methods work by collecting oriented samples at measured intervals throughout core sequences and measuring the polarity of fine-grained magnetic minerals <17 μm that fall through the water column, orient themselves with Earth's magnetic field, and preserve this orientation within the sediments. These orientations can be measured through a sedimentary sequence and the resultant time series compared with independently dated reference curves, which include marked excursions and reversals in the Earth's magnetic field, in order to create a chronology. The first application of relative paleomagnetic intensity (RPI) dating in the subgla-cial sediments of Hodgson Lake [*Hodgson et al.*, 2009a] identified at least one major excursion that was linked to a reference excursion at subaerial Mono Lake, providing a proof of concept for RPI dating methods in subglacial lake settings. A further benefit of RPI dating is its relationship to variations in cosmogenic isotopes (especially ^{10}Be) in ice cores [e.g., *Leduc et al.*, 2006; *Zimmerman et al.*, 2006]. Meteoric ^{10}Be concentrations measured in ice cores are inversely related to RPI because the increased production of cosmogenic nuclides, such as ^{10}Be, in the atmosphere has been shown to increase at times of reduced solar irradiance and hence weakened geomagnetic field [*Beer et al.*, 1990; *Bond et al.*, 2001]. Relative paleomagnetic intensity and polarity excursion data from otherwise undateable subglacial lake sediments could therefore potentially allow global or at least intrahemispheric correlation of ice, marine, and subglacial lake sedimentary records. In subglacial settings where marine sediments have been identified, the concentration of meteoric ^{10}Be, which adheres to marine particles in the surface waters and is subsequently deposited on the seafloor, can also provide an approximate age for the marine detritus [Scherer et al., 1998] and can be compared with the concentrations measured in Antarctic ice [Raisbeck et al., 1987]. Finally, if marine sediments are encountered in subglacial sediments from beneath the marine-based WAIS, and microfossils are present, then biostratigraphic methods may help establish a chronology (e.g. see Scherer et al. [1998]). These rely on assigning relative ages to the sediment units by correlating the fossil assemblages contained within them to dated geological deposits elsewhere.

6.2. Detecting Past Ice Sheet Configurations and Paleoenvironmental Change

Sediments from subglacial lakes, especially where not extensively reworked or sheared by overriding ice, have the potential to provide a record of the evolution of the EAIS and WAIS, particularly the transition to full ice sheet conditions in the Miocene. For the marine-based WAIS, the presence of marine deposits beneath the ice sheet would provide evidence of previous periods of WAIS collapse [cf. Scherer et al. 1998], and changes in the mineralogy and grain size composition may provide evidence of the succession of glacial interglacial cycles because subglacial hydrology is fundamentally tied to the geometry, flow rates, and the thermal state of the ice sheet. The latter has been demonstrated in models and in Lake Hodgson, where a change from fine-grained sediments together with sands, gravels, and small clasts to finer grained sediments characteristic of lower energy delivery marks the transition from a glacial to an interglacial sedimentation regime [Hodgson et al., 2009a]; (Figure 4). Some of the methods that could be applied to dating and identifying changes in ice sheet configuration are discussed by Siegert et al. [2006] and listed in Table 1.

6.3. Detecting Evidence of Life

A lot of research has been carried out on life in subglacial environments in general, but none has been carried out in the sediments of deep continental subglacial lakes, and very little has been carried out on sediments in the more recently sampled and accessible subglacial lakes at the margins of the ice sheet. As discussed elsewhere in this volume, the evidence of biological activity comes in two forms: direct evidence such as microscopic identification, DNA, or other molecular biological fingerprints, and indirect evidence such as changes in sediment geochemistry that can be attributed to biological activity (Table 1). Preliminary studies of this nature were carried out on the subglacial sediments from Lake Hodgson [Hodgson et al., 2009a]. Light microscopy revealed no microfossil evidence of life: a few diatom frustules were present, but these were at levels consistent with the very low diatom concentrations (~34 cells L^{-1}) found in filtered glacial ice from Antarctica [Kellogg and Kellogg, 1996; Scherer et al., 1998] or incorporation from catchment soils (possibly during a previous interglacial) eroded into the lake under the glaciers. Further electron microscopy and DNA fingerprinting of Lake Hodgson sediments are under way. Of the geochemical markers measured, total organic carbon varied from 0.2% to 0.6%, and there were some changes in $\delta^{13}C$ and C/N. These measurements were consistent with local reference measurements of catchment gravel and fine-grained sediments. Thus, these measurements of the carbon chemistry could not be unequivocally linked to in situ biological activity as some of the signal could simply be a result of the incorporation of catchment soils and gravels and possibly old CO_2 in meteoric ice. Even if life were present in these samples, the carbon measurements suggest that it would be in extremely low concentrations [cf. Christner et al. 2006].

Nevertheless, it is anticipated that the application of molecular biological methods (see Table 1 and Skidmore [this volume]) will reveal a subglacial lake biota including instances of, for example, chemoautotrophic bacteria, proteobacteria (anaerobes commonly found in rocks), and small autotrophic eukaryotes such as those present in glacial ice [Lanoil et al., 2004]. A range of direct methods is available for their detection, together with indirect studies of sediment geochemistry. For example, geochemical interactions such as microbial acquisition of dissolved organic carbon from bedrock [Hodson et al., 2008] and the identification of potential biogeochemical pathways that could use this carbon to support heterotrophic metabolisms, for example, that use O_2, NO_3^-, SO_4^{2-}, S_0, or Fe^{2m} as electron acceptors, can reveal what forms of life might be present (Table 1) [Priscu et al., 2008]. Efforts will concentrate particularly on analyzing evidence of life at the sediment-water interface and in the surface sediments of modern subglacial environments where metabolic activity and bacterial diversity are predicted to be focused. This is among the most likely sites for life on account of the biogeochemical interactions at the sediment-water interface that can supply the necessary carbon, nutrients, and electron acceptors. There are several recent reviews on life detection technologies in subglacial environments [Siegert et al., 2006; Hodson et al., 2008] and within accreted subglacial ice [Christner et al., 2006].

7. CONCLUSIONS

Subglacial sediments have long been recognized as a potential source of paleoenvironmental and biological information beneath ice sheets and possibly extending this information to timescales beyond those of ice cores. Physical characteristics of subglacial lake sediments can be inferred from geophysical surveys of lakes beneath the Antarctic ice sheet and from the accretion ice above Lake Vostok. We have also demonstrated the potential of using sedimentary sequences deposited from subglacial lakes inferred beneath the former Northern Hemisphere ice sheets as analogs, but sound a cautionary note on using such sediments until they have been independently demonstrated to have formed in subglacial lake settings. Former subglacial lakes, emerging from or outside the present margin of the Antarctic ice sheet, have been sampled by coring and provide useful data on potential sedimentary environments. We have identified five sediment pathways for subglacial lakes and discussed their likely relative contributions in terms of composition, distribution, and volume: Sediment derived from subglacial meltwater is likely to be dominant in many subglacial lake settings, and advection of subglacial sediment may be important for sub-ice stream lakes such as Whillans. Melt-out of basal debris may be locally important especially on the up-flow side of lakes. Supraglacial rain-out and authigenic chemical sedimentation will likely be volumetrically much less significant, but these components could contain important information such as extraterrestrial dust flux or information on lake water chemistry. From the sedimentary analogs, emerging subglacial lakes, and a review of glacio-lacustrine processes, we have developed conceptual models for sedimentation in lakes Vostok, Whillans, and Ellsworth. These are aimed at providing hypotheses that can be tested by direct sampling and detailed geophysical surveys. Landforms associated with subglacial lakes and their margins have largely been ignored in the literature to date, but we have described here some likely sediment-landform assemblages and suggested how they might record changes in the lakes through time. In particular, we have discussed the possible effects of lake drainage and filling cycles and that such variability is likely to be recorded by sediment erosion around the littoral zone of lakes and by pulsed sedimentation events in the deeper, distal parts of lakes.

The nature of the sediments partly dictates the most appropriate coring technology: variants of percussion corers are being developed for use in the forthcoming lake access experiments at Ellsworth and Whillans. When sediments are finally sampled from such lakes, their analysis will include significant challenges, especially determining a full chronology. The most effective chronological technique is likely to be geomagnetic dating. Following the establishment of a chronology, there is a range of analyses that can be applied to understand the environmental history of the lake and complement the biological, physical, and geochemical measurements on lake water.

Acknowledgments. M.B. thanks Stuart Lane and Alex Densmore for information on the relationships between sediment discharge and catchment size in subaerial catchments and Martyn Tranter for discussion on the potential for chemical sedimentation in subglacial lakes. The Design and Imaging Unit at University of Durham drafted Figure 5. Chris Clark and an anonymous reviewer provided excellent reviews that helped improve the paper and pointed us toward some relevant literature that we were unaware of. This is a contribution of the Lake Ellsworth Consortium, funded by the U.K. Natural Environment Research Council. S.T. performed this research as part of his participation in the WISSARD project (Whillans Ice Stream Subglacial Access Research Drilling) with major funding provided by the National Science Foundation.

REFERENCES

Alley, R. B., D. D. Blankenship, C. R. Bentley, and S. T. Rooney (1987), Till beneath Ice Stream-B .3. Till deformation—Evidence and implications, *J. Geophys. Res.*, *92*(B9), 8921–8929.

Alley, R. B., D. D. Blankenship, S. T. Rooney, and C. R. Bentley (1989), Sedimentation beneath ice shelves—The view from Ice Stream-B, *Mar. Geol.*, *85*(2–4), 101–120.

Alley, R. B., T. K. Dupont, B. R. Parizek, S. Anandakrishnan, D. E. Lawson, G. J. Larson, and E. B. Evenson (2006), Outburst flooding and the initiation of ice-stream surges in response to climatic cooling: A hypothesis, *Geomorphology*, *75*(1–2), 76–89.

Alley, R. B., S. Anandakrishnan, T. K. Dupont, B. R. Parizek, and D. Pollard (2007), Effect of sedimentation on ice-sheet grounding-line stability, *Science*, *315*(5820), 1838–1841.

Anandakrishnan, S., and R. B. Alley (1997), Stagnation of Ice Stream C, West Antarctica by water piracy, *Geophys. Res. Lett.*, *24*(3), 265–268.

Anandakrishnan, S., D. D. Blankenship, R. B. Alley, and P. L. Stoffa (1998), Influence of subglacial geology on the position of a West Antarctic ice stream from seismic observations, *Nature*, *394*(6688), 62–65.

Anandakrishnan, S., G. A. Catania, R. B. Alley, and H. J. Horgan (2007), Discovery of till deposition at the grounding line of Whillans Ice Stream, *Science*, *315*(5820), 1835–1838.

Ashley, G. M., J. Shaw, and N. D. Smith (1985), *Glacial Sedimentary Environments*, 246 pp., Soc. of Econ. Paleontol. and Mineral., Tulsa, Okla.

Augustin, L., et al. (2004), Eight glacial cycles from an Antarctic ice core, *Nature*, *429*(6992), 623–628.

Bamber, J. L., J. L. Gomez-Dans, and J. A. Griggs (2009), A new 1km digital elevation model of the Antarctic derived from combined satellite radar and laser data—Part 1: Data and methods, *Cryosphere*, *3*, 101–111.

Beer, J., et al. (1990), Use of Be-10 in polar ice to trace the 11-year cycle of solar-activity, *Nature*, *347*(6289), 164–166.

Bell, R. E., M. Studinger, A. A. Tikku, G. K. C. Clarke, M. M. Gutner, and C. Meertens (2002), Origin and fate of Lake Vostok water frozen to the base of the East Antarctic Ice Sheet, *Nature*, *416*(6878), 307–310.

Bell, R. E., M. Studinger, M. A. Fahnestock, and C. A. Shuman (2006), Tectonically controlled subglacial lakes on the flanks of the Gamburtsev Subglacial Mountains, East Antarctica, *Geophys. Res. Lett.*, *33*, L02504, doi:10.1029/2005GL025207.

Benn, D. I., and D. J. A. Evans (1998), *Glaciers and Glaciation*, 734 pp., Edward Arnold, London, U. K.

Best, J. L., R. A. Kostaschuk, J. Peakall, P. V. Villard, and M. Franklin (2005), Whole flow field dynamics and velocity pulsing within natural sediment-laden underflows, *Geology*, *33*(10), 765–768.

Bindschadler, R. A., M. A. King, R. B. Alley, S. Anandakrishnan, and L. Padman (2003), Tidally controlled stick-slip discharge of a West Antarctic ice stream, *Science*, *301*(5636), 1087–1089.

Blankenship, D. D., C. R. Bentley, S. T. Rooney, and R. B. Alley (1986), Seismic measurements reveal a saturated porous layer beneath an active Antarctic ice stream, *Nature*, *322*, 54–57.

Blankenship, D. D., C. R. Bentley, S. T. Rooney, and R. B. Alley (1987), Till beneath Ice Stream B: 1. Properties derived from seismic travel times, *J. Geophys. Res.*, *92*(B9), 8903–8911.

Bond, G., B. Kromer, J. Beer, R. Muscheler, M. N. Evans, W. Showers, S. Hoffmann, R. Lotti-Bond, I. Hajdas, and G. Bonani (2001), Persistent solar influence on north Atlantic climate during the Holocene, *Science*, *294*(5549), 2130–2136.

Bougamont, M., S. Tulaczyk, and I. Joughin (2003a), Numerical investigations of the slow-down of Whillans Ice Stream, West Antarctica: Is it shutting down like Ice Stream C?, *Ann. Glaciol.*, *37*, 239–246.

Bougamont, M., S. Tulaczyk, and I. Joughin (2003b), Response of subglacial sediments to basal freeze-on: 2. Application in numerical modeling of the recent stoppage of Ice Stream C, West Antarctica, *J. Geophys. Res.*, *108*(B4), 2223, doi:10.1029/2002JB001936.

Bouma, A. H. (1962), *Sedimentology of Some Flysch Deposits: A Graphic Approach to Facies Interpretation*, 168 pp., Elsevier, Amsterdam, Netherlands.

Bromwich, D. H., E. R. Toracinta, H. L. Wei, R. J. Oglesby, J. L. Fastook, and T. J. Hughes (2004), Polar MM5 simulations of the winter climate of the Laurentide Ice Sheet at the LGM, *J. Clim.*, *17*(17), 3415–3433.

Bromwich, D. H., E. R. Toracinta, R. J. Oglesby, J. L. Fastook, and T. J. Hughes (2005), LGM summer climate on the southern margin of the Laurentide Ice Sheet: Wet or dry?, *J. Clim.*, *18*(16), 3317–3338.

Burckle, L. H. (1993), Is there direct evidence for late Quaternary collapse of the West Antarctic Ice Sheet, *J. Glaciol.*, *39*(133), 491–494.

Carter, S. P., D. D. Blankenship, M. E. Peters, D. A. Young, J. W. Holt, and D. L. Morse (2007), Radar-based subglacial lake classification in Antarctica, *Geochem. Geophys. Geosyst.*, *8*, Q03016, doi:10.1029/2006GC001408.

Christner, B. C., G. Royston-Bishop, C. M. Foreman, B. R. Arnold, M. Tranter, K. A. Welch, W. B. Lyons, A. I. Tsapin, M. Studinger, and J. C. Priscu (2006), Limnological conditions in subglacial Lake Vostok, Antarctica, *Limnol. Oceanogr.*, *51*(6), 2485–2501.

Christoffersen, P., and S. Tulaczyk (2003), Thermodynamics of basal freeze-on: Predicting basal and subglacial signatures of stopped ice streams and interstream ridges, *Ann. Glaciol.*, *36*, 233–243.

Christoffersen, P., S. Tulaczyk, N. J. Wattrus, J. Peterson, N. Quintana-Krupinski, C. D. Clark, and C. Sjunneskog (2008), Large subglacial lake beneath the Laurentide Ice Sheet inferred from sedimentary sequences, *Geology*, *36*(7), 563–566.

Clarke, G. K. C. (2003), Hydraulics of subglacial outburst floods: New insights from the Spring-Hutter formulation, *J. Glaciol.*, *49* (165), 299–313.

Crookshanks, S., and R. Gilbert (2008), Continuous, diurnally fluctuating turbidity currents in Kluane Lake, Yukon Territory, *Can. J. Earth Sci.*, *45*, 1123–1138.

Cuffey, K. M., H. Conway, A. M. Gades, B. Hallet, R. Lorrain, J. P. Severinghaus, E. J. Steig, B. Vaughn, and J. W. C. White (2000), Entrainment at cold glacier beds, *Geology*, *28*(4), 351–354.

Das, S. B., I. Joughin, M. D. Behn, I. M. Howat, M. A. King, D. Lizarralde, and M. P. Bhatia (2008), Fracture propagation to the base of the Greenland Ice Sheet during supraglacial lake drainage, *Science*, *320*(5877), 778–781.

Domack, E., D. Amblas, R. Gilbert, S. Brachfeld, A. Camerlenghi, M. Rebesco, M. Canals, and R. Urgeles (2006), Subglacial morphology and glacial evolution of the Palmer deep outlet system, Antarctic Peninsula, *Geomorphology*, *75*(1–2), 125–142.

Doran, P. T., J. C. Priscu, W. Berry Lyons, R. D. Powell, D. T. Andersen, and R. J. Poreda (2004), Paleolimnology of extreme cold terrestrial and extraterrestrial environments, Long-term Environmental Change in Arctic and Antarctic Lakes8in *Developments in Palaeoenvironmental Research*, edited by R. Pienitz, M. S. V. Douglas, and J. P. Smol, pp. 475–507, Springer, Dordrecht, Netherlands.

Dowdeswell, J. A., and M. J. Siegert (2003), The physiography of modern Antarctic subglacial lakes, *Global Planet. Change*, *35* (3–4), 221–236.

Drewry, D. (1986), *Glacial Geologic Processes*, 276 pp., Edward Arnold, London, U. K.

Dyke, A. S., J. T. Andrews, P. U. Clark, J. H. England, G. H. Miller, J. Shaw, and J. J. Veillette (2002), The Laurentide and Innuitian ice sheets during the Last Glacial Maximum, *Quat. Sci. Rev.*, *21* (1–3), 9–31.

Engelhardt, H., N. Humphrey, B. Kamb, and M. Fahnestock (1990), Physical conditions at the base of a fast moving Antarctic ice stream, *Science*, *248*(4951), 57–59.

Evans, D. J. A. (1990), The effect of glacier morphology on surficial geology and glacial stratigraphy in a high Arctic mountainous terrain, *Z. Geomorphol.*, *34*(4), 481–503.

Evans, D. J. A., B. R. Rea, J. F. Hiemstra, and C. Ó Cofaigh (2006), A critical assessment of subglacial mega-floods: A case study of glacial sediments and landforms in south-central Alberta, Canada, *Quat. Sci. Rev.*, *25*(13–14), 1638–1667.

Evatt, G. W., A. C. Fowler, C. D. Clark, and N. R. J. Hulton (2006), Subglacial floods beneath ice sheets, *Philos. Trans. R. Soc. A*, *364*(1844), 1769–1794.

Eyles, N., C. H. Eyles, and A. D. Miall (1983), Lithofacies types and vertical profile models; an alternative approach to the description and environmental interpretation of glacial diamict and diamictite sequences, *Sedimentology*, *30*, 393–410.

Filina, I., V. Lukin, V. N. Masolov, and D. D. Blankenship (2007), Unconsolidated sediments at the bottom of Lake Vostok from seismic data, in *10th International Symposium on Antarctic Earth Sciences*, edited by A. K. Cooper and C. R. Raymond, *Open File Rep., 2007-1047, Short Res. Pap. 031*, 5 pp., U.S. Geol. Surv., Santa Barbara, Calif.

Filina, I. Y., D. D. Blankenship, M. Thoma, V. V. Lukin, V. N. Masolov, and M. K. Sen (2008), New 3D bathymetry and sediment distribution in Lake Vostok: Implication for pre-glacial origin and numerical modeling of the internal processes within the lake, *Earth Planet. Sci. Lett.*, *276*(1–2), 106–114.

Fricker, H. A., and T. Scambos (2009), Connected subglacial lake activity on lower Mercer and Whillans ice streams, West Antarctica, 2003–2008, *J. Glaciol.*, *55*(190), 303–315.

Fricker, H. A., T. Scambos, R. Bindschadler, and L. Padman (2007), An active subglacial water system in West Antarctica mapped from space, *Science*, *315*(5818), 1544–1548.

Fyfe, G. J. (1990), The effect of water depth on ice-proximal glaciolacustrine sedimentation: Salpausselkä I, southern Finland, *Boreas*, *19*(2), 147–164.

Gilbert, R., and S. Crookshanks (2009), Sediment waves in a modern high-energy glacilacustrine environment, *Sedimentology*, *56*, 645–659.

Gjessing, J. (1960), The drainage of the deglaciation period, its tends and morphogenetic activity in northern Atnedalen, with comparative studies from northern Gudbrandsdalen and northern Österdalen (east-central southern Norway), in *Ishavsmeltningstidens drenering, dens forløb pg formdannende virkning Nordre Atnedalen, Ad Novas*, pp. 441–489, Norw. Geogr. Soc., Oslo.

Gray, L., I. Joughin, S. Tulaczyk, V. B. Spikes, R. Bindschadler, and K. Jezek (2005), Evidence for subglacial water transport in the West Antarctic Ice Sheet through three-dimensional satellite radar interferometry, *Geophys. Res. Lett.*, *32*, L03501, doi:10.1029/2004GL021387.

Hambrey, M. (1994), *Glacial Environments*, 296 pp., Univ. Coll. London Press, London, U. K.

Hay, W. W. (1998), Detrital sediment fluxes from continents to oceans, *Chem. Geol.*, *145*(3–4), 287–323.

Hodder, K. R., and R. Gilbert (2007), Evidence for flocculation in glacier-fed Lillooet Lake, British Columbia, *Water Res.*, *41*, 2748–2762.

Hodgson, D. A., E. Verleyen, A. H. Squier, K. Sabbe, B. J. Keely, K. M. Saunders, and W. Vyverman (2006), Interglacial environ-ments of coastal east Antarctica: Comparison of MIS 1 (Holocene) and MIS 5e (Last Interglacial) lake-sediment records, *Quat. Sci. Rev.*, *25*(1–2), 179–197.

Hodgson, D. A., S. J. Roberts, M. J. Bentley, E. L. Carmichael, J. A. Smith, E. Verleyen, W. Vyverman, P. Geissler, M. J. Leng, and D. C. W. Sanderson (2009a), Exploring former subglacial Hodgson Lake. Paper II: Palaeolimnology, *Quat. Sci. Rev.*, *28*, 2310–2325.

Hodgson, D. A., et al. (2009b), Exploring former subglacial Hodgson Lake. Paper I: Site description, geomorphology and limnology, *Quat. Sci. Rev.*, *28*, 2295–2309.

Hodson, A. J., A. M. Anesio, M. Tranter, A. G. Fountain, A. M. Osborn, J. Priscu, J. Laybourn-Parry, and B. Sattler (2008), Glacial ecosystems, *Ecol. Monogr.*, *78*, 41–67.

Hoffman, P. F. (1987), Continental transform tectonics: Great Slave Lake shear zone (ca.1.9 Ga), northwest Canada, *Geology*, *15*(9), 785–788.

Houbolt, J. J. H. C., and J. B. M. Jonker (1968), Recent sediments in the eastern part of the lake of Geneva (Lac Leman), *Geol. Mijnbouw*, *47*, 131–148.

Joughin, I., and L. Padman (2003), Melting and freezing beneath Filchner-Ronne Ice Shelf, Antarctica, *Geophys. Res. Lett.*, *30*(9), 1477, doi:10.1029/2003GL016941.

Joughin, I., and S. Tulaczyk (2002), Positive mass balance of the Ross Ice Streams, West Antarctica, *Science*, *295*(5554), 476–480.

Joughin, I., S. Tulaczyk, R. Bindschadler, and S. F. Price (2002), Changes in west Antarctic ice stream velocities: Observation and analysis, *J. Geophys. Res.*, *107*(B11), 2289, doi:10.1029/2001JB001029.

Joughin, I., S. Tulaczyk, D. R. MacAyeal, and H. Engelhardt (2004), Melting and freezing beneath the Ross ice streams, Antarctica, *J. Glaciol.*, *50*(168), 96–108.

Joughin, I., et al. (2005), Continued deceleration of Whillans Ice Stream, West Antarctica, *Geophys. Res. Lett.*, *32*, L22501, doi:10.1029/2005GL024319.

Jouzel, J., J. R. Petit, R. Souchez, N. I. Barkov, V. Y. Lipenkov, D. Raynaud, M. Stievenard, N. I. Vassiliev, V. Verbeke, and F. Vimeux (1999), More than 200 meters of lake ice above subglacial lake Vostok, Antarctica, *Science*, *286*(5447), 2138–2141.

Karl, D. M., D. F. Bird, K. Bjorkman, T. Houlihan, R. Shackelford, and L. Tupas (1999), Microorganisms in the accreted ice of Lake Vostok, Antarctica, *Science*, *286*(5447), 2144–2147.

Kellogg, D. E., and T. B. Kellogg (1996), Diatoms in South Pole ice: Implications for eolian contamination of Sirius Group deposits, *Geology*, *24*(2), 115–118.

King, E. C., R. C. A. Hindmarsh, and C. R. Stokes (2009), Formation of mega-scale glacial lineations observed beneath a West Antarctic ice stream, *Nat. Geosci.*, *2*(8), 585–588.

Knighton, A. D. (1999), Downstream variation in stream power, *Geomorphology*, *29*(3–4), 293–306.

Lajeunesse, P., and G. St-Onge (2008), The subglacial origin of the Lake Agassiz–Ojibway final outburst flood, *Nat. Geosci.*, *1*, 184–188.

Lambert, A. (1982), Turbidity currents from the Rhine river on the bottom of Lake Constance, *Wasserwirtschaft*, *72*(4), 1–4.

Lambert, F., B. Delmonte, J.-R. Petit, M. Bigler, P. R. Kaufmann, M. A. Hutterli, T. F. Stocker, U. Ruth, J. P. Steffensen, and V. Maggi (2008), Dust-climate couplings over the past 800,000 years from the EPICA Dome C ice core, *Nature*, *452*, 616–619.

Lamoureux, S., and R. Gilbert (2004), Physical and chemical properties and proxies of high laititude lake sediments, in *Long-term Environmental Change in Arctic and Antarctic Lakes*, edited by R. Pienitz, M. S. V. Douglas, and J. P. Smol, pp. 53–87, Springer, Dordrecht, Netherlands.

Lanoil, B., M. Skidmore, S. Han, W. Foo, and D. Bui (2004), A microbial community in sediments beneath the Western Antarctic Ice Sheet, Ice Stream C (Kamb), *Eos Trans. AGU*, *85*(47), Fall Meet. Suppl., Abstract B23C-04.

Larter, R. D., and L. E. Vanneste (1995), Relict subglacial deltas on the Antarctic Peninsula outer shelf, *Geology*, *23*, 33–36.

Leduc, G., N. Thouveny, D. L. Bourles, C. L. Blanchet, and J. T. Carcaillet (2006), Authigenic Be-10/Be-9 signature of the Laschamp excursion: A tool for global synchronisation of paleo-climatic archives, *Earth Planet. Sci. Lett.*, *245*(1–2), 19–28.

Lewis, T., R. Gilbert, and S. F. Lamoureux (2002), Spatial and temporal changes in sedimentary processes at proglacial Bear Lake, Devon Island, Nunavut, Canada, *Arct. Antarct. Alp. Res.*, *34*(2), 119–129.

Lister, G. S. (1984), Deglaciation of the Lake Zurich area: A model based on the sedimentological record, in *Quaternary Geology of Lake Zurich: An Interdisciplinary Investigation by Deep-Lake Drilling*, edited by K. J. Hsu and K. R. Kelts, *Contrib. Sedimentol.*, *13*, 117–185.

Lowe, A. L., and J. B. Anderson (2003), Evidence for abundant subglacial meltwater beneath the paleo–ice sheet in Pine Island Bay, Antarctica, *J. Glaciol.*, *49*(164), 125–138.

Lythe, M. B., and D. G. Vaughan (2001), BEDMAP: A new ice thickness and subglacial topographic model of Antarctica, *J. Geophys. Res.*, *106*(B6), 11,335–11,351.

Masolov, V. N., S. Popov, V. Lukin, A. N. Sheremetyev, and A. Popov (2006), Russian geophysical studies of Lake Vostok, Central East Antarctica, in *Antarctica—Contributions to Global Earth Sciences. Proceedings of the 9th ISAES*, edited by D. Fütterer, pp. 135–140, Springer, Berlin.

Mayer, C., K. Grosfeld, and M. J. Siegert (2003), Salinity impact on water flow and lake ice in Lake Vostok, Antarctica, *Geophys. Res. Lett.*, *30*(14), 1767, doi:10.1029/2003GL017380.

McCabe, A. M., and C. Ó. Cofaigh (1994), Sedimentation in a subglacial lake, Enniskerry, eastern Ireland, *Sediment. Geol.*, *91*(1–4), 57–95.

Middleton, G. V., and M. A. Hampton (1976), Subaqueous sediment transport and deposition by sediment gravity flows, in *Marine Sediment Transport and Environmental Management*, edited by D. J. Stanley and D. J. P. Swift, pp. 197–218, John Wiley, New York.

Munro-Stasiuk, M. J. (2003), Subglacial Lake McGregor, south-central Alberta, Canada, *Sediment. Geol.*, *160*(4), 325–350.

Nemec, A. (1990), Aspects of sediment movement on steep delta slopes, in *Coarse-Grained Deltas*, edited by A. Colella and D. B. Prior, *Spec. Publ. Int. Assoc. Sedimentol.*, *10*, 29–73.

Nürnberg, D., I. Wollenburg, D. Dehtleff, H. Eicken, H. Kassens, T. Letzig, E. Reimnitz, and J. Thiede (1994), Sediments in Arctic sea ice: Implications for entrainment, transport and release, *Mar. Geol.*, *119*(3–4), 185–214.

Nye, J. F. (1973), Water at the bed of a glacier, in *Symposium on the Hydrology of Glaciers*, *IAHS AISH Publ.*, *95*, 189–194.

Nye, J. F. (1976), Water flow in glaciers: Jokullhlaups, tunnels and veins, *J. Glaciol.*, *17*, 181–207.

Oswald, G. K. A., and G. D. Q. Robin (1973), Lakes beneath Antarctic ice sheet, *Nature*, *245*(5423), 251–254.

Pattyn, F. (2008), Investigating the stability of subglacial lakes with a full Stokes ice-sheet model, *J. Glaciol.*, *54*(185), 353–361.

Pattyn, F., B. De Smedt, and R. Souchez (2004), Influence of subglacial Vostok lake on the regional ice dynamics of the Antarctic ice sheet: A model study, *J. Glaciol.*, *50*(171), 583–589.

Peltier, W. R. (2002), Global glacial isostatic adjustment: Palaeo-geodetic and space-geodetic tests of the ICE-4G (VM2) model, *J. Quat. Sci.*, *17*(5–6), 491–510.

Peters, L. E., S. Anandakrishnan, R. B. Alley, and A. M. Smith (2007), Extensive storage of basal meltwater in the onset region of a major West Antarctic ice stream, *Geology*, *35*(3), 251–254.

Peters, L. E., S. Anandakrishnan, C. W. Holland, H. J. Horgan, D. D. Blankenship, and D. E. Voigt (2008), Seismic detection of a subglacial lake near the South Pole, Antarctica, *Geophys. Res. Lett.*, *35*, L23501, doi:10.1029/2008GL035704.

Pickrill, R. A., and J. Irwin (1982), Predominant headwater inflow and its control of lake-river interactions in Lake Wakatipu, *N. Z. J. Mar. Freshwater Res.*, *16*, 201–213.

Priscu, J. C., S. Tulaczyk, M. Studinger, M. C. Kennicutt, II, B. C. Christner, and C. M. Foreman (2008), Antarctic subglacial water: Origin, evolution and ecology, in *Polar Lakes and Rivers—Limnology of Arctic and Antarctic Aquatic Ecosystems*, edited by W. F. Vincent and J. Laybourn-Parry, pp. 119–135, Oxford Univ. Press, Oxford, U. K.

Raisbeck, G., F. Yiou, D. Bourles, C. Lorius, J. Jouzel, and N. I. Barkov (1987), Evidence for two intervals of enhanced 20Be deposition in Antarctic ice during the last glacial period, *Nature*, *326, 273–277*.

Rebesco, M., A. Camerlenghi, L. De Santis, E. Domack, and M. Kirby (1998), Seismic stratigraphy of Palmer Deep: A fault-bounded late Quaternary sediment trap on the inner continental shelf, Antarctic Peninsula Pacific margin, *Mar. Geol.*, *151*(1–4), 89–110.

Ross, N., et al. (2011), Ellsworth Subglacial Lake, West Antarctica: A review of its history and recent field campaigns, in *Antarctic Subglacial Aquatic Environments*, *Geophys. Monogr. Ser.*, doi: 10.1029/2010GM000936, this volume.

Royston-Bishop, G., J. C. Priscu, M. Tranter, B. Christner, M. J. Siegert, and V. Lee (2005), Incorporation of particulates into accreted ice above subglacial Vostok lake, Antarctica, *Ann. Glaciol.*, *40*, 145–150.

Russell, H. A. J., R. W. C. Arnott, and D. R. Sharpe (2003), Evidence for rapid sedimentation in a tunnel channel, Oak Ridges Moraine, southern Ontario, Canada, *Sediment. Geol.*, *160*(1–3), 33–55.

Sawagaki, T., and K. Hirakawa (1997), Erosion of bedrock by subglacial meltwater, Soya Coast, East Antarctica, *Geogr. Ann., Ser. A*, *79*(4), 223–238.

Scherer, R. (1993), There is direct evidence for Pleistocene collapse of the West Antarctic Ice Sheet, *J. Glaciol.*, *39*(133), 716–722.

Scherer, R. P., A. Aldahan, S. Tulaczyk, G. Possnert, H. Engelhardt, and B. Kamb (1998), Pleistocene collapse of the West Antarctic Ice Sheet, *Science*, *281*(5373), 82–85.

Shepherd, A., A. Hubbard, P. Nienow, M. King, M. McMillan, and I. Joughin (2009), Greenland Ice Sheet motion coupled with daily melting in late summer, *Geophys. Res. Lett.*, *36*, L01501, doi:10.1029/2008GL035758.

Shoemaker, E. M. (1991), On the formation of large subglacial lakes, *Can. J. Earth Sci.*, *28*(12), 1975–1981.

Shreve, R. L. (1972), Movement of water in glaciers, *J. Glaciol.*, *11*, 205–214.

Siegert, M. J. (2000), Antarctic subglacial lakes, *Earth Sci. Rev.*, *50*(1–2), 29–50.

Siegert, M. J., J. C. Ellis-Evans, M. Tranter, C. Mayer, J. R. Petit, A. Salamatin, and J. C. Priscu (2001), Physical, chemical and biological processes in Lake Vostok and other Antarctic subglacial lakes, *Nature*, *414*(6864), 603–609.

Siegert, M. J., R. Hindmarsh, H. Corr, A. Smith, J. Woodward, E. C. King, A. J. Payne, and I. Joughin (2004), Subglacial Lake Ellsworth: A candidate for in situ exploration in West Antarctica, *Geophys. Res. Lett.*, *31*, L23403, doi:10.1029/2004GL021477.

Siegert, M. J., S. Carter, I. Tabacco, S. Popov, and D. D. Blankenship (2005), A revised inventory of Antarctic subglacial lakes, *Antarct. Sci.*, *17*(3), 453–460.

Siegert, M. J., et al. (2006), Exploration of Ellsworth Subglacial Lake: A concept paper on the development, organisation and execution of an experiment to explore, measure and sample the environment of a West Antarctic subglacial lake, *Rev. Environ. Sci. Biotechnol.*, *6*(1–3), 161–179.

Skidmore, M. (2011), Microbial communities in Antarctic subglacial aquatic environments, in *Antarctic Subglacial Aquatic Environments*, *Geophys. Monogr. Ser.*, doi: 10.1029/2010 GM000995, this volume.

Skidmore, M., M. Tranter, S. Tulaczyk, and B. Lanoil (2010), Hydrochemistry of ice stream beds—Evaporitic or microbial effects?, *Hydrol. Processes*, *24*, 517–523.

Smith, A. M. (1997), Basal conditions on Rutford Ice Stream, West Antarctica, from seismic observations, *J. Geophys. Res.*, *102*(B1), 543–552.

Smith, A. M., and T. Murray (2009), Bedform topography and basal conditions beneath a fast-flowing West Antarctic ice stream, *Quat. Sci. Rev.*, *28*(7–8), 584–596.

Smith, A. M., T. Murray, K. W. Nicholls, K. Makinson, G. Aoalgeirsdottir, A. E. Behar, and D. G. Vaughan (2007), Rapid erosion, drumlin formation, and changing hydrology beneath an Antarctic ice stream, *Geology*, *35*(2), 127–130.

Smith, A. M., J. Woodward, N. Ross, M. J. Siegert, H. F. J. Corr, R. C. A. Hindmarsh, E. C. King, D. G. Vaughan, and M. A. King (2008), Physical conditions in Subglacial Lake Ellsworth, *Eos Trans. AGU*, *89*(53), Fall Meet. Suppl., Abstract C11A-0467.

Smith, B. E., H. A. Fricker, I. R. Joughin, and S. Tulaczyk (2009), An inventory of active subglacial lakes in Antarctica detected by ICESat (2003–2008), *J. Glaciol.*, *55*, 573–595.

Smith, N. D. (1978), Sedimentation processes and patterns in a glacier-fed lake with low sediment input, *Can. J. Earth Sci.*, *15*, 741–756.

Souchez, R., J. R. Petit, J. L. Tison, J. Jouzel, and V. Verbeke (2000), Ice formation in subglacial Lake Vostok, Central Antarctica, *Earth Planet. Sci. Lett.*, *181*(4), 529–538.

Studinger, M., et al. (2003), Ice cover, landscape setting, and geological framework of Lake Vostok, East Antarctica, *Earth Planet. Sci. Lett.*, *205*(3–4), 195–210.

Studinger, M., R. E. Bell, and A. A. Tikku (2004), Estimating the depth and shape of subglacial Lake Vostok's water cavity from aerogravity data, *Geophys. Res. Lett.*, *31*, L12401, doi:10.1029/2004GL019801.

Sugden, D. E., G. H. Denton, and D. R. Marchant (1991), Subglacial meltwater channel systems and ice sheet overriding, Asgard Range, Antarctica, *Geogr. Ann., Ser. A*, *73*(2), 109–121.

Thoma, M., K. Grosfeld, and C. Mayer (2007), Modelling mixing and circulation in subglacial lake Vostok, Antarctica, *Ocean Dyn.*, *57*(6), 531–540.

Thoma, M., K. Grosfeld, and C. Mayer (2008a), Modelling accreted ice in subglacial Lake Vostok, Antarctica, *Geophys. Res. Lett.*, *35*, L11504, doi:10.1029/2008GL033607.

Thoma, M., C. Mayer, and K. Grosfeld (2008b), Sensitivity of subglacial Lake Vostok's flow regime on environmental parameters, *Earth Planet. Sci. Lett.*, *269*(1–2), 242–247.

Thoma, M., K. Grosfeld, C. Mayer, and F. Pattyn (2010), Interaction between ice sheet dynamics and subglacial lake circulation: A coupled modelling approach, *Cryosphere*, *4*(1), 1–12.

Tikku, A. A., R. E. Bell, M. Studinger, and G. K. C. Clarke (2004), Ice flow field over Lake Vostok, East Antarctica inferred by structure tracking, *Earth Planet. Sci. Lett.*, *227*(3–4), 249–261.

Tulaczyk, S., W. B. Kamb, and H. F. Engelhardt (2000), Basal mechanics of Ice Stream B, West Antarctica 2. Undrained plastic bed model, *J. Geophys. Res.*, *105*(B1), 483–494.

Tulaczyk, S., R. P. Scherer, and C. D. Clark (2001), A ploughing model for the origin of weak tills beneath ice streams: A qualitative treatment, *Quat. Int.*, *86*, 59–70.

van Rensbergen, P., M. de Batist, C. Beck, and E. Chapron (1999), High-resolution seismic stratigraphy of glacial to interglacial fill of a deep glacigenic lake: Lake Le Bourget, Northwestern Alps, France, *Sediment. Geol.*, *128*(1–2), 99–129.

Vaughan, D. G., A. Rivera, J. Woodward, H. F. J. Corr, J. Wendt, and R. Zamora (2007), Topographic and hydrological controls on subglacial Lake Ellsworth, West Antarctica, *Geophys. Res. Lett.*, *34*, L18501, doi:10.1029/2007GL030769.

Williams, M. J. M. (2001), Application of a three-dimensional numerical model to Lake Vostok: An Antarctic subglacial lake, *Geophys. Res. Lett.*, *28*(3), 531–534.

Winberry, J. P., S. Anandakrishnan, and R. B. Alley (2009), Seismic observations of transient subglacial water-flow beneath MacAyeal Ice Stream, West Antarctica, *Geophys. Res. Lett.*, *36*, L11502, doi:10.1029/2009GL037730.

Wingham, D. J., M. J. Siegert, A. Shepherd, and A. S. Muir (2006), Rapid discharge connects Antarctic subglacial lakes, *Nature*, *440* (7087), 1033–1036.

Woodward, J., A. M. Smith, N. Ross, M. P. A. Thomassen, H. F. J. Corr, E. C. King, M. A. King, K. Grosfeld, M. Tranter, and M. J. Siegert (2010), Location for direct access to subglacial Lake Ellsworth: An assessment of geophysical data and modelling, *Geophys. Res. Lett.*, *37*, L11501, doi:10.1029/2010GL042884.

Wright, A. P., M. J. Siegert, A. M. Le Brocq, and D. B. Gore (2008), High sensitivity of subglacial hydrological pathways in Antarc-

tica to small ice-sheet changes, *Geophys. Res. Lett.*, *35*, L17504, doi:10.1029/2008GL034937.

Wuest, A., and E. Carmack (2000), A priori estimates of mixing and circulation in the hard-to-reach water body of Lake Vostok, *Ocean Modell.*, *2*, 29–43.

Zimmerman, S. H., S. R. Hemming, D. V. Kent, and S. Y. Searle (2006), Revised chronology for late Pleistocene Mono Lake sediments based on paleointensity correlation to the global reference curve, *Earth Planet. Sci. Lett.*, *252*(1–2), 94–106.

Zwally, H. J., W. Abdalati, T. Herring, K. Larson, J. Saba, and K. Steffen (2002), Surface melt-induced acceleration of Greenland Ice-Sheet flow, *Science*, *297*(5579), 218–222.

M. J. Bentley and A. M. Le Brocq, Department of Geography, University of Durham, South Road, Durham DH1 3LE, UK. (M.J. Bentley@durham.ac.uk)

P. Christoffersen, Department of Geography, University of Cambridge, Downing Place, Cambridge CB2 3EN, UK.

D. A. Hodgson and A. M. Smith, British Antarctic Survey, High Cross, Madingley Road, Cambridge CB3 0ET, UK.

S. Tulaczyk, Department of Earth and Planetary Sciences, University of California, Santa Cruz, Santa Cruz, CA 95064, USA.

The Geomorphic Signature of Massive Subglacial Floods in Victoria Land, Antarctica

David R. Marchant

Department of Earth Sciences, Boston University, Boston, Massachusetts, USA

Stewart S. R. Jamieson

Department of Geography, Durham University, Durham, UK

David E. Sugden

School of GeoSciences, University of Edinburgh, Edinburgh, UK

We describe a landscape in the Dry Valleys area consisting of channels, potholes, plunge pools, stripped bedrock, and scabland created by subglacial outburst floods on a scale similar to that of the massive outbursts of Lake Missoula in Washington, United States. The features are dated to the mid-Miocene and occurred when a thicker East Antarctic ice sheet overrode this part of the Transantarctic Mountains. Ice sheet modeling is used to reconstruct potential source areas for meltwater discharge. Results suggest that meltwater from Lake Vostok, or other interior basins, would have bypassed the Dry Valleys area and instead flowed out beneath the troughs of David and Byrd glaciers, north and south of the Dry Valleys, respectively. A more likely source might have been lakes on the inner flank of the Transantarctic Mountains, with periodic outbursts associated with local breaches in the cold-based rim of the ice sheet. The efficacy of erosion by such outbursts is remarkable. The most concentrated effects were in the labyrinth on the bed of the former Wright outlet glacier. Such outbursts are likely to contribute to the erosion of present-day outlet glacier troughs.

1. INTRODUCTION

The aim of this article was to examine the geomorphology of meltwater landforms formed by massive outburst floods beneath a formerly more extensive Miocene Antarctic ice sheet in Victoria Land. The scale of the features is of the same order of magnitude as the ice-dammed Lake Missoula outbursts that created the channeled scablands of east-central Washington [*Bretz*, 1923, 1969]. A difference is that, unlike the Missoula floods, the Antarctic features were formed subglacially.

The geomorphology of subglacial meltwater features is one approach to the study of water flow at the base of ice sheets. Theoretical work, backed up by observations mainly in the Northern Hemisphere, has shown that subglacial meltwater is fundamental to the dynamics of ice flow. Meltwater can form major subglacial rivers with predictable networks and routes [*Shreve*, 1972]. Moreover, meltwater controls the

Antarctic Subglacial Aquatic Environments
Geophysical Monograph Series 192
Copyright 2011 by the American Geophysical Union
10.1029/2010GM000943

111

of Greenville Valley, Battleship Promontory itself, and parts of the labyrinth could be classified as scabland. However, the dominance of scalloped bedrock basins and buttes on the upland areas seems distinctive.

3. CHRONOLOGY

The meltwater features have been dated to the mid-Miocene using two independent methods. In the first method, $^{40}Ar/^{39}Ar$ dates of ~12.5 and 12.44 Ma on pristine ashfall that stratigraphically overlies terrain scoured by ice and meltwater in the western Asgard and western Olympus ranges provide minimum ages for meltwater incision [*Marchant et al.*, 1993b; *Lewis et al.*, 2007]. In addition, pristine drifts that drape across scoured terrain likewise provide minimum ages for meltwater incision (Plate 8). Further, if one assumes that meltwater incision occurred during glacial overriding, then a date of 14.8 Ma, on an in situ ash layer found beneath till drawn out from a bedrock hollow in Nibelungen Valley [*Marchant et al.*, 1993a, 1993b], provides a maximum age for meltwater incision. Taken together, the radiometric dates bracket meltwater incision at high elevations in the Dry Valleys to sometime

between 14.8 and ~12.5 Ma. A closer fix on the timing of meltwater incision comes from radiometric ages on pristine and reworked tephra in the labyrinth. A date of 12.37 ± 0.66 Ma on pristine ashfall on ice-scoured bedrock provides a minimum age for meltwater incision [*Lewis et al.*, 2006]. A date of 14.36 ± 0.43 Ma on reworked tephra found within stratified deposits provides a maximum age for at least one meltwater event. Taken together, the radiometric data call for at least one meltwater flood in Wright Valley trough sometime between ~14.3 and ~12.3 Ma [*Lewis et al.*, 2006].

The second independent method was the use of cosmogenic ^{3}He to date the dolerite flood deposits at an elevation of 2000 m in the Coombs Hills [*Margerison et al.*, 2005]. Three dolerite clasts within 15 m of each other were selected from a flat corrugated sandstone surface well distant from any possible slope activity. The clasts were undisturbed since deposition in that their upper surfaces were pockmarked with 4 cm deep solution holes due to chemical weathering, a conclusion subsequently confirmed by comparing analyses of their upper and lower surfaces. Assuming scaling factors appropriate for Antarctica and no erosion, the analysis showed that the three clasts had been exposed to cosmic rays for at least 8.6 to

Plate 6. Oblique aerial view of the sandstone escarpment at Battleship Promontory. View is to the west. The sandstone escarpment is 200 m high. Huge plunge pools occur at the base of the cliff. The area just above the cliff is pockmarked with giant potholes. Battleship Promontory makes up part of an extensive 34 km long channel-and-pothole system that begins at a summit divide (partially seen in background) and drops 1000 m in elevation.

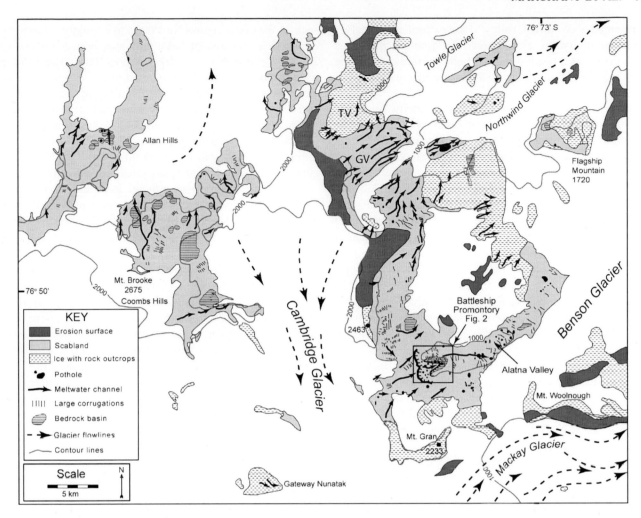

Figure 3. Geomorphic map highlighting scabland topography, major channels, and large potholes in the Convoy Range; also shown is the location of Battleship Promontory and Figure 2. GV, Greenville Valley; TV, Towle Valley. Adapted from *Sugden and Denton* [2004].

10.4 Ma. Allowing for an erosion rate of ~0.03 m Ma^{-1}, calculated from other samples and a rate typical for the region [e.g., *Summerfield et al.*, 1999; *Staiger et al.*, 2006; *Schäfer et al.*, 1999; *Balco et al.*, 2008], then an exposure age of ~14 Ma is likely for the oldest clast. The analysis supports the hypothesis that the clasts were deposited by a flood ~14 Ma ago and have remained undisturbed ever since.

4. INTERPRETATION

Following the arguments outlined in the work of *Denton and Sugden* [2005], we interpret the stripped bedrock surfaces, channels, potholes, plunge pools, and scabland as the result of high-magnitude subglacial floods. The continuity, sinuosity, and confluence pattern of the channels and their association with potholes are features best explained by the action of water. There are several characteristics that demonstrate the features were cut subglacially under hydrostatic pressure. These include irregular, convex, long profiles of channels, their anastomosing pattern cut in bedrock, and their overall orientation parallel to ice flow, often across the grain of the underlying topography. In the latter case, many channels and other features reflect flow northeastwardly across the mountains, a direction indicated by streamlined landforms of ice erosion. This orientation is reflected in the northeastward flow of channels in scabland on the upper plateau and, in areas of more relief, in the location of channels in saddles in the underlying topography. The Convoy Range channel systems originate in saddles in the main mountain crest and are a good example of the latter relationship. In deeper valleys, such as Wright and Taylor valleys, the channel systems are more constrained by topography and are oriented E-W down valley. This pattern is typical of subglacial meltwater flow driven by ambient pressures within the ice as a result of primarily the surface slope of the overriding ice and secondarily the shape of the underlying topography.

It seems as though the ice between the meltwater forms was cold-based. This is exemplified by the widespread preservation of preexisting regolith and till patches immediately adjacent to the meltwater features. The preservation of undisturbed premeltwater volcanic ash in the Asgard Range [*Marchant et al.*, 1993b; see also *Lewis et al.*, 2007, 2008] demonstrates near-perfect preservation of antecedent sediments outside scour zones. In the Allan Hills, Sirius Group and related deposits, thought to reflect the earliest glaciation of this part of the Transantarctic Mountains, have been preserved in patches immediately adjacent to flood ripples (Plate 2). Indeed, the corrugations have actually exposed the deposit in places. Cosmogenic isotope analysis has shown that cold-based ice can override clasts without removing them [*Briner et al.*, 2006; *Sugden et al.*, 2005], and thus, it is reasonable to suggest that the overriding ice was cold-based, at least over the high mountain rim, and may have been in response to cooling in the mid-Miocene [*Lewis et al.*, 2007, 2008]. Thus, the juxtaposition of such deposits and flood features is best explained by a subglacial flood breaching a cold-based ice sheet covering the mountains.

An unusual feature of the meltwater landforms in the Dry Valleys area is the relative lack of sediment. The Missoula flood routes are associated with sediment deposition at certain locations, and the undulations on the bed may consist of giant ripples topped by dunes. Sediment in the Dry Valleys area is confined to boulders and patches of coarse gravel in parts of the labyrinth and to crescentic ripples topping some of the bedrock corrugations in otherwise stripped bedrock areas. What is missing is evidence of eskers, valley trains, and outwash fans that are so typical of the beds of former Northern Hemisphere ice sheets. Perhaps much of the sediment froze on to the bottom of the ice sheet and was transported offshore by normal processes of ice flow. The location of distinct spatulate-shaped patches of stripped bedrock downstream of channels, as in the case of the Asgard Range and Battleship Promontory, could indicate such a process [*Marchant et al.*, 1993b]. Perhaps the sediments were transported offshore into the Ross Sea embayment by meltwater and have been deposited as deltas near the ice edge. In support of such an idea is the recording of mid-Miocene gravel deltas and channels within the till sequences on the floor of the Ross Sea [*Chow and Bart*, 2003; *Haywood et al.*, 2008].

There are three arguments to suggest that the meltwater features were cut by megafloods. First, the sheer scale of the landforms is similar to those known from other huge floods, such as the floods across Northwest Washington, caused by the sudden drainage of Lake Missoula with a volume of 2600 km³ through its ice dam [*Baker*, 2009]. The 150 km width of the mountain front affected by flood features in the Dry Valleys area compares to an equivalent figure of 50 to 100 km for the Missoula floods. Moreover, individual channels in both areas are several kilometers long and hundreds of meters deep. The plunge pools in the Convoy Range are kilometers across with backing cliffs of 100 to 200 m and are the equivalent of the similarly sized coulees in Washington.

Second, the landform assemblage is similar. *Bretz* [1923] coined the word scabland to describe areas of bedrock from which the loess cover had been stripped by water. In Washington, the "usual development is small channels and rock basins surrounding buttes and mesas with a typical relief of 30 to 100 m. The rock basins range in scale from shallow saucers or deep potholes, 10 to 100 m in width, to Rock Lake, a huge inner channel that is 11 km long and 30 m deep" [*Baker*, 2009]. Such words are excellent descriptions of the scabland in Antarctica. Furthermore, the association of Washington scabland with potholes, cataracts or plunge pools, longer channels, and transverse bedforms, 2 to 3 m in amplitude and a spacing of up to 60 m, all have counterparts in Antarctica. Such features are typical of separated flow at high flow velocities in rock channels [*Hancock et al.*, 1998; *Baker*, 1978]. The concentration of potholes on lee slopes, as in the Dry Valleys area, is characteristic. The transverse corrugations are linked to standing waves typical of high velocities [*Allen*, 1971].

The third argument in favor of a massive outburst flood is that even approximate calculations suggest the discharges are in excess of 1 million m³ s⁻¹ and thus can be classified as megafloods. In the case of Antarctica, minimum estimates on the basis of (1) the 1.8 × 1.8 m face dimensions of transported dolerite blocks in the labyrinth and (2) the assumption that the flood did not overfill the channels yield discharges of 1.6 to 2.2 million m³ s⁻¹ [*Lewis et al.*, 2006]. If, as is possible in view of potholes existing on channel interfluves, the flood overfilled the channels, then the discharge could have been much higher. Working in a location where the upper flood surface can be estimated, discharges of ~17 million m³ s⁻¹ have been calculated for some Washington flood peaks [*Waitt*, 1985; *O'Connor and Baker*, 1992].

5. GLACIOLOGICAL MODELING

Radio echo sounding has revealed that there are many subglacial lakes beneath the East Antarctic ice sheet [*Siegert*, 2000]. Lake Vostok near the center of the East Antarctic ice sheet beneath Dome C is comparable in size to the Great Lakes of North America and measures 260 × 80 km and has a depth of over 500 m. Lake Ellsworth in West Antarctica lies in an overdeepened trough and is around 10 km long and at least tens of meters deep [*Siegert et al.*, 2004]. Observations from past and present ice sheets show that such lakes can drain suddenly. *Evatt et al.* [2006] calculate that Antarctic subglacial lakes can repeatedly drain suddenly, and satellite observations of ice-surface elevation changes as a result of one subglacial lake draining and flowing rapidly down glacier to another subglacial lake basin have demonstrated the process in action [*Wingham et al.*, 2006]. The route of the flood follows the ambient pressures within the ice and thus

reflects the dominant role of ice-surface gradient and lesser role of subglacial topography. The interval between floods may be hundreds to thousands of years and depends on such factors as the size of the lake catchment and the rate of basal melting. Under the present ice sheet, meltwater from Lake Vostok with a total volume of 6×10^{11} m^3 would find its way to the Ross Sea embayment via the Byrd glacier trough, which breaches the Transantarctic Mountains 350 km south of Taylor Valley [*Evatt et al.*, 2006].

The meltwater features in the wider Dry Valleys area formed beneath a former expanded ice sheet, and *Sugden*

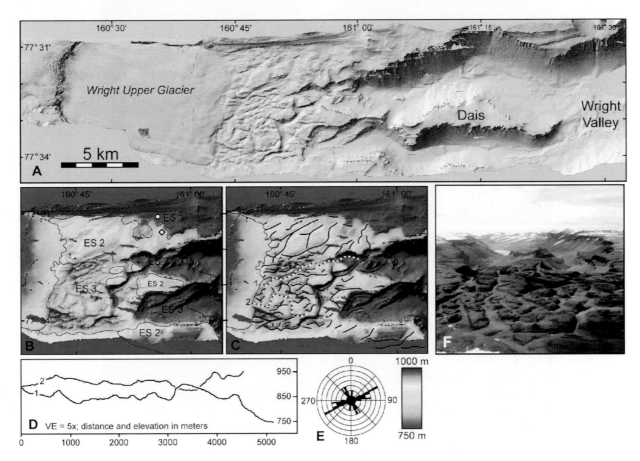

Plate 7. Digital elevation models (DEMs) of the labyrinth; data from a joint NSF/NASA/USGS effort, with basic processing at the Byrd Polar Research Center, The Ohio State University; north is to the top, and scale bar is applicable to all DEMs. (a) Shaded relief DEM of Wright Upper Glacier, the labyrinth, and central Wright Valley (see Figure 1 for location). (b) Erosion surfaces (ES 1, ES 2, and ES 3) comprising the labyrinth; white circle and diamond show locations for in situ and reworked ashfall, respectively, that provide chronologic control for the most recent flood event between 12.37 ± 0.66 and 14.36 ± 0.43 Ma (see text). (c) Channels, depicted as black lines, and location of longitudinal profiles (1 and 2) depicted as dashed lines. Individual channels feature potholes at tributary junctions, with plunge pools, blind terminations, and reverse gradients. (d) Longitudinal profiles for channels 1 and 2 as indicated in (c); adverse slopes, up to 30 m in relief, occur within some of the largest channels. North Fork and South Fork basins, on opposing sides of the Dais, are as much as 800 m deep and carved by glaciers during headword erosion of Wright Valley. (e) Joint orientations in bedrock displayed on rose diagram (note similarity with overall channel pattern within the labyrinth). (f) Photograph of channels in the labyrinth; view to the east. VE, vertical exaggeration. Adapted from *Lewis et al.* [2006].

and Denton [2004] showed a model of such an ice sheet. They pointed out that the surface flow lines ran from Vostok across the area and wondered whether the source of the meltwater could indeed have been Lake Vostok. In such a case, the water might have flowed progressively and slowly down glacier, perhaps in a number of stages, until it eventu-

ally reached the cold-based rim over the mountain axis. Once it was able to lift the overlying ice and breach the dam, the water drained catastrophically to form the flood features in the lee of the mountain crest. Alternatively, the adverse slopes along the interior flank of the Transantarctic Mountains would most likely have provided the requisite

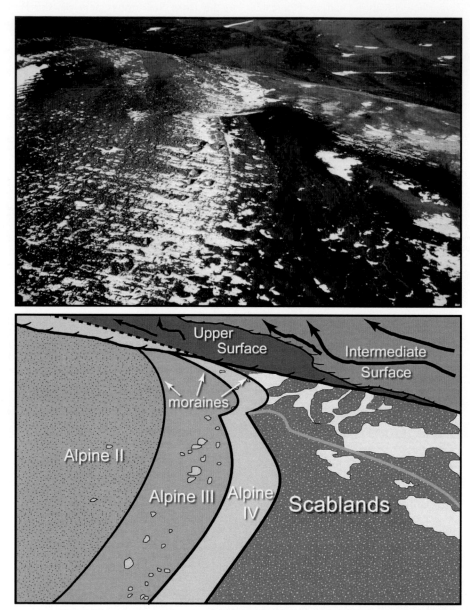

Plate 8. (top) Oblique aerial view (looking southeast) showing pristine alpine drifts (<1 m thick) that drape across scabland topography in the Olympus Range (foreground) and the upper erosion surface of the labyrinth (background). (bottom) Interpretive sketch highlighting cross-cutting relations and relative chronology. The alpine drifts, undated at present, clearly postdate meltwater incision. In addition, the pristine nature of the drift sheets and moraines suggest that aeolian processes in recent times have not been sufficient to remove thin regolith and expose underlying sandstone bedrock. The latter is consistent with a nonuniformitarian process, e.g., massive floods, as a likely cause for the observed scabland features. Field of view at base of photograph is ~300 m.

glaciological conditions for subglacial meltwater ponding [*Clarke et al.*, 2005]. Particularly favorable locations would be the troughs on the inner flank of the mountains revealed by radar remote sensing [*Calkin*, 1974], which were probably cut by local mountain ice caps at an early stage of Antarctic glaciation [*Jamieson and Sugden*, 2008]. The cross-cutting channels and erosion surfaces of the labyrinth call for repeated failure of such an ice dam and multiple discharge events. We explore this possibility in general terms using an ice sheet model.

In order to analyze subglacial drainage in the region of the Dry Valleys, we use models of hydraulic potential at the base of both the present-day ice sheet and an ice sheet that is ~1000 m thicker over the inland end of Wright Valley. BED-MAP data from *Lythe et al.* [2001] provides the topographic and surface elevations for the modern analysis. The thicker ice sheet configuration is based on modeled data used to reconstruct past basal processes and water flow over wider regions of Antarctica [*Jamieson and Sugden*, 2008; *Jamieson et al.*, 2010]. The thicker ice is generated using the GLIM-MER community ice sheet model with the assumption that the slopes of the bedrock and of the ice surface are shallow

and therefore that longitudinal stress components are negligible [*Rutt et al.*, 2009]. The geometry of the modeled ice mass is determined by patterns of ice flow, accumulation, ablation, and basal melting. A 20 km resolution, isostatically rebounded version of the BEDMAP bedrock data provides the initial boundary condition, and it is assumed that precipitation patterns are similar to present but that intensity is dependent on elevation. This enhances snowfall and helps to produce a thickened steady state ice sheet over the Dry Valleys region.

The hydraulic potential is calculated for both the present-day and thicker ice sheets, assuming, as others have done [*Evatt et al.*, 2006], that basal water pressures are equal to ice overburden pressure. The resultant surface allows tracking of dominant subglacial water pathways with flow perpendicular to the pressure gradient. Hydraulic potential and therefore paths of subglacial flow are most strongly controlled by ice-surface gradient with the influence of bed slope being felt to a lesser extent unless at a local scale [*Shreve*, 1972]. Thus, the locations of potential lakes, where hydraulic gradients converge inward, are often, although not always, aligned with topographic overdeepenings. The locations of these

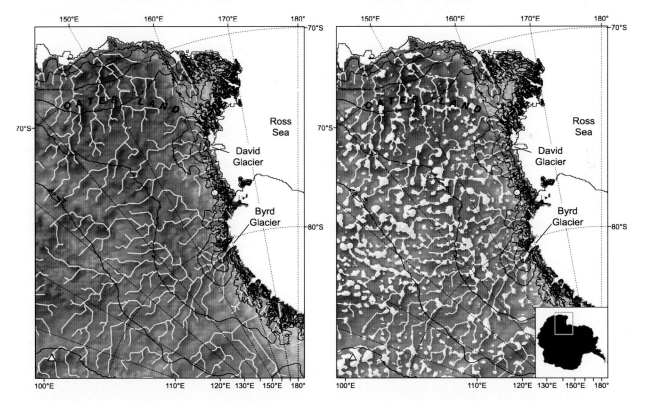

Figure 4. Predicted subglacial pathways (white lines) (left) without and (right) with potential lakes (white areas) beneath the present-day East Antarctic ice sheet. Hydraulic potential was calculated using BEDMAP ice-surface and bed data [*Lythe et al.*, 2001]. White circle, labyrinth; white triangle, Lake Vostok; gray shading, hill-shaded subglacial bed topography; black lines, 500 m ice-surface contours.

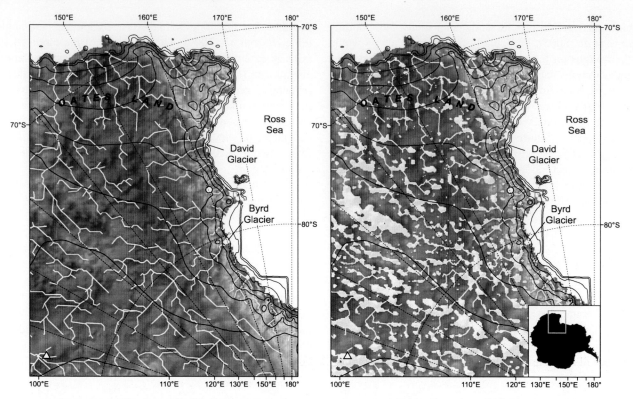

Figure 5. Reconstruction of subglacial pathways (white lines) (left) without and (right) with potential lakes (white areas) beneath a thickened East Antarctic ice sheet. Hydraulic potential was calculated using ice-surface and bed data from an ice sheet model that predicts ice 1000 m thicker than present over the Dry Valleys region [*Jamieson and Sugden*, 2008]. White circle, labyrinth; white triangle, Lake Vostok; gray shading, hill-shaded subglacial bed topography; black lines, 500 m ice-surface contours.

potential lakes are robust. However, their extents are particularly sensitive to the smoothness of the ice surface and, in this case, are maximum estimates.

Figure 4 shows predicted subglacial meltwater flow paths, with and without potential lake basins, beneath a large sector of the present East Antarctic ice sheet, and covers the Dry Valleys and Lake Vostok and the area in between. The pattern is of radiating dendritic flow from the central dome of the ice sheet near Vostok (Dome C). At a macroscale, it is evident that the Transantarctic Mountain blocks concentrate both ice and subglacial water flow into the Byrd and David glacier troughs that breach the barrier at the present day. In the north, there is freer and more direct flow toward the coast of Oates Land. The Dry Valleys area emerges as a zone of divergence with meltwater from the ice sheet interior being diverted north and south around the mountain blocks. This is expected in view of the fact that the ice sheet is not sufficiently thick to cover the mountains except for a few outlets such as Ferrar and Mackay glaciers. The dendritic flow connects chains of potential subglacial lake sites, where the equipotentials are zero or reversed.

Figure 5 shows a modeled thicker ice sheet with surface elevations over the Dry Valleys area 1000 m higher at around 2500 to 3000 m. Encouragingly, the direction of ice flow, perpendicular to the surface contours, shows ice flowing across the mountains in a northeasterly direction. This is in good agreement with the evidence of ice molding and the orientation of subglacial meltwater flow. The broad pattern of flow of subglacial meltwater is similar to that of the present-day reconstruction in that it radiates out from Dome C. However, in this case, it is more focused on outlet glaciers breaching the Transantarctic Mountains, and there is a larger area draining toward the Oates coast. Moreover, Lake Vostok, if it drains, is likely to go to the Wilkes coast due to the size and position of a palaeo-Dome C. A persistent feature is that the Dry Valleys area still encourages interior water to diverge around the mountain blocks.

We carried out sensitivity tests in order to identify whether a lack of detail in the basal topography could be important in understanding the flow routing in the region of the Dry Valleys. Although BEDMAP data were collected at a relatively high resolution over the adjacent Taylor Dome, it is possible that the detail of key topographic ridges, hollows, or troughs may be poorly represented in the model. The sensitivity tests made topographic conditions as favorable as possible to the hypothesis that water draining from the interior of East Antarctica could be routed through the Dry Valleys. Tests included lowering the mountains of the Dry Valleys area by 500 m to simulate possible tectonic activity or ice thickening and blocking the surrounding Byrd and David troughs at the elevation of the adjacent mountains to cover the possibility that they had not been fully excavated by the mid-Miocene. In none of these experiments did we induce meltwater from interior East Antarctica to cross the mountains in the Dry Valleys area. However, in these experiments, the original modeled ice-surface elevation is retained, and given the importance of ice-surface gradient in driving hydraulic potential, we cannot rule out the hypothesis that water may have been directed through the Dry Valleys if the Byrd and David breaches were blocked. Nevertheless, there is a strong tendency for water to drain preferentially through the vicinity of the Byrd and David glaciers due to their adjacent mountain elevation, size, depth, significant inland catchment areas, and the influence on ice-surface gradients. Not only do these troughs and their hinterlands offer natural funnels, but also they force ice-surface gradients to such an extent that ice flow and hydraulic potential diverge behind the Dry Valleys block.

So we are left with a puzzle. The evidence tells of a huge lake draining subglacially and catastrophically over the mountains in the area, and yet there is no obvious route from the deep interior. So what are the possible explanations for an apparent mismatch between the model output and the field evidence? First, there is the question of whether the resolution of the model and of the topography is sufficient to pick out a relatively narrow meltwater route across the mountains. This is a distinct possibility because the model simulates ice sheet behavior on a much coarser resolution (20 km) than the BEDMAP data (5 km), and this will have the effect of smoothing the topography and reducing high and low points. Also, the model generalizes the surface gradients of the ice and assumes an equilibrium state. During glacial cycles, there would have been periods of thinner ice over the mountains, and in such a case, ice-surface gradients could have been steeper. Further, there is the observational evidence that Wright outlet glacier had sufficient ice flux to excavate an overdeepened trough 800 m deep and 7 km wide in Wright Valley (Plate 7), which shows that the glaciological conditions and topography did permit focused ice flow across the mountain rim. Such ice flow would favor subglacial meltwater flow in the same direction.

A second possibility is that the topography, especially the Byrd and David troughs, has been modified since the mid-Miocene. It is easy to imagine that outlet glaciers breaching

the mountains are continuing to deepen their troughs and are thus changing the subglacial hydrology. Such an idea underlay the idea of infilling in the troughs in one sensitivity test and requires further work to explore fully. Nonetheless, the experiment showed that it was the overall geometry and elevation of the mountain rim rather than the depth of the troughs that were more significant. Moreover, it is likely that the troughs exploited preexisting river valleys that had already cut down to sea level by the Miocene and that a low level route to the sea already existed [Sugden and Denton, 2004]. At present, this seems an unlikely explanation.

Perhaps a third and the most likely explanation is that there is a local source for the lake on the inland slopes of the mountains. Experience has shown that even extremely large topographic features can remain hidden below the ice [Le Brocq et al., 2008] and that perhaps a trough cut by local ice might exist and offer a location for a massive lake to accumulate. In such a case, the ice gradients on the peripheries of the ice sheet would be the main drivers of the subglacial flow across the mountain rim. Perhaps the lake outbursts could be influenced by mid-Miocene glacial cycles and relate to periods when the overlying ice thins and ice-surface gradients steepen. Such a local source avoids the problem of divergence suggested by the model for meltwater flow from the deep ice sheet interior.

These three possible explanations all relate to our knowledge of past and present topography. Given the importance of ice-surface gradients on hydraulic potential, any deficiency will be magnified by ice surfaces generated using ice sheet models. Therefore, it seems likely that one or more of these alternatives holds the key to explaining the source and routing of the meltwater outbursts and landforms in the Dry Valleys area. At present, our preferred hypothesis is that meltwater accumulated subglacially in a bedrock basin on the inner flanks of the Transantarctic Mountains and that on more than one occasion, and possibly on many occasions, it breached the cold-based ice sheet rim and flowed beneath the ice as a megaflood toward the northeast and the sea.

6. CONCLUSIONS

1. We describe a landscape in the Dry Valleys area consisting of channels, potholes, plunge pools, stripped bedrock, and scabland created by subglacial outburst floods on a scale similar to that of the massive outbursts of Lake Missoula in Washington, United States.

2. The features are dated to the mid-Miocene and formed when a thicker East Antarctic ice sheet overrode this part of the Transantarctic Mountains.

3. Ice sheet modeling suggests that meltwater from Lake Vostok or other interior basins would have bypassed the Dry

Valleys area and instead would flow out the troughs beneath the Byrd and David glaciers.

4. It is likely that the source of the massive subglacial lake is a trough excavated by local glaciers on the inner flank of the mountains and that periodically the lake breached the cold-based rim of the ice sheet and drained suddenly.

5. The efficacy of erosion by such outbursts is remarkable. The most concentrated effects were in the labyrinth on the bed of the former Wright outlet glacier. Such outbursts are likely to contribute to the erosion of present-day outlet glacier troughs.

Acknowledgments. This research was supported by the Division of Polar Programs of the U.S. National Science Foundation and by the U.K. Natural Environment Research Council. We are indebted to George Denton for many stimulating and creative discussions in the field and subsequently. We thank Jeff Kargel and an anonymous reviewer for helpful comments on the initial manuscript.

REFERENCES

Ackert, R. P., Jr. (1990), Surficial geology and geomorphology of the western Asgard Range, Antarctica; implications for late Tertiary glacial history, unpublished M.Sc. thesis, 147 pp., Univ. of Maine, Orono.

Allen, J. R. L. (1971), Transverse erosional marks of mud and rock: Their physical basis and geological significance, *Sediment. Geol.*, *5*, 167–385.

Baker, V. R. (1978), Paleohydraulics and hydrodynamics of Scabland floods, in *The Channelled Scabland*, edited by V. R. Baker and D. Nummedal, pp. 59–79, NASA, Washington, D. C.

Baker, V. R. (2009), The Channelled Scabland: A retrospective, *Annu. Rev. Earth Planet. Sci.*, *37*, 393–411.

Balco, G., J. O. Stone, N. A. Lifton, and T. J. Dunai (2008), A complete and easily accessible means of calculating surface exposure ages or erosion rates from 10Be and 26Al measurements, *Quat. Geochronol.*, *3*, 174–195.

Boulton, G. S. (1996), The origin of till sequences by sediment deformation beneath mid-latitude ice sheets, *Ann. Glaciol.*, *22*, 75–84.

Bretz, J. H. (1923), The Channelled Scabland of the Columbia Plateau, *J. Geol.*, *31*, 617–649.

Bretz, J. H. (1969), The Lake Missoula floods and the Channelled Scabland, *J. Geol.*, *77*, 505–543.

Briner, J. P., G. H. Miller, P. Davis, and R. C. Finkel (2006), Cosmogenic radionuclides from fjord landscapes support differential erosion by overriding ice sheets, *Geol. Soc. Am. Bull.*, *118*, 406–420.

Calkin, P. E. (1974), Subglacial geomorphology surrounding the ice-free valleys of southern Victoria Land, Antarctica, *J. Glaciol.*, *13*, 415–429.

Clarke, G. K. C., D. W. Leverington, J. T. Teller, and A. S. Dyke (2005), Fresh arguments against the Shaw megaflood hypothesis:

A reply to comments by David Sharpe on "Paleohydraulics of the last outburst flood from glacial Lake Agassiz and the 8200 B.P. cold event", *Quat. Sci. Rev.*, *24*, 1533–1541, doi:10.1016/j.quascirev.2004.12.003.

Chow, J. M., and P. J. Bart (2003), West Antarctic Ice Sheet grounding events on the Ross Sea outer continental shelf during the Middle Miocene, *Palaeogeogr. Palaeoclimatol. Palaeoecol.*, *198*, 169–186.

Denton, G. H., and D. E. Sugden (2005), Meltwater features that suggest Miocene ice-sheet overriding of the Transantarctic Mountains in Victoria Land, Antarctica, *Geogr. Ann., Ser. A*, *87*, 67–85.

Denton, G. H., M. L. Prentice, D. E. Kellogg, and T. B. Kellogg (1984), Late Tertiary history of the Antarctic Ice Sheet: Evidence from the Dry Valleys, *Geology*, *12*, 263–267.

Denton, G. H., D. E. Sugden, D. R. Marchant, B. L. Hall, and T. I. Wilch (1993), East Antarctic ice sheet sensitivity to Pliocene climatic change from a Dry Valleys perspective, *Geogr. Ann., Ser. A*, *75*, 155–204.

Domak, E., D. Amblas, R. Gilbert, S. Brachfeld, A. Camerlenghi, M. Rebesco, M. Canals, and R. Urgeles (2006), Subglacial morphology and glacial evolution of the Palmer deep outlet system, Antarctic Peninsula, *Geomorphology*, *75*, 125–142.

Evatt, G. W., A. C. Fowler, C. D. Clark, and N. Hulton (2006), Subglacial floods beneath ice sheets, *Philos. Trans. R. Soc. A*, *364*, 1769–1794.

Goodwin, I. D. (1988), The nature and origin of a jökulhlaup near Casey Station, Antarctica, *J. Glaciol.*, *34*, 95–101.

Graham, A. G. C., R. D. Larter, K. Gohl, C.-D. Hillenbrand, J. A. Smith, and G. Kuhn (2009), Bedform signature of a West Antarctic paleo-ice stream reveals a multi-temporal record of flow and substrate control, *Quat. Sci. Rev.*, *28*, 2774–2793.

Hancock, G. S., R. S. Anderson, and K. X. Whipple (1998), Beyond power: Bedrock river incision processes and form, in *Rivers and Rock, Geophys. Monogr. Ser.*, vol. 107, edited by K. J. Tinkler and E. E. Wohl, pp. 35–60, AGU, Washington, D. C.

Haywood, A. M., et al. (2008), Middle Miocene to Pliocene history of Antarctica and the Southern Ocean, *Dev. Earth Environ. Sci.*, *8*, 401–463, doi:10.1016/S1571-9197(08)00010-4.

Iken, A., H. Röthlisberger, A. Flotron, and W. Haeberli (1983), The uplift of Unteraargletscher at the beginning of the melt season — A consequence of water storage at the bed?, *J. Glaciol.*, *29*, 28–47.

Iverson, N. R. (1991), Potential effects of subglacial water-pressure fluctuations on quarrying, *J. Glaciol.*, *37*, 27–36.

Jamieson, S. S. R., and D. E. Sugden (2008), Landscape evolution of Antarctica, in *Antarctica: A Keystone in a Changing World — Proceedings of the 10th International Symposium on Antarctic Earth Sciences*, edited by A. K. Cooper et al., pp. 39–54, Natl. Acad. Press, Washington, D. C.

Jamieson, S. S. R., D. E. Sugden, and N. R. J. Hulton (2010), The evolution of the subglacial landscape of Antarctica, *Earth Planet. Sci. Lett.*, *203*, 1–27.

Le Brocq, A. M., A. Hubbard, M. J. Bentley, and J. L. Bamber (2008), Subglacial topography inferred from ice surface terrain analysis reveals a large un-surveyed basin below sea level in East Antarctica, *Geophys. Res. Lett.*, *35*, L16503, doi:10.1029/2008GL034728.

Lewis, A. R., D. R. Marchant, D. E. Kowalewski, S. L. Baldwin, and L. E. Webb (2006), The age and origin of the Labyrinth, western Dry Valleys, Antarctica: Evidence for extensive middle Miocene subglacial floods and water freshwater discharge into the Southern Ocean, Geology, 34, 513–516.

Lewis, A. R., D. R. Marchant, A. C. Ashworth, S. R. Hemming, and M. L. Machlus (2007), Major middle Miocene global climate change: Evidence from East Antarctica and the Transantarctic Mountains, Geol. Soc. Am. Bull., 119(11/12), 1449–1461.

Lewis, A. R., et al. (2008), Mid-Miocene cooling and the extinction of tundra in continental Antarctica, Proc. Natl. Acad. Sci. U. S. A., 105(31), 10,676–10,689, doi:10.1073/pnas.0802501105.

Lowe, A. L., and J. B. Anderson (2003), Evidence for abundant subglacial meltwater beneath the paleo-ice sheet in Pine Island Bay, Antarctica, J. Glaciol., 49, 125–138.

Lythe, M., D. G. Vaughan, and B. Consortium (2001), BEDMAP: A new ice thickness and subglacial topographic model of Antarctica, J. Geophys. Res., 106, 11,335–11,352.

Marchant, D. R., G. H. Denton, and C. C. Swisher, III (1993a), Miocene-Pliocene-Pleistocene glacial history of Arena Valley, Quartermain Mountains, Antarctica, Geogr. Ann., Ser. A, 75, 269–302.

Marchant, D. R., G. H. Denton, D. E. Sugden, and C. C. Swisher, III (1993b), Miocene glacial stratigraphy and glacial landscape evolution of the western Asgard Range, Geogr. Ann., Ser. A, 75, 303–330.

Margerison, H. R., W. M. Phillips, F. M. Stuart, and D. E. Sugden (2005), Cosmogenic ³He concentrations in ancient flood deposits from the Coombs Hills, northern Dry Valleys, East Antarctica: Interpreting exposure ages and erosion rates, Earth Planet. Sci. Lett., 230, 163–175.

Nienow, P. W., A. L. Hubbard, B. P. Hubbard, D. M. Chandler, D. W. F. Mair, M. J. Sharp, and I. C. Willis (2005), Hydrological controls on diurnal ice flow variability in valley glaciers, J. Geophys. Res., 110, F04002, doi:10.1029/2003JF000112.

Nye, J. F. (1976), Water flow in glaciers: Jökulhlaups, tunnels, and veins, J. Glaciol., 17, 181–207.

Ó Cofaigh, C., C. J. Pudsey, J. A. Dowdeswell, and P. Morris (2002), Evolution of subglacial bedforms along a paleo-ice stream, Antarctic Peninsula continental shelf, Geophys. Res. Lett., 29(8), 1199, doi:10.1029/2001GL014488.

O'Connor, J. E., and V. R. Baker (1992), Magnitudes and implications of peak discharge from Lake Missoula, Geol. Soc. Am. Bull., 104, 267–279.

Röthlisberger, H. (1972), Water pressure in intra- and subglacial channels, J. Glaciol., 11, 177–203.

Rutt, I. C., M. Hagdorn, N. R. J. Hulton, and A. J. Payne (2009), The Glimmer community ice sheet model, J. Geophys. Res., 114, F02004, doi:10.1029/2008JF001015.

Schäfer, J. M., S. Ivy-Ochs, R. Wieler, I. Leya, H. Baur, G. H. Denton, and C. Schlüchter (1999), Cosmogenic noble gas studies in the oldest landscape on Earth: Surface exposure ages of the Dry Valleys, Antarctica, Earth Planet. Sci. Lett., 167, 215–226.

Shreve, R. L. (1972), Movement of water in glaciers, J. Glaciol., 11, 205–214.

Siegert, M. J. (2000), Antarctic subglacial lakes, Earth Sci. Rev., 50, 29–50.

Siegert, M. J., R. Hindmarsh, H. Corr, A. Smith, J. Woodward, E. C. King, A. J. Payne, and I. Joughin (2004), Subglacial Lake Ellsworth: A candidate for in situ exploration in West Antarctica, Geophys. Res. Lett., 31, L23403, doi:10.1029/2004GL021477.

Smith, B. E., H. A. Fricker, I. R. Joughin, and S. Tulaczyk (2009), An inventory of active subglacial lakes in Antarctica detected by ICESat (2003–2008), J. Glaciol., 55, 573–595.

Sugden, D. E., and G. H. Denton (2004), Cenozoic landscape evolution of the Convoy Range to Mackay Glacier area, Transantarctic Mountains: Onshore to offshore synthesis, Geol. Soc. Am. Bull., 116, 840–857, doi:10.1130/B25356.1.

Sugden, D. E., and B. S. John (1971), Raised marine features and phases of glaciation in the South Shetland Islands, Br. Antarct. Surv. Bull., 24, 45–111.

Sugden, D. E., G. H. Denton, and D. R. Marchant (1991), Subglacial meltwater channel systems and ice sheet overriding, Asgard Range, Antarctica, Geogr. Ann., Ser. A, 73, 109–121.

Sugden, D. E., M. A. Summerfield, G. H. Denton, T. I. Wilch, W. C. McIntosh, D. R. Marchant, and R. H. Rutford (1999), Landscape development in the Royal Society Range, southern Victoria Land, Antarctica: Stability since the mid-Miocene, Geomorphology, 28, 181–200.

Sugden, D. E., G. Balco, S. G. Cowdery, J. O. Stone, and L. C. Sass, III (2005), Selective glacial erosion and weathering zones in the coastal mountains of Marie Byrd Land, Antarctica, Geomorphology, 67, 317–334, doi:10.1016/j.geomorph.2004.10.007.

Staiger, J. W., D. R. Marchant, J. M. Schaefer, P. Oberholzer, J. V. Johnson, A. R. Lewis, and K. M. Swanger (2006), Plio-Pleistocene history of Ferrar Glacier, Antarctica: Implications for climate and ice sheet stability, Earth Planet. Sci. Lett., 243, 489–503.

Summerfield, M. A., F. M. Stuart, H. A. P. Cockburn, D. E. Sugden, G. H. Denton, T. Dunai, and D. R. Marchant (1999), Long-term rates of denudation in the Dry Valleys region of the Transantarctic Mountains, southern Victoria Land based on in-situ produced cosmogenic ²¹Ne, Geomorphology, 27, 113–129.

Waitt, R. B. (1985), Case for periodic, colossal jökulhlaups from Pleistocene glacial Lake Missoula, Geol. Soc. Am. Bull., 96, 1271–1286.

Wingham, D. J., M. J. Siegert, A. P. Shepherd, and A. S. Muir (2006), Rapid discharge connects Antarctic subglacial lakes, Nature, 440, 1033–1036.

Weertman, J. (1957), On the sliding of glaciers, J. Glaciol., 3, 33–38.

S. S. R. Jamieson, Department of Geography, Durham University, Science Laboratories, South Road, Durham DH1 3LE, UK.

D. R. Marchant, Department of Earth Sciences, Boston University, Boston, MA 02215, USA. (marchant@bu.edu)

D. E. Sugden, School of GeoSciences, University of Edinburgh, Edinburgh EH8 9XP, UK.

Subglacial Environments and the Search for Life Beyond the Earth

Charles S. Cockell,[1] Elizabeth Bagshaw,[2] Matt Balme,[1] Peter Doran,[3] Christopher P. McKay,[4]
Katarina Miljkovic,[1] David Pearce,[5] Martin J. Siegert,[6] Martyn Tranter,[2] Mary Voytek,[7] and Jemma Wadham[2]

One of the most remarkable discoveries resulting from the robotic and remote sensing exploration of space is the inferred presence of bodies of liquid water under ice deposits on other planetary bodies: extraterrestrial subglacial environments. Most prominent among these are the ice-covered ocean of the Jovian moon, Europa, and the Saturnian moon, Enceladus. On Mars, although there is no current evidence for subglacial liquid water today, conditions may have been more favorable for liquid water during periods of higher obliquity. Data on these extraterrestrial environments show that while they share similarities with some subglacial environments on the Earth, they are very different in their combined physicochemical conditions. Extraterrestrial environments may provide three new types of subglacial settings for study: (1) uninhabitable environments that are more extreme and life-limiting than terrestrial subglacial environments, (2) environments that are habitable but are uninhabited, which can be compared to similar biotically influenced subglacial environments on the Earth, and (3) environments with examples of life, which will provide new opportunities to investigate the interactions between a biota and glacial environments.

1. INTRODUCTION

The robotic exploration of the solar system has revealed an increasing number of glacial environments. They include, for example, glacial deposits on Mars [*Head and Marchant*,

[1]Planetary and Space Sciences Research Institute, Open University, Milton Keynes, UK.
[2]School of Geographical Sciences, University of Bristol, Bristol, UK.
[3]Department of Earth and Environmental Sciences, University of Illinois at Chicago, Chicago, Illinois, USA.
[4]NASA Ames Research Center, Mountain View, California, USA.
[5]British Antarctic Survey, Cambridge, UK.
[6]School of GeoSciences, University of Edinburgh, Edinburgh, UK.
[7]NASA Headquarters, Washington, D. C., USA.

Antarctic Subglacial Aquatic Environments
Geophysical Monograph Series 192
Copyright 2011 by the American Geophysical Union
10.1029/2010GM000939

2003], which were originally thought to be restricted to the polar regions but have now been identified at lower latitudes; oceans under ice covers on moons orbiting Jupiter, including Europa, Ganymede, and Callisto [*Carr et al.*, 1998; *Baker et al.*, 2005]; and moons orbiting Saturn, including Enceladus [*Parkinson et al.*, 2008] and, potentially, Titan [*Grindrod et al.*, 2008].

Of the known extraterrestrial glacial environments, in one case, the subglacial environment has been sampled. The icy plumes produced in the southern polar regions of Enceladus were sampled in a flyby by the Cassini spacecraft in 2005 and found to contain water ice, CO, CO_2, N_2, CH_4, organics [*Waite et al.*, 2006, 2009; *Matson et al.*, 2007], and more recently, NH_3 has been interpreted [*Waite et al.*, 2009]. It is not known at what depth these plumes emanate.

At the time of writing, liquid water has only been indirectly measured in some of these subglacial environments. The most substantial liquid water body associated with an extraterrestrial subglacial environment is the ocean of Europa [*Carr et al.*, 1998], which, based on the volume of the moon, contains more water than the terrestrial oceans. Here we review the various subglacial environments (any environments under ice sheets or glaciers) that are likely to exist in the solar system.

Scientific interest in extraterrestrial subglacial environments stems from two motivations. The first is the interest in understanding the major physical and chemical processes that drive surface and subsurface evolution on other planetary bodies. The melting, refreezing, and movement of ices, which entrain volatiles and salts, are the result of, or result in, geologically active processes. The investigation of these environments provides insights into the early conditions in the solar system. The second motivation is the search for life. Liquid water is presumed to be a basic requirement for life (in addition to many other requirements, such as an energy source). The discovery of glacial environments on other planetary bodies that may harbor subglacial liquid water suggests promising possibilities for the search for extraterrestrial life. However, the habitability of these environments can only be properly assessed when the presence of liquid water has been confirmed and their physical and chemical conditions have been determined.

Extraterrestrial environments might yield three new types of subglacial environments for study:

1. These environments could be completely different from those known on the Earth and could be uninhabitable. Many extraterrestrial subglacial environments are likely to have very different physical and chemical conditions to subglacial environments on the Earth. Different redox environments, different absolute and fluctuating conditions of radiation (ionizing and UV radiation), temperature, and pH will yield different environmental conditions. These environments will expand the known physical and chemical parameter space of subglacial environments and improve knowledge of the known boundaries for habitability.

2. These environments could be habitable but uninhabited. A habitable subglacial environment in a planetary location where there is no life to take advantage of it (a plausible example could be a localized and transient impact melting of glacial ice on Mars) would allow biogeochemists to examine how geochemical cycles operate in glacial environments without a biota. This research might therefore provide "control" environments in which only geochemical cycles occur without the influence of a biota, which can be compared to terrestrial environments to understand better the role of a biota in shaping glacial environments on the Earth.

3. These environments could be inhabited. These environments would provide new data points to study the interaction of a biota and its subglacial environment and to investigate new examples of life.

These three possibilities are conceptually illustrated in Figure 1.

In this chapter, our objectives are to (1) briefly review what we know about subglacial environments on the Earth and their major physical, chemical, and biological characteristics, (2) review the current state of knowledge of some of the major extraterrestrial subglacial environments that have been examined to date, (3) discuss parallels between analog environments on the Earth and extraterrestrial subglacial environments, and (4) summarize current plans for the future exploration of extraterrestrial subglacial environments.

2. DISTRIBUTION AND BIOLOGICAL POTENTIAL OF SUBGLACIAL HABITATS ON THE EARTH

Ice covers between 11% and 18% of the Earth's surface during Quaternary glacial cycles and may have been even more widespread in ancient periods of the Earth's history such as the Neoproterozoic [*Schrag and Hoffman*, 2001]. The Antarctic Ice Sheet (>80% of world glacier area) and Greenland Ice Sheet (10% of world glacier area) currently dominate the distribution of subglacial environments on Earth.

In contrast to other parts of the biosphere, the composition and function of microbial communities in deep, cold environments is poorly understood, since they were once believed to be devoid of life, and direct access is hampered by the overlying ice cover. The temperature profile and substrate of the basal environment of glaciers and ice sheets has a major bearing on the rates and pathways of microbial activity. The most biologically active subglacial environments are those where liquid water is present. Here physical erosion of the bedrock may also promote the accumulation of reactive debris, which acts as a substrate for microbes, in addition to a source of energy, organic carbon, and nutrients [*Tranter et al.*, 2005; *Wadham et al.*, 2010]. Temperate valley glaciers have ice at the pressure melting point throughout and possess dynamic subglacial hydrological systems. Here significant concentrations (10^4–10^7 cells/mL) of active microorganisms have been reported [*Sharp et al.*, 1999; *Skidmore et al.*, 2000, 2005; *Botrell and Tranter*, 2002; *Foght et al.*, 2004]. Colder polar systems may present a more challenging environment for microbial communities, since a proportion of glacier bed ice is below the pressure melting point, and there are restrictions on the availability of liquid water. Small, thin polar glaciers are entirely composed of "cold" ice, and there is little or no free liquid water at the glacier bed. The larger "polythermal" systems display a layer of cold ice at the surface and around the margins [*Paterson*, 1994]. Here microbial activity prevails in the "warm" core of the glacier, where ice at the pressure melting point and liquid water is present [*Skidmore et al.*, 2000, 2005; *Wadham et al.*, 2004].

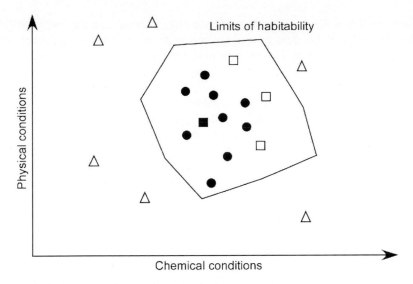

● Terrestrial subglacial environments (inhabited)
△ Extraterrestrial subglacial environments that are uninhabitable
■ Extraterrestrial subglacial environments that are inhabited
□ Extraterrestrial subglacial environments that are uninhabited, but habitable

Figure 1. Conceptual illustration of the contribution of the study of extraterrestrial environments to subglacial studies. Extraterrestrial subglacial environments might be so extreme as to be uninhabitable and lie outside the boundaries of habitable conditions, providing new insights into the physical and chemical conditions suitable for life. They might provide new examples of extreme but habitable subglacial environments. If they are inhabited, then they will provide new examples of subglacial life to study. If they are habitable but are uninhabited, then they might be used as "control" environments to study geochemical cycles in habitable environments without the influence of life.

In certain cases, the basal liquid water can be brine, supporting microbial communities, as at Taylor Glacier, Antarctica [*Mikucki and Priscu*, 2007; Mikucki et al., 2009]. Less is known about microbial communities in sub-ice sheet environments, despite their large areal extent on Earth. Satellite and geodetic data have begun to reveal the nature of sub-ice sheet hydrological environments, demonstrating the widespread presence of liquid water at the ice sheet bed. For example, there is rapid drainage of surface meltwater to the bed in Greenland [*Zwally et al.*, 2002; *Das et al.*, 2008], and the transfer of kilometer cubed volumes of meltwater between Antarctic subglacial lakes [*Wingham et al.*, 2006; *Fricker et al.*, 2007]. A mounting body of data demonstrates the existence of microbial communities beneath ice sheets [*Miteva and Brenchley*, 2005; *Lanoil et al.*, 2009], though further information regarding their composition, distribution, and function awaits direct access of subglacial aquatic environments (e.g., Antarctic subglacial lakes) [*Priscu et al.*, 2005; *Lake Ellsworth Consortium*, 2007] via deep drilling campaigns. A major control on the diversity of these microbiological communities and the biogeochemical processes they

are involved with is the prevalent redox environment at the bed.

Redox conditions at glacier beds, as in any aquatic environment at or near the Earth surface, are controlled by the rate of supply of oxidants versus the rate of oxidation [*Tranter*, 2004]. The types of oxidants that can be supplied at glacier beds include O_2, Fe(III), Mn(IV), NO_3^-, and SO_4^{2-}. The main electron donors are often sulfides and organic matter [*Wadham et al.*, 2008].

Oxidants are derived from any surface meltwaters, which reach the glacier bed, basal ice melt, and comminuted rock flour. Oxygenated water flows from the surface during periods of snow and ice melting through moulins, crevasses, and other englacial channels to the bed in many smaller ice masses and near the margins of most of the larger ice masses [*Tranter et al.*, 2005]. The oxygen content of the waters depends on the altitude from which the waters were sourced and any overpressuring via interactions with entrained air that may occur during descent to the bed. The type of drainage system receiving these waters at the bed is usually a channelized or low-pressure drainage system, which is characterized by relatively short residence times (hours to days) and low rock-water ratios (<1 g/L). Water flow rates are also relatively high (~1 m/s), and so much of the fine sediment to which the biomass is attached is suspended in the water column. Hence, waters flowing through channelized drainage systems are usually oxic because of the high oxygen supply relative to the potential for oxidation [*Tranter et al.*, 2002]. Lower oxygen levels may be found in the channel marginal zone that flanks the main channels. Here water floods the subglacial till during rising water levels and drains out during falling levels. Hence, the water residence time is higher, the rock-water ratio is higher. Reduced compounds such as organic matter, Fe (II), and Mn (II) on the surfaces of comminuted mineral grains and sulfide minerals, such as pyrite, become depleted over time by microbial activity, although surface organic matter may be washed in during rising water levels [*Tranter et al.*, 2005].

The channelized or low-pressure drainage system and its marginal zone (similar to the hyporeic zone in streams and rivers) are flanked by the much more pervasive distributed, or high-pressure, drainage system. Water flows approximately in the direction of ice flow in the low-pressure drainage system, whereas it flows across the direction of ice flow in a high-pressured drainage system. Waters flow quite slowly in the distributed drainage system (<0.1 m/s), and so rock-water contact times are higher. Glaciers and ice sheets often have areas of the bed that are draped with subglacial till, and where this is unfrozen, flow within the water-laden till is part of the distributed drainage system. Hence, rock-water ratios are also higher in the distributed system, and so too is the biomass as a consequence. Oxidation of organic matter and sulfide minerals with molecular oxygen is the most thermodynamically favorable redox reaction to occur in till-rich environments within the distributed drainage system. Should organic matter and sulfide minerals, which are reducing agents, be plentiful, microbially catalyzed reactions may deplete the oxygen along the water flow path to such an extent that all the waters become anoxic and other oxidizing agents, such as NO_3^-, Fe(III), Mn (IV), and SO_4^{2-}, are utilized to oxidize organic matter and sulfide minerals. Should sufficient reactive organic matter be present, for example, in the form of overridden paleosoils, then methanogenesis may occur [*Tranter et al.*, 2002, 2005; *Wadham et al.*, 2008].

Sulfides are common components of many rocks, and Fe(II) is a common component of many primarily silicate minerals. Hence, glacial comminution of bedrock produces a ready supply of reducing agents. Organic matter is also found in many rocks. Beneath the interior of ice sheets, where no surface meltwater reaches the bed, it is highly probable that the supply of O_2 (and other oxidizing agents such as NO_3^- and SO_4^{2-}) from the melting of basal ice, either as a consequence of geothermal heating, frictional/deformational heating or regelation, is less than the supply of reducing agents from subglacial comminution. Hence, it is likely that water flowing through subglacial till beneath thick ice becomes anoxic and that the biomass is dominated by communities capable of existing at low Eh conditions [*Wadham et al.*, 2008]. Types of anoxic environments might include the till beneath ice streams [*Tulaczyk et al.*, 2000] and in the areas between hydrologically connected subglacial lakes, which may transmit water periodically and at least partially freeze between connection events [*Wingham et al.*, 2006]. By contrast, the larger, hydrologically closed subglacial lakes, such as Vostok Subglacial Lake, are more likely to be oxic, since the input of comminuted glacial debris is more limited, and there is continual oxic recharge of the lake with meteoric ice melt [*Siegert et al.*, 2003], although biotic and abiotic sinks for O_2 do not rule out anoxia even in these systems. Hence, ice sheet beds are likely to display a wide spectrum of redox conditions, that are connected to the type of drainage system and the quantity and nature of the till present. Consequently, ice sheet beds are also likely to be colonized by a diverse spectrum of microorganisms [*Miteva and Brenchley*, 2005; *Skidmore et al.*, 2005; *Tung et al.*, 2005; *Christner et al.*, 2008; *Lanoil et al.*, 2009; *Mikucki et al.*, 2009; *Skidmore et al.*, 2010].

3. EXTRATERRESTRIAL SUBGLACIAL ENVIRONMENTS

3.1. Mars

3.1.1. Introduction. Mars is the fourth planet from the Sun with an equatorial radius of 3397 km. Mars is rocky with a thin (~6 mbar; about 1/200th that of the Earth) with an atmosphere comprised primarily of CO_2. The absence of a thick atmosphere means that Martian surface temperatures are highly variable: daytime temperatures can be higher than 20°C at the equator, but nighttime temperatures are tens of degrees Celsius below zero. A similar latitudinal control of temperature exists as the Earth, with polar regions being the coldest. At about 25.1°, Mars' obliquity is similar to that of the Earth, meaning that Mars also experiences seasonal climate variations.

In many ways, Mars is the most similar planet in the solar system to Earth, and decades of research have allowed a detailed picture of Mars' geological history to be constructed. A particular focus of Mars missions has been tracing the history of water, and Mars is thought to have once been "warmer and wetter" than today, as demonstrated by observations of ancient valley networks [e.g., *Fanale et al.*, 1992; *Mangold et al.*, 2004], outflow channels [*Baker*, 1982], possible deltas [*Pondrelli et al.*, 2008], and in situ identifications of minerals interpreted to have formed by groundwater processes [*Squyres et al.*, 2004]. Mars might also have once possessed a frozen ocean [*Parker et al.*, 1989; *Taylor Perron et al.*, 2007] that occupied its low-standing northern hemisphere, although debate over the existence of such an extensive, long-lived body of water is still ongoing [e.g., *Carr and Head*, 2003].

The upper few kilometers of the Martian crust contains large amounts of water-ice [*Squyres et al.*, 1992]. At high latitudes, the surface regolith can contain more than 50% ice [*Boynton et al.*, 2002; *Feldman et al.*, 2004] by volume and is covered by only a few centimeters of ice-rich dust [*Smith et al.*, 2009]. Models (summarized by *Squyres et al.* [1992]) suggest that such ice persists to depths of several kilometers, at which point the melting isotherm (the depth at which the geothermal temperatures are high enough to melt ice) causes the ice to become liquid water. Nearer the equator, the top of the ice table is driven deeper, but the base of the ice table is shallower, because of warmer year-round surface temperatures [*Fanale*, 1976].

The most recent high-resolution imaging data from Mars have shown that the action of liquid water on the surface has not been confined to the ancient past. Images of fluvial-like gullies [e.g., *Malin and Edgett*, 2000], geologically recent outflow channels [*Burr et al.*, 2002b], and low-latitude periglacial landforms [*Balme and Gallagher*, 2009; *Page*, 2007] have demonstrated the action of liquid water at the surface in the past few millions years. Further studies [e.g., *Costard et al.*, 2002] have linked gully formation with changes in the Martian climate, which are driven, in turn, by changes in Mars' obliquity [*Head et al.*, 2003a; *Kieffer and Zent*, 1992; *Schorghofer*, 2007]. The Martian obliquity varies periodically by more than 20° in an ~125,000-year cycle [*Laskar et al.*, 2004]. This suggests that recent cycles of deposition, removal, and even perhaps thaw of ice have controlled the Martian surface environment over the past few million years. Recently, a variety of hydrated minerals have been identified on Mars that provide further evidence for groundwater [*Gendrin et al.*, 2005; *Poulet et al.*, 2005].

3.1.2. Physicochemical conditions and the prospects for life. Mars hosts a range of terrains and environments that, in a similar way to the Earth, have evolved over geological time, often leaving only morphological or geological traces of their existence. Therefore, it is impossible to summarize all the possible habitats that might have come and gone over the planet's history. Chemically, the Martian crust is poorly documented, for a blanket of fine dust, dominated by silicates and iron, calcium, aluminum, and magnesium oxides, drapes much of the surface and makes remote sensing studies difficult.

Martian environments under ice covers, such as lake ice covers, have for a long time been recognized to be potential habitats for life, based on studies of analogous ice-covered habitats on Earth [*McKay et al.*, 1985]. One subsurface candidate for a stable subglacial environment is the location of the melting isotherm, several kilometers beneath the surface of the Martian cryosphere. This perhaps provides the environment most conducive for life, for water here could remain liquid for geologically significant time periods. Second, Mars' extensive, kilometer-thick perennial polar caps [*Phillips et al.*, 2008] are mainly water-ice, and it has been suggested that pressure-induced melting or geothermal action could lead to the formation of pockets of liquid at the base of the ice [*Clifford*, 1987], in a similar way to the preservation of subglacial lakes in the Antarctic. Indeed, it has also been suggested that Chasma Boreale, a large reentrant and valley system within the north polar cap, was formed by catastrophic flooding from just such a subglacial liquid reservoir [*Clifford*, 1987; *Fishbaugh and Head*, 2003]. Although simulations suggest that pressure melting is unlikely [*Greve et al.*, 2004], the presence of salts that depress the melting point of water (such as perchlorates, recently discovered at the Phoenix Landing site) [*Hecht et al.*, 2009] could allow melt to form. If this were the case, then the margins of the north polar cap would form an attractive target for study of past subglacial environments.

Radar studies of massive ice deposits on Mars have not revealed liquid water [e.g., *Holt et al.*, 2008] It is plausible that the Martian water table is hidden and that attenuating material within ice might hide aquifers beneath ice deposits [*Farrell et al.*, 2009]. However, at the time of writing, it seems that most Martian ice deposits are more similar to cold terrestrial polar glaciers with little, if any, water in the subglacial environment. However, thin layers of water at the base of glaciers would not be easily visible to radar analysis.

Near-surface candidate subglacial environments can be split into either (1) locations beneath extant surface ice or (2) regions in which ice was recently present at the surface but has since been removed. Although there is good evidence for extant surface ice (usually dust or debris covered) in the form of tropical mountain glaciers [*Head and Marchant*, 2003], ice-rich and glacier-like flows [*Lucchitta*, 1981; *Pierce and Crown*, 2003], midlatitude ice-rich dust mantles [*Mustard et al.*, 2001], and high-latitude patterned ground [e.g., *Mangold*, 2005], most authors have found no evidence for liquid water in these systems. For example, the Martian glaciers seen today are inferred to be cold-based with no basal melting [*Head and Marchant*, 2003; *Shean et al.*, 2005], similar to cold polar glaciers on the Earth (section 3).

The evidence for recent wet-based glaciers in Mars' past is also somewhat equivocal. Observations of possible eskers (ridge-like landforms caused by deposition of sediments in sub- or intraglacial fluvial channels) and associated landforms in highlands surrounding the Argyre and Hellas impact basins point to possible wet-based glaciation [e.g., *Banks et al.*, 2008; *Kargel and Strom*, 1991], although it is likely that these features are a billion years old or more in age.

Another example of a surface morphology that might indicate a recent subglacial environment comes from the recent description of an equatorial "frozen sea" in the Elysium Planitia region of Mars [*Murray et al.*, 2005]. These deposits occur at the end of what is probably the youngest, large-scale outflow channel on Mars (Athabasca Vallis), which might have been active only a few million years ago [*Burr et al.*, 2002b]. Although some authors have argued that the "sea" is now occupied by flood lavas, the pattern of interlinked basins and channels in which it sits strongly suggest that there was a water-filled basin here at some point [*Balme et al.*, 2010]. Like many other flood channels, Athabasca Vallis was carved by liquids emanating from a deep tectonic fracture [*Burr et al.*, 2002a; *Head et al.*, 2003b]. Given that the source of the floods is probably long-lived, subsurface aquifers, the slowly freezing water that occupied the basin at the termination of the channel could plausibly have provided a transient subglacial habitat for any organisms that were once present in the aquifer, deep below ground.

The only geologically recent environments in which ice and water are likely to coexist near the surface appear to be periglacial, rather than subglacial. Evidence for very recent thaw of ground ice on Mars is amassing [e.g., *Balme and Gallagher*, 2009; *Balme et al.*, 2009; *Costard et al.*, 2002; *Levy et al.*, 2009; *Page*, 2007; *Soare et al.*, 2008] and includes both the well-known gullies and also more controversial features such as thermokarst, pingos, and sorted patterned ground. Again, though, these environments probably only contain(ed) transient liquid water, and in many cases, the actual amount of water was likely to be very small. Periods of higher obliquity in the past may also have caused glacial melting [*Jakosky et al.*, 2003]. Although these environments may not be conducive to life today, they might be plausible sites to search for past life on Mars.

Currently, there are no extant subglacial Martian environments that can be easily observed. The astrobiologically most promising subglacial habitats are at the base of the polar ice caps and deep within the crust, at the base of the cryosphere. It is likely that deep drilling will be required to analyze either of these environments. Debris-rich basal ice in the marginal zones of the ice caps, especially the North Polar cap, are also promising from an astrobiological perspective and do not require deep drilling [*Skidmore et al.*, 2000].

Assuming that melting can, or could, occur in Martian subglacial environments, then they are likely to be anaerobic, consistent with Martian atmospheric composition. Therefore, the closest terrestrial analogs in terms of available redox couples are likely to be anaerobic zones in subglacial environments (section 3). A range of electron acceptors found in terrestrial subglacial environments (section 3) are available on Mars including Fe(III) and probably Mn(IV), both from comminuted oxidized basaltic rocks. Despite the probable lack of photosynthesis to create a ready supply of other oxidized elements for use as electron acceptors, there are plentiful supplies of sulfate in salts detected across the Martian surface [*Clark et al.*, 1982; *Rieder et al.*, 1997; *Gendrin et al.*, 2005; *Langevin et al.*, 2005; *Squyres et al.*, 2006]. Perchlorate, identified in the Martian soil by the Phoenix Lander [*Hecht et al.*, 2009], is also a microbial electron acceptor. Electron donors in subglacial environments could plausibly include Fe(II), again produced from comminuted Martian basalts, and possibly organic material delivered exogenously in meteorites and either directly delivered into the subsurface through glaciers or leached there by melting in the past.

Other elements required for life are likely to be present in Martian subglacial environments, including trace elements such as Zn, Cu, Ni, and other elements found in glacially comminuted basaltic rocks and phosphorus from apatite. More uncertain is the source of nitrogen to sustain a Martian subglacial biota. Without a biological nitrogen cycle, fixed nitrogen will be produced in low abundance on the present-day subsurface or surface. However, fixed nitrogen could have been produced by volcanic or impact processing in the early history of the planet [*Segura and Navarro-Gonzalez*, 2005; *Summers and Khare*, 2007; *Manning et al.*, 2009], with nitrate-containing minerals subsequently made available by glacial comminution.

The limited, or lack of, supraglacial melting today would limit the movement of water through glaciers to generate the flow paths of nutrients observed in present-day subglacial settings on the Earth (section 3), meaning that subglacial environments on present-day Mars are more likely to reach chemical equilibrium and have unfavorable conditions for the persistence or replenishment of redox couples for life.

These considerations show that Martian subglacial environments would be favorable places for life at any point in time at which melting could occur, driving fluid movement to supply nutrients and generate chemical disequilibria in analogy to present-day subglacial environments on the Earth. Although these conditions cannot be ruled out today, they are more likely to have occurred during geologically recent obliquity changes or in the more distant past history of the planet. Thus, the search for extant life in Martian subglacial environments is a valid objective, but subglacial environments are also particularly favorable locations to search for past life on Mars.

3.2. Europa

3.2.1. Introduction. At least three of the Jovian moons may harbor liquid water oceans (Callisto, Ganymede, and Europa) [*Baker et al.*, 2005]. The extent of any putative oceans or their state (frozen or liquid) in Ganymede or Callisto is not known [*Spohn and Schubert*, 2003]. Greatest attention has been given to Europa, which will be discussed in detail here. Europa is the sixth moon of Jupiter and has a radius of 1550 km. The moon was discovered in 1610 by Galileo Galilei. It is the smallest of the four Galilean moons. Europa has a similar bulk composition to the terrestrial planets. The surface has a high albedo caused by water ice and few surface impact craters, suggesting a young reworked surface, possibly of 20 to 180 million years old.

Europa has a variety of features, which suggest a relatively active geology. The moon's surface is crisscrossed with lines (lineae), which are dark streaks across its surface (Figure 2). The lines appear to be cracks in the ice on either side of which sheets of ice move relative to one another. Cross-sections of these features reveal a ridge-like morphology. Chaos terrain, differently interpreted as the result of diapirism or melting of the ice, is observed on the surface [*Riley et al.*, 2000; *Greeley et al.*, 2004].

Europa is thought to host a liquid water ocean beneath its icy crust, a supposition resulting from three observations: (1) the presence of an induced magnetic field as the moon passes through Jupiter's magnetic field, detected by the Galileo spacecraft provides evidence for a conducting medium [*Khurana et al.*, 1998; *Kivelson et al.*, 2000], (2) the active

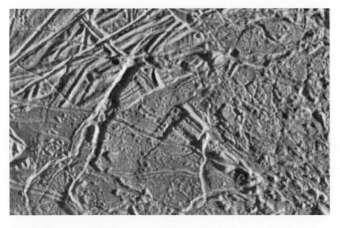

Figure 2. Conamara Chaos on Europa. Visible are cracks in the ice of the Europan ice shell. The image is about 100 km across.

geological nature of its surface, which suggests a mobile medium beneath the ice [*Carr et al.*, 1998], and (3) the asynchronous rotation of Europa, which suggests a decoupling of its silicate core and icy surface. The extent of the ocean remains debated. The ice layer above it may be kilometers to tens of kilometers thick and the ocean about 80–170 km deep [*Anderson et al.*, 1998; *Turtle and Pierazzo*, 2001; *Greeley et al.*, 2004]. Nevertheless, even with lower estimates, the ocean would still contain substantially more water than the Earth's oceans.

3.2.2. Physicochemical conditions and the prospects for life. The physical conditions at the surface of the Europan ice are better constrained than the ocean. It is comprised primarily of water ice and has temperatures of 86–132 K. Particle bombardment from Jupiter's magnetosphere delivers H, S, and O, and also drives complex chemistry resulting in the formation of compounds such as O_2, SO_2, and H_2O_2 [*Kargel*, 1998; *Carlson et al.*, 1999, 2002; *Hand et al.*, 2006], which have been observed. The radiolysis of water contributes to the formation of oxygen at the surface of Europa. The entrapment of oxygen within the surface ice and its exposure to radiation may also generate O_3, but the spectral feature of this compound has remained controversial [*Johnson et al.*, 2003] because it may be mixed with other absorption features such as those caused by –OH or organics [*Johnson et al.*, 2003].

Models suggest that sulfuric acid, hydrated salts, and other compounds should be present. Different lines of evidence support the presence of sulfates on Europa and in its oceans. The infrared signatures on Europa's surface can be explained with sulfates [*McCord et al.*, 2002], and sulfate might be produced radiolytically [*Johnson et al.*, 2003]. The sulfate concentrations of Europa's oceans were modeled by *McKinnon and Zolensky* [2003]. They derive an upper limit of 10%. Other chemical parameters of the Europan ocean are poorly constrained. For example, the pH of the ocean is unknown; it may have a low pH [*Kargel et al.*, 2000].

In addition to salts, simple organic molecules should also be produced on the surface, and the dark lineations on the surface of the ice may well contain more complex organic chemistry. Meteoritic and cometary infall to the surface would be expected to deliver organic molecules [*Pierazzo and Chyba*, 1999, 2002].

The presence of a subsurface ocean on Europa has made the moon the focus of many astrobiological investigations as a potential habitat for life [*Reynolds et al.*, 1983; *Jakosky and Shock*, 1998; *Gaidos et al.*, 1999; *Greenberg et al.*, 2000; *Chyba and Hand*, 2001; *Chyba and Phillips*, 2001; 2002; *Schulze-Makuch and Irwin*, 2002; *Pierazzo and Chyba*, 2002]. The various potential habitats that could exist in and under the Europan ice sheet have led to a proposed taphonomy of Europa, with suggestions on the best locations to search for preserved life [*Lipps and Rieboldt*, 2005].

Assessing the ocean as an abode for life depends critically on knowledge of the physicochemical properties and their many factors, important for biochemistry, which are still poorly constrained. A number of studies have attempted to assess potential sources of energy in a Europan ocean. Energy might come from the surface in the form of organics or oxidants produced during interactions of radiation with the surface [*Chyba*, 2000; *Chyba and Hand*, 2001; *Cooper et al.*, 2001, 2003]. These forms of energy would require an active connection between the surface and the Europan ocean.

In the ocean, phototrophy would be unlikely, since solar-derived light will be reduced to below the minimum required for photosynthesis in the first few meters of the Europan ice layer [*Cockell*, 2000], although phototrophy could plausibly occur using geothermal energy from hydrothermal vent-like environments [*Beatty et al.*, 2005].

Depending on the oxidation state of the ocean, life might be able to use a variety of other redox couples available in the Europan ocean. There are a diversity of plausible candidates including H_2 and CO_2 used in methanogenesis. H_2 would be derived from serpentinization of ultramafic rocks in the Europan silicate core and CO_2 derived from the primordial inventory. Fe^{3+} and H_2 could act as a redox couple for iron reduction, with Fe^{3+} derived from the silicate core [*Schulze-Makuch and Irwin*, 2002].

McKinnon and Zolensky [2003] point out the critical lack of information on sulfur evolution in Europa, which has implications for the assessment of its habitability. The oxidation state of the sulfur in the ocean and its concentration will have an important influence on the extent to which the ocean is in direct circulatory contact with a silicate core. For example, if sulfur is present as thick beds of sulfur at the bottom of the ocean, these beds would impede any possibility of life analogous to hydrothermal vents in the Earth's deep oceans.

Schultze-Makuch and Irwin [2002] considered speculative organisms other than those that use traditional redox couples as possible inhabitants of a Europan ocean. Their study was an investigation of unconventional energy acquisition pathways that organisms might evolve in an environment where chemical energy is limited. They considered organisms using thermal energy, kinetic energy, osmotic gradients, magnetic fields, and gravitational energy. Of these various potential energy sources, they concluded that kinetic energy and osmotic energy might be the most promising candidates [*Schultze-Makuch and Irwin*, 2002].

A critical parameter still not well constrained is the temperature within the ocean. *Marion et al.* [2003] investigated temperature and salinity in model Europan oceans and suggested that temperatures within the ocean might be too low for life (<253 K) and salinity high. Thermal diapirs within the ice crust might yield more favorable environments for life [*Ruiz et al.*, 2007].

All of these studies illustrate that at the current time, there is insufficient data on the ocean composition and that future missions will dramatically improve the basis with which to assess the habitability of Europa.

Accessing the subglacial environment on Europa to search for life will be hugely challenging on account of the need to penetrate the ice layer that may be many kilometers thick. However, if ocean-surface connection exists in lineae and chaotic terrains, then biosignatures might be sought on the surface of Europa [*Dalton et al.* 2003].

3.3. Enceladus

3.3.1. Introduction. Enceladus is a small Saturnian moon with a mean radius of 252 km. It was first observed in 1789 by William Herschel. The first investigations of the moon were carried out by the Voyager robotic craft, which determined it had a high albedo, probably caused by water ice, and an association with the Saturnian E ring. Voyager 2 also determined that the surface of the moon was comprised of terrains of different ages.

The Cassini spacecraft, which first visited the moon in 2005, revealed the presence of multiple gas plumes emanating from the south polar terrain of the moon, which coalesce into a giant plume over 80 km from the surface of the moon. It was this phenomenon that heightened astrobiological interest in the moon.

The region from which the plumes are ejected has characteristic "stripes" (named "tiger stripes," Figure 3), several hundred meters wide and hundreds of kilometers long. They are morphologically analogous to the ridges observed in the ice sheet of Europa. The plumes are associated with an anomalous source of heat suggesting temperatures near the plumes in the near surface of 190 K and ~6 GW of energy [*Spencer et al.*, 2006], implying even higher temperatures (250–273 K) [*Parkinson et al.*, 2008] in the deeper subsurface. The plume material is responsible for the formation of the Saturnian E ring.

3.3.2. Physicochemical conditions and the prospects for life. Active geological turnover within the plume-generating region could occur. If the tiger stripes are formed in a mechanism analogous to the spreading at terrestrial mid-ocean ridges, then a regional heat flux sufficient to generate the observed thermal anomaly could be created (~250 mW m^{-2}), with recycling of the crust on a 1 to 5 Ma time scale,

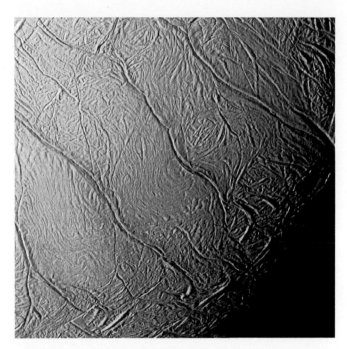

Figure 3. Tiger stripes in the southern polar region of Enceladus.

in which case, the heat source would occur from within the ice. An alternative hypothesis is that the heat is generated within a silicate core [*Castillo-Rogez et al.*, 2007], and it might be driven by tidal heating of the moon by Saturn. At the current time, the exact source of the heat and its geological history is not well understood. The extent of differentiation of the moon is also not fully known. The core might be ~150 km in diameter [*Parkinson et al.*, 2008].

The composition of the plumes provides tantalizing insights into the possible composition of Enceladus' interior (Figure 4). The plumes contain water ice and vapor (91%), which may come from a subsurface (subglacial) ocean, or it may be ice entrained from the surface of the moon [*Porco et al.*, 2006; *Kieffer et al.*, 2006].

Intriguing is the observation of N_2 (~4%) and CH_4 (~1.6%) in the plume. There are many plausible explanations for the presence of these gases [*Waite et al.*, 2006, 2009; *Matson et al.*, 2007; *Glein et al.*, 2008]. The gas production may be linked to ancient sources including primordial thermal degradation of ammonia into N_2 and organics into CH_4, which would be consistent with the presence of propane and acetylene in the plumes, although the latter could also be formed by photolysis of methane [*Parkinson et al.*, 2008]. Alternatively, these gases are primary in source but are trapped within the ice as clathrates and are steadily released in the plumes. Although there were originally no reports of ammonia in the present-day plumes (Figure 4), new studies of the data suggest it is present [*Waite et al.*, 2009].

From an astrobiological point of view, a significant observation is the lack of salts in the plumes, since that might imply a lack of available cations and anions required as nutrients in life. However, as *McKay et al.* [2008] point out, this may reflect preferential retention of salts in deeper water bodies with ices nearer the surface remaining "fresher," analogous to perennially ice-covered lakes in Antarctica, whose ice covers are low in salt but whose deeper layers contain salts. The same situation has also been proposed for Vostok Subglacial Lake [*McKay et al.*, 2003], where gas buildup in the lake may occur as a consequence of gassed water input to the lake via meteoric ice and the removal of water as pure ice into the ice base accretion [*Siegert et al.*, this volume].

Many of the discussions on habitability elaborated for Europa apply to Enceladus. At the time of writing, there are many key unknowns about the moon that preclude accurate modeling of the conditions for life on Enceladus. Key among these questions are the following: (1) Is there exchange between the surface of Enceladus and its subsurface? (2) Are there salts in the deeper layers of ice or in deep water bodies? (3) How can point 2 be reconciled with the presence of a silicate core, and is a core linked to any putative subsurface water body? (4) What are the sources and sinks of redox couples?

McKay et al. [2008] discuss a range of possible terrestrial analogous ecosystems for Enceladus. The production of methane from CO_2 and H_2 by methanogens is one plausible reaction scheme. The CO_2 would be derived from primordial CO_2 entrained within the ice (which is observed in the plumes) [*Matson et al.*, 2007] and the H_2 from serpentinization reactions occurring in Enceladus' core. Ultimately, the recycling of these redox couples would be achieved by thermal degradation of the methane produced.

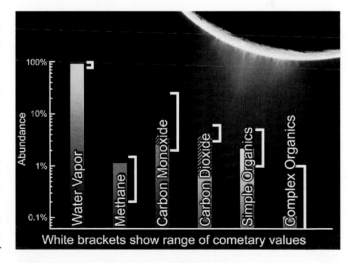

Figure 4. Composition of the plumes of Enceladus showing range of cometary values (N_2 is not included).

Indeed the discussions by *McKay et al.* [2008] underline, as with all subglacial environments, how critical the link is between subglacial liquid water and silicate rocks for life. Silicate cores open other possibilities for the generation of redox couples [*Parkinson et al.*, 2007], including the radiolytic production of H_2 from H_2O [*Lin et al.*, 2006], which can be used in methanogenesis as above or in sulfate reduction (Figure 5).

Aside from redox couples, silicate rocks provide a range of other elements (Mg, Ca, K, etc.) required by all organisms on the Earth and trace elements (Ni, Cu, Zn, Mo, etc.), which are used by organisms. It is beyond the scope of this paper to speculate on what "other" life would use among the range of available elements, but it can be assumed that at least some subset of them is required to do any meaningful complex biochemistry.

If the plumes on Enceladus are directly emanating from subsurface liquid water reservoirs, then the lack of observed salts suggests that at least the water observed is not directly in contact with the silicate core and is depauperate in biologically useful elements. However, organisms can use extremely low levels of elements. Further, many organisms can extract essential nutrients from silicate grains. Even if the silicate content of the plumes was less than 1% [*Parkinson et al.*, 2008], then they would provide a source of many elements. However, if the grains were covered in almost pure water ice, then silicate grains would be difficult to detect. A thorough investigation of habitability awaits a more complete and detailed inventory of elements within the plumes and their component ices/grains.

Unlike Europa, the supraglacial environment of Enceladus is not as heavily processed by radiation. However, *Parkinson et al.* [2008] propose a mechanism whereby water in the Saturnian E ring (itself from Enceladus) would be affected by UV radiation and ionizing radiation to generate oxidants such as H_2O_2, which would then be swept up by the moon on its surface. Whether these compounds would have any astrobiological significance would depend on their circulation into the interior of the moon.

One important factor that distinguishes the environment of Enceladus from terrestrial subglacial environments is the possible presence of ammonia in the ice at over 10% mass [*Squyres et al.*, 1983]. Ammonium is used as an effective biologically available source of N by terrestrial organisms, and in some subglacial environments, nitrate is ultimately produced by N fixation and nitrification in the supraglacial environment (section 3). However, the effects of long-term concentrations of high ammonia concentrations to biochemical systems are not well known.

The ejection of plumes into outer space from Enceladus offers remarkable possibilities for future astrobiology missions, since they provide an opportunity for technically "easy" sampling of an extraterrestrial subglacial environment with the minimum chance for forward contamination of the environment (see below). Already, the data we have on these plumes was achieved by a flyby with the Cassini spacecraft, which was not designed for this task, showing the readily achieved sampling of the plumes. Future missions dedicated to astrobiology might study the isotopic composition of the CH_4 in the plumes to attempt to determine a biotic or abiotic source [*McKay et al.*, 2008]. An investigation of higher carbon compounds would also yield information on pathways of organic complexification in the subsurface of Enceladus.

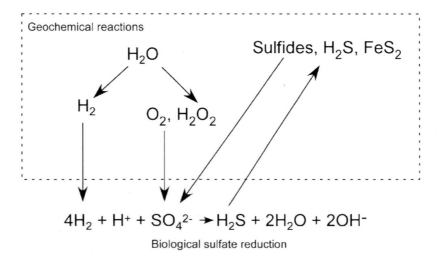

Figure 5. Hypothetical scheme for sulfate reduction in a subglacial environment using the radiolysis of water to generate hydrogen as an electron donor and oxidants to regenerate sulfate [after *McKay et al.*, 2008].

3.4. Titan

3.4.1. Introduction. Titan is the largest Saturnian moon with a radius of 2575 km. It was first observed in 1655 by Christian Huygens. Titan is composed primarily of rock and water ice and hosts lakes of liquid hydrocarbons, the only body other than the Earth known to possess stable liquid on its surface. The atmosphere of the moon is 98.4% nitrogen, the remainder is comprised of methane and trace gases including ethane, diacetylene, methylacetylene, acetylene, and propane [*Nieman et al.*, 2005]. The surface pressure is ~1.5 bar. The atmosphere is opaque to many wavelengths, and a complete reflectance spectrum of its surface cannot be obtained from present space measurements.

3.4.2. Physicochemical conditions and the prospects for life. *Fortes* [2000] presents a model for a Titan with a water-ammonia ocean early in its history. The formation of a thick N_2 atmosphere combined with loss of other atmospheric components would cause this ocean to roof over with frozen volatiles. The depth of any putative ocean is uncertain; estimates have ranged from an initial depth of 50 km to a present-day depth of 200 km [*Grasset and Sotin*, 1996]. The temperature of this putative ocean may be ~235 K, set by the cutectic temperature of the H_2O–NH_3 system. It is not clear that if an ocean was present, it would be in contact with a silicate core. Titan is thought to have such a core, but the ocean could be separated from it by a thick layer of H_2O–NH_3 ice.

Fortes [2000] examines a range of physical and chemical parameters that might be found in a Titan subglacial ocean. These considerations are similar to those that have more recently been applied to Europa and Enceladus. Nutrients are presumed to be made available from early chondritic input, including the input of major elements and organic carbon. The latter was suggested to provide a potential source of energy for heterotrophs.

Insofar as the reaction of acetylene, ethane, and organic solids with hydrogen is thermodynamically favorable, these substrates could also provide a redox couple for any proposed Titan biosphere [*McKay and Smith*, 2005]. The detection of life on Titan could focus on the use of carbon isotopic analysis of organic material or methane [e.g., *Fortes*, 2000] or the anomalous depletion of gases in the atmosphere or on the surface [*McKay and Smith*, 2005].

4. TERRESTRIAL ANALOGS: PARALLELS AND LIMITATIONS

Finding analogs on the Earth with which to assess the habitability of extraterrestrial subglacial environments is se-verely limited by a lack of comprehensive information on the physicochemical conditions in extraterrestrial subglacial environments. *Marion et al.* [2003] consider the case of Europa and split their discussion into three sections: the ice layer, the brine ocean, and the seafloor environment. They point out that the choice of analog environment depends upon the biological factor under consideration, suggesting that deep ice cores might be good analogs for understanding the preservation of biosignatures in Europan ice, but that ecosystem structure and constraints to life in ice are best investigated in the perennially ice-covered lakes of Antarctica. Assuming that salinity, acidity, and temperature will be three of the most important factors limiting life in aqueous environments, *Marion et al.* [2003] list a number of analogs for understanding different types of brine systems and their influence on a biota. Deep brine basins such as Orca Basin, the Gulf of Mexico, and Mediterranean deep brine basins are suggested as possible analog environments for investigating Europan ocean seafloor environments.

Analog environments can provide a set of observations concerning life that can be tested by exploring extraterrestrial subglacial environments, and they can also be used as technology test beds for developing methods to be used to explore extraterrestrial environments.

Greatest attention has been given to the most extreme terrestrial environments as analogs. Subglacial lakes have previously been investigated as analogs for extraterrestrial subglacial water bodies. For example, Vostok Subglacial Lake, located ~4 km deep in the East Antarctic ice sheet and isolated from the atmosphere for ~15 Ma (although ancient atmospheric gases are continuously being delivered to the lake through basal pressure melting) has previously been discussed as an analog for extraterrestrial subglacial environments [e.g., *Karl et al.*, 1999; *Lipps and Rieboldt*, 2005; *Siegert et al.*, this volume]. Deep ice core drilling has yielded organisms and environmental data from the accretion ice just above the lake. The accretion ice had two- to sevenfold higher bacterial numbers than the ice above it, suggesting that the lake is a source of bacteria. Members of the beta, gamma, and delta subdivisions of "Proteobacteria" were identified [*Christner et al.*, 2006]. In addition to a focus on biological data, the study of Vostok Subglacial Lake has focused on many of its physical characteristics, which influence the biota [*Wells and Wettlaufer*, 2008]. For example, the extent of mixing is important for determining the nutrients and redox couples available to life. The lake also receives mineral and biological input from the Antarctic ice. Models developed for understanding the mixing regimen in subglacial lakes will prove valuable for calculating the mixing regimens in extraterrestrial water bodies when sufficient data exists to constrain them [see *Siegert et al.*, this volume].

The drilling of a subglacial volcanic lake in Iceland (under the Vatnajokull ice cap) provided evidence of a community containing a diversity of chemolithotrophs using sulfide, sulfur, or hydrogen as electron donors or oxygen, sulfate, or CO_2 as electron acceptors [*Gaidos et al.*, 2004]. The lake water is fresh and slightly acidic, and the lake chemistry is dominated by glacial melt. Between 1922 and 1991, 78% of the lake water is estimated to have been supplied by basal melting of the ice sheet. Microbial numbers in the water column were 2×10^4 mL^{-1}. The microbial diversity was explained by the mixing of sulfidic and oxygenated water. Some of these redox couples could provide analogies to extraterrestrial redox couples. However, insofar as the oxygenated water is linked to the surface aerobic environment, the lake chemistry and its biota is strongly coupled to the terrestrial aerobic biosphere and is likely to limit the analog [*Gaidos et al.*, 1999].

It is also clear that the best analogs for extraterrestrial subglacial environments may not be subglacial environments on the Earth. *Prieto-Ballesteros et al.* [2003] describe Tirez Lake, a briny lake in La Mancha, Spain. The lake contains a $Mg-Na-SO_4-Cl$ brine. Although the lake contains a phototrophic population, which is not considered relevant to Europa, it also contains a population of sulfate reducers and methanogens, which the authors discuss as a potential analog for the anaerobic use of the sulfate or methanogenesis in a Europan ocean.

It is possible that no environment on the Earth can truly represent analogous conditions to any extraterrestrial subglacial environments (Table 1). Apart from putative ecosystems operating entirely independently from the surface photosynthetic biosphere, on the Earth, even anaerobic subglacial environments are linked to the availability of redox couples, e.g., sulfate and nitrate, generated, in large part, in the aerobic biosphere (section 3). Understanding fluxes of solutes into and out of extraterrestrial subglacial environments can only be properly achieved by directly taking measurements in situ; only then can their astrobiological potential be assessed.

A case in point is again the Europan ocean. *Gaidos et al.* [1999] discuss the biotic potential of the ocean. Many of the redox couples associated with terrestrial subglacial environments may be difficult or impossible to sustain. For example, more reducing conditions in the silicate core of Europa may mean that carbon is primarily outgassed as methane rather than carbon dioxide, preventing methanogenesis (although other substrates such as acetate and methanol can support methanogenesis. Furthermore, methane is itself a substrate for methanotrophy). Similarly, sulfur, produced as sulfides, might deny life a source of oxidized electron acceptors. If turnover with the icy surface of the moon, where radiation bombardment could produce electron acceptors such as peroxide, is insufficient, then the Europan ocean may provide an extremely energy-poor environment for life. The turnover within the ocean will also determine the extent to which any redox couples reach equilibrium or whether geochemical disequilibria can be maintained over geologically long time scales to provide energy and nutrients for life. It is not yet clear to what extent this is the case for many of the extraterrestrial subglacial environments discussed here.

5. RISK OF CONTAMINATION

Concern about the contamination of extraterrestrial bodies that might be capable of sustaining life has led to substantial considerations on measures to prevent forward contamination. Apart from the threat to an indigenous biota, it is conceivable that extraterrestrial subglacial environments might locally have habitable conditions for life, but these environments are not inhabited because life has not originated on that body, been transferred to it, or had the opportunity to move into highly localized conditions for habitability (e.g., transient liquid water). Contaminating these sites might prejudice an ability to study habitable but abiotic environments.

For Mars, detailed consideration has been given to the environmental parameters that would define regions as "special regions," i.e., locations in which conditions might be suitable for the replication of terrestrial organisms. To address what these parameters are, an extensive workshop was held to synthesize our existing knowledge of the limits of terrestrial life with a special focus on low temperature and water activity limits [*Kminek et al.*, 2010]. It was concluded that any region experiencing temperatures $>-25°C$ for a few hours a year and a water activity >0.5 can potentially allow the replication of terrestrial microorganisms. These constraints were based on the addition of a 5°C buffer to a range of data that suggest that metabolic activity, let along reproduction, does not occur below $\sim-20°C$ and the known limits of water activity in microorganisms, which generally cannot grow at water activities below ~0.6. Physical features on Mars that can be interpreted as meeting these conditions constitute a Mars Special Region. Based on current knowledge of the Martian environment and the conservative nature of planetary protection, these regions include gullies and bright streaks associated with them, pasted-on terrain, deep subsurface, dark streaks only on a case by case basis, and others to be determined. The deep subsurface could include subglacial environments.

Insofar as this study considered available knowledge on the low temperature and water activity limits for life, then its conclusions apply to other extraterrestrial subglacial environments, accepting, of course, that life needs much more than

Table 1. Some Differences and Similarities Between Terrestrial and Extraterrestrial Subglacial Environments[a]

Factor	Comments	Consequences for Life	Example Subglacial Environment	Reference
		Differences		
Ammonia	High ammonia concentrations in some extraterrestrial subglacial environments	High concentrations of ammonia potentially toxic	Enceladus and Titan	*Fortes* [2000], *Mitri et al.* [2008], and *McKay et al.* [2008]
Highly reducing conditions	Lack of source of oxidants	Shortage of electron acceptors	Europa	*Gaidos et al.* [1999] and *Schulze-Makuch and Irwin* [2002]
Lack of organics	No biological input of exogenous organics	Heterotrophic modes of production limited	Potentially many extraterrestrial subglacial environments	*Gaidos et al.* [1999]
Thick ice covers	Reduced geochemical exchange with supraglacial environment and reduced light penetration to liquid water below	Limitation in supply of useful compounds, e.g., meteoritic organics and radiation-formed oxidants; prevention of phototrophy	Enceladus and Europa	*Anderson et al.* [1998] and *Turtle and Pierazzo* [2001]
Low temperatures	Temperatures may be well below limit of metabolic activity in organisms	Low-temperature limitation to life	All extraterrestrial subglacial environments	*Marion et al.* [2003]
Salt concentrations	Briny conditions may create water activity below minimum in which metabolic activity can occur	Water activity limit to life	Europa	*Marion et al.* [2003]
		Similarities		
Connection to silicate surfaces	Silicate minerals weather to produce redox couples and nutrients/trace elements	Large range of elements/mineral available to drive redox reactions and biochemistry	Mars, Europa, and Enceladus	*Parkinson et al.* [2008]
Geological activity	Turnover in environments with subglacial oceans in contact with a silicate core	Prevents system running to equilibrium with respect to redox couples	Europa and Enceladus	*Parkinson et al.* [2008]
Organic input	Meteoritic input in extraterrestrial environments	Source of basic organic compounds and source of electron donors	Potentially all extraterrestrial subglacial environments	*Pierazzo and Chyba* [1999, 2002]
Liquid water	Indirect evidence for presence of liquid water in many extraterrestrial subglacial environments	Required for biochemical reactions	Potentially all extraterrestrial subglacial environments; Mars at higher obliquity than today	*Carr et al.* [1998], *Parkinson et al.* [2008], and *McKay et al.* [2008]

[a]For Encaladus, Titan, and Europa, the table refers to the ocean environment.

just high water activity and liquid water above $-25°C$ to grow. In the absence of detailed information on the chemistry of the Europan ocean or the source of the plumes of Enceladus, then initial missions to characterize these moons that involve direct contact with the surfaces will require stringent planetary protection protocols to prevent forward contamination.

Perhaps one of the most useful contributions of subglacial exploration on the Earth is the development of technologies that will improve the ability to access and explore extraterrestrial subglacial environments. *Doran and Vincent* [this volume], discuss the growing procedures and guidelines developed for responsible stewardship of subglacial environments on Earth.

E. Bagshaw, M. Tranter, and J. Wadham, School of Geographical Sciences, University of Bristol, Bristol BS8 1SS, UK.

M. Balme, C. S. Cockell, and K. Miljkovic, Planetary and Space Sciences Research Institute, Open University, Milton Keynes MK7 6AA, UK. (C.S.Cockell@open.ac.uk)

P. Doran, Department of Earth and Environmental Sciences, University of Illinois at Chicago, Chicago, IL 60607, USA.

C. P. McKay, NASA Ames Research Center, Mountain View, CA 94035, USA.

D. Pearce, British Antarctic Survey, Madingley Road, Cambridge CB3 0ET, UK.

M. J. Siegert, School of GeoSciences, University of Edinburgh, Edinburgh EH9 3JW, Edinburgh, UK.

M. Voytek, NASA Headquarters, Washington, DC 20546, USA.

Environmental Protection and Stewardship of Subglacial Aquatic Environments

Peter T. Doran

Earth and Environmental Sciences, University of Illinois at Chicago, Chicago, Illinois, USA

Warwick F. Vincent

Centre for Northern Studies and Department of Biology, Laval University, Québec, Québec, Canada

Environmental stewardship is a guiding principle of the Antarctic Treaty System. Efforts began in the 1990s to generate specific guidelines for stewardship of many terrestrial environments, including surface lakes and rivers. The relatively recent documentation of widespread subglacial aquatic environments, and planning for acquiring samples from them, has generated a need for stewardship guidelines for these environments. In response to a request from the U.S. National Science Foundation, the National Research Council of the National Academies of Sciences (NAS) created the Committee on the Principles of Environmental and Scientific Stewardship for the Exploration and Study of Subglacial Environments. The committee made 13 recommendations and a decision tree as a framework and flow chart for environmental management decisions. The committee report was also largely the basis of a Code of Conduct (CoC) for the exploration of subglacial environments formulated by a Scientific Committee on Antarctic Research Action Group. Both the NAS report and CoC have been used as guidance, to varying degrees, by subglacial research currently in progress.

1. INTRODUCTION

Antarctic subglacial aquatic environments (SAEs) have been documented for some time using remote sensing, geophysical techniques, but only very recently have there been plans devised and implemented to sample and study these environments directly. The long lead up to the sampling of these lakes is largely related to the logistical difficulty of penetrating their thick ice caps, but also due to the cautious approach warranted by the pristine nature of the environments and their almost completely unknown capacity to sustain viable ecosystems.

Environmental stewardship is one of the central principles of the Antarctic Treaty System. Although the initial treaty did not deal with environmental issues directly, the "Protocol on" Environmental Protection to the Antarctic Treaty (also known as the Madrid Protocol) adopted in 1991, extends wide environmental protection over the continent. The protocol codified all the many recommendations on environmental management and proposed specific approaches for various tools like environmental impact assessments. The protocol also provides for the establishment of a Committee for Environmental Protection (CEP) to advise the Antarctic Treaty Consultative Meeting (ATCM), requires the development of contingency plans to respond to environmental emergencies and provides for the elaboration of rules relating to liability for environmental damage (details at the Antarctic Treaty System website: http://www.ats.aq/e/ep.htm).

In this chapter we first review the leading initiatives to protect surface aquatic environments in continental and

Antarctic Subglacial Aquatic Environments
Geophysical Monograph Series 192
Copyright 2011 by the American Geophysical Union
10.1029/2010GM000947

3.7 Annex V of the Protocol allows areas to be designated as Antarctic Specially Protected Areas (ASPA), either to manage areas for research purposes or to conserve them as pristine exemplars for future generations. Subglacial lakes used as research sites should therefore be demarcated ASPAs to protect their long term scientific value, to regulate activities at these sites, and to formalize the requirements for full documentation and information exchange. In this way, each lake researched will have a known history of usage that later researchers can take into account. Once more direct information is available about the characteristics of subglacial lakes, attention should also be given to selecting and designating exemplar subglacial aquatic environments as ASPAs for long term conservation, in accordance with Article 3 of Annex V of the Protocol.

4. Drilling and SAE-entry

4.1 Unless there is site-specific evidence to the contrary, drilling to the base of Antarctic ice sheets should assume that the basal ice is underlain by liquid water, and that this water forms part of a subglacial drainage network requiring a high level of environmental protection. In general, downstream sites, particularly those closest to the sea, can be viewed to have lower environmental risk than upstream sites.

4.2 Exploration protocols should also assume that the subglacial aquatic environments contain living organisms, and precautions should be adopted to prevent any permanent alteration of the biology (including introduction of alien species) or habitat properties of these environments.

4.3 Drilling fluids and equipment that will enter the subglacial aquatic environment should be cleaned to the extent practicable, and records should be maintained of sterility tests (e.g., bacterial counts by fluorescence microscopy at the drilling site). As a provisional guideline for general cleanliness, these objects should not contain more microbes than are present in an equivalent volume of the ice that is being drilled through to reach the subglacial environment. This standard should be re-evaluated when new data on subglacial aquatic microbial populations become available.

4.4 The concentrations of chemical contaminants introduced by drill fluids and sampling equipment should be documented, and clean drilling technologies (e.g., hot-water) should be used to the full extent practicable.

4.5 The total amount of any contaminant added to these aquatic environments should not be expected to change the measurable chemical properties of the environment.

4.6 Water pressures and partial pressures of gases in lakes should be estimated prior to drilling in order to avoid downflow contamination or destabilisation of gas hydrates respectively. Preparatory steps should also be taken for potential blow-out situations.

5. Sampling and instrument deployment

5.1 Sampling plans and protocols should be optimized to ensure that one type of investigation does not accidentally impact other investigations adversely, that sampling regimes plan for the maximum interdisciplinary use of samples, and that all information is shared to promote greater understanding.

5.2 Protocols should be designed to minimize disrupting the chemical and physical structure and properties of subglacial aquatic environments during the exploration and sampling of water and sediments.

5.3 Sampling systems and other instruments lowered into subglacial aquatic environments should be meticulously cleaned to ensure minimal chemical and microbiological contamination, following recommendations under point 4.3.

5.4 Certain objects and materials may need to be placed into subglacial aquatic environments for monitoring purposes. This may be to measure the long term impacts of human activities on the subglacial environment and would be defined in the project's environmental impact assessment, or it may be for scientific purposes; e.g., long term monitoring of geophysical or biogeochemical processes. These additions should follow the microbiological constraints in 4.3, and for scientific uses should include an analysis of environmental risks (e.g., likelihood and implications of lack of retrieval) versus scientific benefits in the environmental assessment documents.

Members of the SCAR Action Group:

Irina Alekhina (Russia)

Peter Doran (USA)

Takeshi Naganuma (Japan)

Guido di Prisco (Italy)

Bryan Storey (New Zealand)

Warwick Vincent (Canada), chair

Jemma Wadham (United Kingdom)

David Walton (United Kingdom)

5. ONGOING DEVELOPMENTS

At the time of writing this chapter, there are three projects underway with the goal of exploring SAEs: (1) the Russians have plans to penetrate into Lake Vostok; (2) a U.K.-led international consortium is planning to enter into subglacial Lake Ellesworth; and (3) the United States has initiated efforts to penetrate and sample beneath the Whillans Ice Stream in West Antarctica. The Russian effort in Lake Vostok went through the CEE process prior to the NAS report and CoC. The UK group at Lake Ellesworth is preparing a CEE and has been making extensive use of both the NRC report and the CoC in that process. They are currently evaluating the levels of cleanliness they will be targeting (R. Clarke, personal communication, 2009). The U.S. group (Whillans Ice Stream Subglacial Access Research Drilling) has also found the NRC report and CoC to be a valuable resource. The United States is preparing an IEE to cover the project in its various phases, rather than deciding ahead of time that the impact will be more than minor or transitory. If they determine that there will be significant impacts, they will work on developing mitigating measures (i.e., engineering controls, management strategies, changing the scope of the operations, etc.) to offset the risks. If the risk(s) cannot be mitigated, then the United States will proceed with a CEE (P. Penhale, personal communication, 2009).

6. CONCLUSIONS

The Protocol on Environmental Protection to the Antarctic Treaty sets general principles and requirements that provide an overarching framework for the stewardship and protection of Antarctic SAEs. Antarctica contains many unusual surface water features, and at several continental and maritime sites, these more readily accessed aquatic environments have been accorded more specific protection through instruments associated with the Protocol. These include detailed management plans, designation of specially protected areas, and the implementation of environmental codes of conduct. There has been little attempt, however, to standardize these protocols across sites, and the sharing of best environmental codes and practices deserves further attention, for example, by the Committee of Managers of National Antarctic Programs.

Building on the NSF-led stewardship of surface aquatic environments and the work of the SCAR SALE group, the NAS Committee report provided steps toward the integrated management of SAEs throughout Antarctica, including specific recommendations and a flow chart for environmental decision making. This, in turn, laid the foundation for a CoC formulated by a SCAR Action Group and now submitted via SCAR to the Antarctic Treaty System. In all of these ongoing efforts, the quality and extent of stewardship depends on input from many disciplines and continued exchange and collaboration among nations.

Exploration of SAEs is still in its infancy, and many fundamental questions remain to be answered about these unique environments. Direct sampling has yet to occur and will need to take place before we get answers to these questions and resolve the ongoing debate about the existence and nature of life in these extreme environments. The potential for there to be a unique flora and fauna in pristine SAEs dictates that extreme caution must be taken in logistics and science planning in order to follow proper guidelines for environmental stewardship. The NRC report and subsequent CoC reviewed here are first and necessary steps in laying the groundwork for proper, long-term environmental management of SAEs.

Acknowledgments. We thank the other members of the NAS Subglacial Aquatic Environments Report: John Hobbie (Chair), Amy Baker, Garry Clarke, David Karl, Barbara Methé, Heinz Miller, Samual Mukasa, Margaret Race, David Walton, James White, as well as the staff of NAS for their expert assistance and guidance throughout the report preparation, review, and publication. We also thank the members of the SCAR Working Group (Irina Alekhina, Peter Doran, Takeshi Naganuma, Guido di Prisco, Bryan Storey, Warwick Vincent (Chair), Jemma Wadham, and David Walton) and also Mahlon Kennicutt, President of SCAR, and Colin Summerhayes, Executive Director of SCAR, for their encouragement and advice.

REFERENCES

Burgess, J. S., and E. Kaup (1997), Some aspects of human impacts on lakes in the Larsemann Hills, Princess Elizabeth Land, eastern Antarctica, in *Ecosystem Processes in Antarctic Ice-Free Landscapes*, edited by W. B. Lyons, C. Howard-Williams, and I. Hawes, pp. 250–264, A. A. Balkema, Rotterdam, Netherlands.

Fricker, H. A., T. Scambos, R. Bindschadler, and L. Padman (2007), An active subglacial water system in West Antarctica mapped from space, *Science*, *315*, 1544–1548.

Gray, L., I. Joughin, S. Tulaczyk, V. B. Spikes, R. Bindschadler, and K. Jezek (2005), Evidence for subglacial water transport in the West Antarctic ice sheet through three-dimensional satellite radar interferometry, *Geophys. Res. Lett.*, *32*, L03501, doi:10.1029/2004GL021387.

Harris, C. M. (1998), Science and environmental management in the McMurdo Dry Valleys, southern Victoria Land, Antarctica, in *Ecosystem Dynamics in a Polar Desert The McMurdo Dry Valleys, Antarctica, Antarct. Res. Ser.*, vol. 72, edited by J. C. Priscu, pp. 337–350, AGU, Washington, D. C.

National Research Council (2007), *Exploration of Antarctic Subglacial Aquatic Environments: Environmental and Scientific Stewardship*, 152 pp., NRC Press, Washington, D. C. (Available at http://www.nap.edu/catalog/11886.html)

Parker, B. C. (Ed.) (1972), *Proceedings of the Colloquium on Conservation Problems in Antarctica*, 356 pp., Allen Press, Lawrence, Kansas.

Priscu, J. C. (Ed.) (1998), *Ecosystem Dynamics in a Polar Desert: The McMurdo Dry Valleys, Antarctica, Antarct. Res. Ser.*, vol. 72, 369 pp., AGU, Washington, D. C.

Toro, M., et al. (2007), Limnological characteristics of the freshwater ecosystems of Byers Peninsula, Livingston Island, in maritime Antarctica, *Polar Biol.*, *30*, 635–649.

Vincent, W. F. (1996), Environmental management of a cold desert ecosystem: The McMurdo Dry Valleys, Antarctica, special publication, 55 pp., Desert Res. Inst., Univ. of Nev., Reno.

Wharton, R. A., and P. T. Doran (1999), McMurdo dry valley lakes: Impacts of research activities, special publication, 54 pp., Desert Res. Inst., Univ. of Nev., Reno.

Wingham, D. J., M. J. Siegert, A. Sheherd, and A. S. Muir (2006), Rapid discharge connects Antarctic subglacial lakes, *Nature*, *440*, 1033–1036.

P. T. Doran, Earth and Environmental Sciences, University of Illinois at Chicago, 845 W. Taylor Street, Chicago, IL 60607, USA. (pdoran@uic.edu)

W. F. Vincent, Centre for Northern Studies (CEN) and Department of Biology, Laval University, Québec, QC G1V 0A6, Canada. (warwick.vincent@bio.ulaval.ca)

Probe Technology for the Direct Measurement and Sampling of Ellsworth Subglacial Lake

Matthew C. Mowlem,[1] Maria-Nefeli Tsaloglou,[1] Edward M. Waugh,[1] Cedric F. A. Floquet,[1] Kevin Saw,[1] Lee Fowler,[1] Robin Brown,[1] David Pearce,[2] James B. Wyatt,[1] Alexander D. Beaton,[3] Mario P. Brito,[1] Dominic A. Hodgson,[2] Gwyn Griffiths,[1] M. Bentley,[4] D. Blake,[2] L. Capper,[2] R. Clarke,[2] C. Cockell,[5] H. Corr,[2] W. Harris,[3] C. Hill,[2] R. Hindmarsh,[2] E. King,[2] H. Lamb,[6] B. Maher,[7] K. Makinson,[2] J. Parnell,[8] J. Priscu,[9] A. Rivera,[10] N. Ross,[11] M. J. Siegert,[11] A. Smith,[2] A. Tait,[2] M. Tranter,[3] J. Wadham,[3] B. Whalley,[12] and J. Woodward[13]

The direct measurement and sampling of Ellsworth Subglacial Lake is a multidisciplinary investigation of life in extreme environments and West Antarctic ice sheet history. The project's aims are (1) to determine whether, and in what form, microbial life exists in Antarctic subglacial lakes and (2) to reveal the post-Pliocene history of the West Antarctic Ice Sheet. A U.K. consortium has planned an extensive logistics and equipment development program that will deliver the necessary resources. This will include hot water drill technology for lake access through approximately 3.2 km of ice, a probe to make measurements with sensors and to collect water and sediment samples, and a percussion corer to acquire an ~3–4 m sediment core. This chapter details the requirements and early stages of design and development of the probe system. This includes the instrumentation package, water samplers, and a mini gravity corer mounted on the front of the probe. Initial design concepts for supporting equipment required at the drill site to deploy and operate the probe are also described. A review of the literature describing relevant technology is presented. The project will implement environmental protection in line with principles set out by the Scientific Committee on Antarctic Research.

[1]National Oceanography Centre, Southampton, UK.

[2]British Antarctic Survey, Cambridge, UK.

[3]School of Geographical Sciences, University of Bristol, Bristol, UK.

[4]Department of Geography, University of Durham, Durham, UK.

[5]Planetary and Space Sciences Research Institute, Open University, Milton Keynes, UK.

[6]Institute of Geography and Earth Sciences, Aberystwyth University, Aberystwyth, UK.

[7]Lancaster Environment Centre, Lancaster University, Lancaster, UK.

[8]School of Geosciences, University of Aberdeen, Aberdeen, UK.

[9]Department of Land Resources and Environmental Sciences, Montana State University, Bozeman, Montana, USA.

[10]Centro de Estudios Científicos, Valdivia, Chile.

[11]School of GeoSciences, University of Edinburgh, Edinburgh, UK.

[12]School of Geography, Queen's University Belfast, Belfast, UK.

[13]School of Built and Natural Environment, Northumbria University, Newcastle upon Tyne, UK.

Antarctic Subglacial Aquatic Environments
Geophysical Monograph Series 192
Copyright 2011 by the American Geophysical Union
10.1029/2010GM001013

This includes application of microbiological control and best practice in protection of pristine environments within a pragmatic and realizable framework. Appropriate environmental protection standards, methods, verification protocols, and technology are being developed by the Lake Ellsworth Consortium. A review of best practice, initial plans, and results is presented.

1. INTRODUCTION

The scientific aims of the direct access and measurement of Ellsworth Subglacial Lake (ESL) are (1) to determine whether, and in what form, microbial life exists in Antarctic subglacial lakes, and (2) to reveal the post-Pliocene history of the West Antarctic Ice Sheet. The project also aims to determine the organic geochemical, hydrochemical, physical, and biological characteristics of the lake. These aims necessitate the collection of water and sediment samples from the lake and favor the use of in situ sensor technology wherever feasible. The project will meet these requirements by deploying a probe consisting of a vehicle, imaging and sonar systems, in situ sensors, water samplers, and a sediment sampler into ESL. The probe (see Figure 1) is heavily negatively buoyant, is tethered to the surface and has only simple maneuverability (depth control via tether and limited rotation). It is ~3.5 m in length, 20 cm in diameter, and consists of two pressure cases separated by water samplers. The lower case houses the instrument package and is tipped with a short sediment sampler (increasing total length to ~4 m).

Microbial control measures will be used throughout to prevent confounding of microbial analysis and to preserve this unique environment. To minimize risk, the probe is being developed using a design approach that specifies and works toward a reliability target.

This chapter begins with a summary of the performance requirements and characteristics of the ESL probe technology as demanded by the aims of the project. We then review the relevant literature describing vehicles, imaging and sonar systems, in situ sensors, sampler systems, microbial control and targeted reliability. In each case, the implications for the ESL probe technology are discussed. The initial designs for the probe and supporting equipment are then detailed.

2. PERFORMANCE REQUIREMENTS FOR ESL PROBE TECHNOLOGIES

The requirements for the ESL probe technologies are based on the most up to date understanding of ESL from geophysical experiments and research on other subglacial habitats. The design is also constrained by the requirement to minimize the environmental impact of lake access, the use of hot water drilling (HWD) to access the lake, the timetable of the program, and nature of the logistics used.

2.1. Microbial Control

The primary requirement for the ESL probe technologies is to ensure environmental protection, particularly by minimizing transfer of exogenous microorganisms to the lake. In addition, it is imperative to avoid contamination of the samples taken from the lake, as this may confound their analysis. The design of the probe and support systems is required to simplify, as far as possible, the process of achieving microbial control and to be resilient to these processes. A review of published best practice and our proposed protection methods are described below.

2.2. Physical Characteristics

The ESL HWD will result in a 360 mm borehole that refreezes at approximately 6 mm h^{-1} [*Siegert et al.*, 2006] resulting in useable diameter of ~200 mm after only a day from first access. Before the hole reduces to this size, both the probe and a sediment corer [see *Bentley et al.*, this volume] must be deployed and retrieved. There will only be sufficient fuel to ream the hole once or twice (dependent on timetable), and once drilling has commenced, there can be no delay. The probe is therefore required to be reliable and to perform on time. To maximize the duration of possible borehole traverse, the probe is required to be <200 mm in diameter. To reduce the time required to complete its deployment, it must be able to descend rapidly (>1 m s^{-1}).

The ESL probe and support equipment will be shipped from the United Kingdom to Chile from August to November 2012 in containers dimensionally identical to standard 20' ISO shipping containers. From Chile, they will be flown to Antarctica by aircraft, which can transport a maximum weight of 17 tons but no more than 10 tons per shipping container. Once in Antarctica, a 2 day overland journey is required to reach the drill site. The probe and all support equipment is therefore required to fit into standard shipping containers (i.e., must be <5.8 m long) and to survive the environmental conditions of this logistics chain.

2.3. Resilience to Harsh Environments

In transport, and in operation, the probe and support systems must withstand harsh environments and some rapid environmental changes. It is possible for us to provide

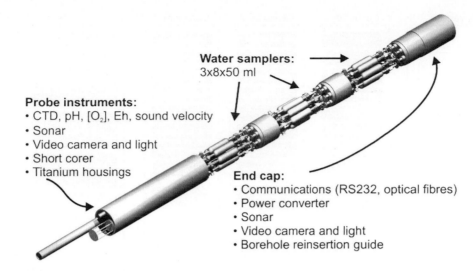

Water samplers:
3x8x50 ml

Probe instruments:
• CTD, pH, [O₂], Eh, sound velocity
• Sonar
• Video camera and light
• Short corer
• Titanium housings

End cap:
• Communications (RS232, optical fibres)
• Power converter
• Sonar
• Video camera and light
• Borehole reinsertion guide

Figure 1. 3D CAD rendering of the Lake Ellsworth probe concept which consist of two gas-filled pressure cases separated by three carousels of water samplers (see Figure 2) all attached to a central core that is attached to the tether. The bottom pressure case houses the majority of the instrument package and is tipped with a short gravity core sediment sampler (increasing total length to ~4 m).

environmental control during transit (by sea, air, and overland in Antarctica), for example, temperature regulation and humidity management. However, it would be simpler and lower risk, if the systems could survive without such a system. For example, if the sample bottles are transported closed, containing sterile water, it would be advantageous if their design would enable them to survive freezing. This would mitigate the risk of failure of the temperature regulation system, while in transit to the drill site. The lowest ambient pressure (<0.7 bar (70 kPA)) will be experienced in flight to Antarctica, while the lowest temperature (<−25°C) will be experienced at the drill site and during overland transport to this location. The atmospheric pressure at the drill site is higher (~0.750–0.780 bar (75–78 kPa), measured at the ice surface over Lake Ellsworth in field season 2007/2008 (J. Woodward, personal communication, 20 August 2010)). In the top (air filled) section of the borehole, the temperature will be close to that of the surrounding ice at moderate depths (~−18°C). On entering the water at approximately 270 m depth in the borehole, the probe and tether will experience a rapid change in external temperature (~−18°C to 2°C in less than 1 s). During the descent of the borehole and lake traverse, the pressure will increase to approximately 30.2 MPa (assuming an ice thickness of 3170 m and a lake depth of 156 m) [see *Ross et al.*, this volume] and temperature will decrease from ~2°C to −5°C. On retrieval, the probe will pass through the water-air interface at the hydrostatic level in the borehole returning to atmospheric pressure and will experience a large thermal gradient (~2°C to −18°C in less than 1 s).

2.4. In Situ Measurements

To maximize the probability of successful characterization of ESL, the probe is required to make measurements of the environment with in situ sensors. The requirements for these sensors are shown in Table 1. These measurements enable the characterization of the lake environment in real-time, providing useful scientific data and informing the sampling strategies.

2.5. Water Samplers

While in situ measurements allow characterization of the environment during the deployment, retrieval of water and sediment samples is a primary aim because of the large number of detailed analyses this enables. The water sampler system is required to enable acquisition of more than 18 individual water samples of volume >50 mL enabling triplicate sampling at six locations. This gives a total of at least 150 mL per sample location, which will be divided between analysis techniques as shown in Table 2. To minimize the deployment duration, and hence maximize the probability of successful probe extraction through the borehole, the water samplers are required to trigger and fill rapidly. It should take less than 1 min to obtain a sample in each bottle. To ease sample processing and distribution to laboratories, the bottles are required to be removable (individually) from the probe and sampling system without compromising the physical seal or introducing contamination. Each bottle should be uniquely identified (e.g., carousel and

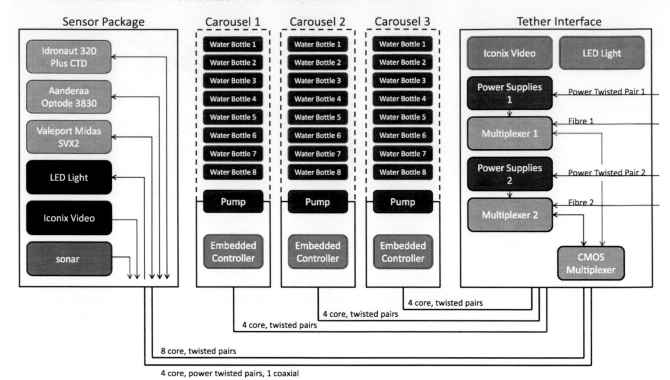

Figure 2. Schematic of the Lake Ellsworth probe highlighting electrical layout, cable connections, and electronic systems. The design depicted does not include backup batteries or an embedded microcontroller (included in design variants) to enable continued operation should the tether power or communications link fail. The current water sample protocol fires one bottle on each of the carousels simultaneously creating three >50 mL samples from each location. The provision of eight bottles allows sampling in six locations and two further locations as a reserve. Carousel 1 will be analyzed for microbiology, carousel 2 will be split between organic geochemistry and microbiology, and carousel 3 will be analyzed for hydrochemistry.

bottle number). One of the scientific aims is to characterize the dissolved gasses present in ESL, and therefore, the sample bottles are required to store water at the pressure it was collected (i.e., gas tight sampling). The design and materials selection of the bottles must minimize contamination with analytes of interest (see Table 2). Iron analysis is required, so stainless steel is not a suitable material (see sampler review below). Gas and organic geochemistry analyses may also be compromised by the use of polymer bottles or liners.

Table 1. Probe Instrumentation Requirements

Sensor	Range	Accuracy	Resolution	Time Constant
Pressure	0–1000 bar (0–100 MPa)	0.01% FS	0.001% FS	15 ms
Temperature	−5°C to −45°C	0.001°C	0.0001°C	50 ms
Conductivity	0–6400 $\mu S\ cm^{-1}$	5 $\mu S\ cm^{-1}$	0.1 $\mu S\ cm^{-1}$	50 ms
Oxygen (electrode)	0–1560 μM	3.1 μM	0.21 μM	3 s
pH	0–14 pH	0.01 pH	0.001 pH	3 s
Redox	−1000 to +1000 mV	1 mV	0.1 mV	3 s
Sound velocity (SV)	1400–1600 m s^{-1} (extended on request)	0.03 m s^{-1}	0.001 m s^{-1}	NC
Temperature (SV system)	−5°C to +35°C	0.01°C	0.005°C	NC
Pressure (SV system)	0–600 bar (0–60 MPa)	0.01% FS	0.001% FS	NC
Oxygen optode	0–500 μM	<8 μM or 5%, whichever is greater	<1 μM	25 s (63%)

Table 2. Analyses to be Applied to Samples Acquired Using the ESL Probe System

Analysis	Sample[a]	Details
Microscopy	Water[b] (25 mL), particles (filtrand[c]), sediment	Range of scales targeted: light microscopy ~0.2 μm, fluorescence ~30–50 nm (using DNA binding stains), SEM ~1–10 nm and TEM ~1 nm, epifluorescence to detect NADH (live organisms), FISH microautoradiography and CARD-FISH to identify specific cells and group and taxon
Molecular biology	Water (25 mL), filtrand[c], sediment	DNA will be used to construct a metagenomic library to screen for novel physiologies 16S rDNA-based community reconstruction PCR and RT-PCR to assess transcription
Microbiological physiology	Water (38 mL)[d], filtrand[c], sediment	Cultures. Incubations (including ^{14}C-labeled) to assess activity. Biomarkers including ATP, PLFAs, lipopolysaccharide, enantiomeric compounds, biogenically precipitated minerals, and stable isotopes
Organic geochemistry	Water (20 mL), filtrand[c]	Single derivatizing agent (BSTFA) and GC-MS[e] analysis to include phenols, alkylphenols, polyaromatic hydrocarbons (PAHs), fatty acids, alcohols, sterols, and amino acids HPLC[f] and (coupled) ICP-MS[g] targeting heteroatoms (e.g., organosulfur and organophosphorus compounds) and organometallic compounds (e.g., porphyrins)
Hydrochemistry	Water (requires minimum 30 mL) [*Siegert et al.*, 2006]	Geochemical reactivity, Cl^-, $\delta^{18}O$-H_2O and δD-H_2O, major cations and anions, DIC, DOC, pH, NO_3^-, NH_4^+, PO_4^{3-}, DON, DOP, O_2, $\delta^{13}C$-DIC, $\delta^{34}S$-SO_4^{2-}, $\delta^{18}O$-SO_4^{2-}, Eh, Mn(II), Fe(II) and other analytes depend on the concentration of the waters examined Gasses: N_2, O_2, CO_2, CH_4, $\delta^{13}C$-CO_2, and $\delta^{13}C$-CH_4
Sedimentology	Sediment	Environmental change: SEM, XRF, XRD, grain size, Nd-Sr, density, magnetic susceptibility, P wave, imaging, gamma radioactivity. Life detection (see microscopy, molecular biology and physiology above) SEM and microfossil analysis. Dating: magnetostratigraphy, cosmogenic isotopes (e.g., 10Be, 3He, etc.), other radioactive isotopes, carbonate isotopes, and U-Th dating

[a]Water volumes listed are the volumes analyzed per sample location of which there will be six.

[b]Water samples will be filtered for microscopic analysis. The filtrate will be retained for further hydrochemistry analysis.

[c]A total of at least 200 L of lake water will be filtered in situ with approximately 100 L filtered at a minimum of two locations. These specific analyses will examine filtrand from approximately 50 L of lake water.

[d]This water sample will be used in part to create cultures which will also be analyzed with microscopy and molecular techniques.

[e]Gas chromatography-mass spectrometry.

[f]High-performance liquid chromatography.

[g]Inductively coupled plasma-MS.

2.6. Particle Samplers

Particle samples are required to investigate the microbial and geochemical characteristics of the lake. The particulate sampling systems is required to acquire at least two samples of filtered particles at use-specified locations each from at least 100 L of lake water and must collect all particles ≥2 μm. A large volume is required to maximize the probability of capturing microorganisms or particles of interest. To enable quantitative studies, the volume filtered must be measured. It may be advisable to stop sampling prior to contact with the sediment when disturbance is likely to cause a small plume, hence triggering control from the surface is required. To ease analysis, the filter should be detachable from the probe and must be easy to isolate from the environment.

2.7. Sediment Sampler

Discussion of the coring operations planned for the ESL experiment is presented by *Bently et al.* [this volume]. The acquisition of a long sediment core (>3 m) will follow retrieval of the probe using a large corer controlled by the same tether as the probe. The main challenges in the design of this large corer are the use of a single tether, operation through a narrow borehole, and achieving sufficient penetration via onboard percussion. For redundancy (e.g., to mitigate the risk that only the probe may be deployed), the probe is also required to capture at least one sample of sediment of >1 cm diameter and >20 cm length. This enables sampling of the sediment-water interface and the top layer of sediment, which are key targets for microbial analysis. A small diameter may be used, as preservation of sediment stratigraphy is not a priority for this sample. This will be investigated in samples from the larger percussion corer. However, disturbance or distortion of the sample should be minimized. It would be desirable to remove the sediment sampler from probe after probe recovery without losing, disturbing, or contaminating the sample.

2.8. Video

Video data is required to inform probe operation during the experiment including triggering and control of sampler systems. Images should be recorded looking downward from the bottom of the probe and upward to image the undersurface of the overlying ice. Color is not essential for imaging of ice and lake bed gross structure but may have uses in this unique environment. Resolution should be standard definition or better. Recording should be performed at the surface only, as in the event of tether failure, there will be insufficient power to run the lights.

2.9. Sonar

Sonar data is required primarily for measuring the distance to the lake floor and imaging the ice ceiling. This must be recorded simultaneously looking upward from the top of the probe and downward from the bottom of the probe. If an upward-looking camera is installed, the upward-looking sonar may be omitted. Sonar data should be recorded at the surface. If used, the upward-looking sonar must have a range of >200 m and a beam angle of >20°. The downward-facing sonar does not need to provide images but must have a range of >200 m. It would be advantageous if the downward-looking sonar enabled measurement of the sub-bottom sediment profile.

2.10. Reliability and Fault Tolerance

A high degree of system reliability is required, as the lake access borehole can only be maintained for approximately 36 h. Little or no repairs can be made on site and in addition, once HWD has commenced, there is little opportunity to alter the timetable of the experiment. The system must work on time. To achieve this reliability, a suitable risk model should be used to enable risk minimization and targeted reliability through design. In addition, testing of

components and systems will provide data to improve reliability estimates. For example, early life failure of components may be reduced by stress testing, for pressure and temperature.

One possible mitigation method is to include fault tolerance, so that the system will continue to operate if one or more systems fail. For example, if power failure on the tether then occurs, the system should allow the probe to continue to communicate with the surface, take samples (water and sediment), and make in situ measurements and record data. If the communications link were to fail, the probe should take water and sediment samples in a predetermined fashion, continue to make all in situ measurements, and record the data. If both power and communications were lost, samples and data should still be acquired. A reduction in the quantity of sample and data returned may be permissible in each of these cases.

The ability to detect and isolate faults may also be necessary. For example, if one of the modules (e.g., a water sampler carousel or an individual pressure vessel) fails or is flooded, it should be possible to turn off the power and shut down this unit without compromising the performance of remaining functional systems.

In addition to the engineered robustness, automation, and fault tolerance described above, risk due to human factors should be mitigated by testing, procedural documentation, checklists, cross-checking, and rehearsal (including the use of simulation).

The final risk mitigation requirement is the construction of two probes, both of which should be delivered to site prior to the experiment.

3. REVIEW OF TECHNOLOGIES AND TECHNIQUES APPLICABLE TO SUBGLACIAL MEASUREMENT AND SAMPLING

3.1. Vehicles

Remotely operated vehicle (ROV) technology has been used successfully to explore the grounding zone at a glacier terminus [*Dawber and Powell*, 1995; *Powell et al.*, 1995] at moderate depths of approximately 100 m using commercial systems. These systems would need considerable adaptation to enable access through a borehole and to enable sterilization. Bespoke ROV systems are in development for sub-ice applications by *Vogel et al.* [2007, 2008]. This Sub-Ice ROV (SIR) is intended for exploration at the base of the Whillans Ice Stream [*Bindschadler et al.*, 2003; *Fricker and Scambos*, 2009; *Wiens et al.*, 2008] downstream of the grounding line. The vehicle transforms between a borehole decent configuration and a sub-ice exploration configuration. On decent, its maximum diameter is 55 cm requiring a 70–75 cm borehole [*Vogel et al.*, 2008]. The tether for this vehicle is integrated with a HWD [*Bentley and Koci*, 2007; *Makinson*, 1993; *Craven et al.*, 2004] hose enabling reaming of the borehole on decent or extraction. In sub-ice configuration, SIR is a capable ROV carrying extensive oceanographic instrumentation, a water sampler, and a sediment corer. ROV operations in the vicinity of glacier-grounding zones have not required microbial control due to the open access to the marine environment in these regions and therefore an assessment of the amenability of the SIR to microbial control has not been published.

Autonomous Underwater Vehicle (AUV) technology has been successfully used under ice in West Lake Bonney, Taylor Valley, Antarctica [*Stone et al.*, 2010] and to survey cenote sinkhole environments [*Gary et al.*, 2008] using multidirectional sonar. The restricted access encountered in these environments is analogous to subglacial applications. The system has autonomy, mapping capability, and sophisticated navigation, including homing to an ice borehole. This vehicle is large (1.9 m diameter) [*Gary et al.*, 2008], has a cruise speed of 0.2 m s^{-1} and has not been designed to enable microbial control. Autosub, a large AUV developed by Natural Environment Research Council (NERC) in the United Kingdom, has successfully mapped ice topography, bathymetry, and hydrographic parameters under ice shelves in Antarctica [*McPhail et al.*, 2009; *Nicholls et al.*, 2006; *Jenkins et al.*, 2010] and under Arctic sea ice [*Wadhams et al.*, 2006]. This vehicle is large (0.9 m diameter) and is not designed for microbial control. *Bruhn et al.* [2005] propose the development of a tethered ROV/AUV for sub-ice applications; however, significant investment would be required to enable this concept to reliably deliver data for the ESL experiment. Under ice, deployments of REMUS AUV (19 cm diameter) [*Plueddemann et al.*, 2008] have been completed off Barrow, Alaska. A docking system has been developed [*Allen et al.*, 2006; *Stokey et al.*, 1997], opening the possibility of untethered lake exploration. Neither the docking system nor the AUV has been designed for microbial control. The GAVIA AUV (20 cm diameter) has been successfully deployed through ice [*Doble et al.*, 2009] though without an automated recovery system for free swimming operations. This vehicle has operated under ice while tethered [*Doble et al.*, 2009] to distances of a few hundred meters. While this has an impact on AUV performance, this is a promising low-risk solution for subglacial applications. The GAVIA AUV is depth limited to 1000 m and has not been designed for microbial control.

Melting probe technology has been used in Antarctic exploration [*Kasser*, 1960; *Ulamec et al.*, 2007; *Aamot*, 1970; *Hansen and Kersten*, 1984; *Kelty*, 1995; *Tüg*, 2003] achieving ice penetration to depths of 1000 m [*Aamot*, 1968]. This technology has also been proposed for astrobiological missions to explore the polar ice caps of Mars [*Zimmerman et al.*, 2002, 2001] and Europa [*Bruhn et al.*, 2005; *Zimmerman et al.*, 2001; *Biele et al.*, 2002; *French et al.*, 2001] where an extensive sub-ice ocean is expected [*Carr et al.*, 1998]. There is some similarity between vehicles designed for Europan and Antarctic exploration [*Ulamec et al.*, 2007; *Biele et al.*, 2002]. Heat generated within the vehicle or probe tip is used to penetrate the ice. Radioisotope thermo(electric) generators are proposed in astrobiological applications [*Zimmerman et al.*, 2001] but would not be permitted in Antarctica. Melt-induced penetration can be problematic in near-vacuum conditions [*Kaufmann et al.*, 2009] or if the ice contains dust or sediment, as this collects at the probe tip forming an impenetrable barrier [*Ulamec et al.*, 2007]. These systems communicate to the surface using a tether that spools out from the topmost part of the vehicle and is refrozen into the ice as the vehicle descends. In this manner, the ice cap seal overlying the water body is not broken at breakthrough into this environment. However, the volume available for the tether spool is limited resulting in either a short, or thin, tether. In deep sub-ice applications, a short tether necessitates a larger taxi vehicle [*Ulamec et al.*, 2007], whereas a thin tether necessitates a probe containing significant onboard power to enable penetration of the ice sheet. Rapid recovery of melting probes (e.g., for sample retrieval) is problematic.

3.2. Implications for the ESL Vehicle (Probe)

The measurement and sampling of ESL requires a vehicle with a unique portfolio of attributes. It must be designed to facilitate microbial control, whereas current ROVs and AUVs are not specifically designed to meet this requirement. Considerable investment would be required to develop protocols for microbial control and validation on these platforms. This development is not without risk. In addition, the combination of stringent requirements for reliability, environmental resilience, and small diameter cannot be met by existing AUV or ROV systems. Melting probes are attractive because of their simplicity, which eases microbial control and enhances reliability. However, problems with drilling caused by debris are well documented, and deep drilling (as required for ESL) is challenging. In addition, a melt probe cannot be rapidly retrieved from the lake, which is incompatible with rapid analysis of water and sediment samples from sub-ice environments, as required for the ESL experiment. It is these unique and unmet requirements that have motivated the development of a custom-made probe.

3.3. Imaging and Sensor Systems

3.3.1. Imagery. Imagery through ice access boreholes (typically formed by HWD) has been used in a number of studies [*Engelhardt et al.*, 1978; *Bruchhausen et al.*, 1979; *Lipps et al.*, 1979; *Craven et al.*, 2005; *Carsey et al.*, 2002; *Behar et al.*, 2001; *Harper et al.*, 2002; *Harrison and Kamb*, 1973]. As is used routinely in ROV technology, video may be streamed in real time via a fiber optic communications link [e.g., *Craven et al.*, 2005]. Borehole imaging systems typically obtain images over short distances and low turbidity is frequently observed in sub-ice environments. Therefore, scattering and optical absorption do not prevent the acquisition of high-quality images [*Craven et al.*, 2005; *Carsey et al.*, 2002; *Harper et al.*, 2002]. Imaging has been achieved with modest separation of the light source and the camera. However, turbidity has been observed as a consequence of thermal/mechanical [*Engelhardt et al.*, 1978] and HWD [*Craven et al.*, 2005] due to suspended glacial flour in drill fluids. Innovations in underwater imaging [*Kocak et al.*, 2008; *Caimi et al.*, 2008; *Kocak and Caimi*, 2005; *Mueller et al.*, 2006] include the extensive use of digital camera technology for still, video, three-dimensional (3-D), and holographic imaging. Digital technology is compact and low-cost with low-light, IR, high-resolution, and high-speed variants available. For subglacial applications, imaging system design must consider the tradeoff between resolution, low-light performance, and frame rate (e.g., CCD still cameras can offer improved resolution or low-light level performance at the expense of frame rate). High-definition video and even higher-resolution formats [*Kocak et al.*, 2008; *Caimi et al.*, 2008] are supported by commercially available camera technologies enabling a good compromise between resolution and frame rate while enabling color recording. Lighting technologies are critical for ensuring adequate range, color representation and minimizing scatter [*Kocak et al.*, 2008; *Caimi et al.*, 2008; *Kocak and Caimi*, 2005; *Jaffe*, 2010, 2005]. LEDs offer robust high-efficiency solutions, while angular separation from the camera and structured light sources dramatically reduce scatter and increase range [*Kocak et al.*, 2008; *Caimi et al.*, 2008; *Jaffe*, 2010, 2005]. Quantitative information, such as object size and location, can be obtained from images using laser spots or with a penalty of increased system size/complexity, with 3-D or holographic imaging systems [*Kocak et al.*, 2008; *Caimi et al.*, 2008; *Mueller et al.*, 2006].

3.3.2. In situ sensors. In situ sensors have been used in glacier and sub-glacial systems to measure temperature [*Engelhardt et al.*, 1990; *Hart et al.*, 2009], pressure [*Engelhardt et al.*, 1990; *Hart et al.*, 2009; *Stone and Clarke*, 1996; *Harrison et al.*, 2004], electrical conductivity [*Hart et al.*, 2009; *Stone and Clarke*, 1996; *Stone et al.*, 1993; *Gordon et al.*, 1998], and turbidity [*Stone and Clarke*, 1996; *Stone et al.*, 1993; *Gordon et al.*, 1998]. In situ oxygen measurements have been made under sea ice [*Kühl et al.*, 2001; *McMinn and Ashworth*, 1998; *Trenerry et al.*, 2002], in laboratory-based sea ice mesocosms [*Mock et al.*, 2002, 2003], and in glacial melt waters and cryoconite holes in Antarctica [*Bagshaw et al.*, 2011]. Larger sensor systems have also been deployed for simultaneous monitoring of carbon dioxide and oxygen concentrations in ice-covered lakes [*Baehr and De-Grandpre*, 2002].

The use of in situ sensor systems is more widespread in an oceanographic setting [*Daly and Byrne*, 2004], where there is often less restricted access, and there is large-scale logistical support (e.g., a research vessel). In this setting conductivity, temperature and pressure (depth) sensing is routine. Commercial sensors exist for many parameters including oxygen [*Tengberg et al.*, 2006; *Falkner et al.*, 2005], chlorophyll, and hydrocarbon fluorescence [*Falkowski and Kiefer*, 1985; *Suggett et al.*, 2009], carbon dioxide [*Degrandpre*, 1993], nutrients/inorganic anions [*Johnson and Coletti*, 2002; *Hanson*, 2005], pH [*Seidel et al.*, 2008], and methane [*Fukasawa et al.*, 2006]. However, sensor performance and suitability for sterilization must be carefully matched to the application to ensure high-quality studies. Larger analytical systems using membrane inlet mass spectroscopy are in the early stages of commercialization [*Camilli and Duryea*, 2007; *Short et al.*, 1999, 2006] and have overcome many of the difficulties in calibration and quantitative analysis [*Bell et al.*, 2007]. Biological and chemical sensor and analytical system development is an active area of research [*Daly and Byrne*, 2004; *Prien*, 2007].

3.4. Implications for Imaging and Sensor Technologies for ESL

The requirement for imaging and in situ sensing in ESL experiment are well matched by commercially available technologies that are proven in similar environments. Some care is required to ensure that these technologies meet the demanding performance specification, and further work is required to prove that microbial control is feasible. The requirement for proven performance and high reliability for any system used in the ESL experiment makes the use of research instrumentation unattractive in this application. While the effects on imaging of turbidity in boreholes formed by HWD are of concern, this phenomena is not anticipated in ESL. During the ESL experiment, HWD fluids will be brought to the surface at a rate equal to the injection rate at the cutting head (~3 L s^{-1}) and filtered (<0.2 μm) before reinjection. This process will limit the accumulation of turbidity in the ESL borehole.

3.5. Sampler Systems

Sampler systems for water, dissolved gasses, and particles have been widely developed for subglacial and oceanographic applications. The requirement for multiple water samples and quantitative gas content analysis in the ESL experiment suggests either gas-tight water sampling or quantitative analysis of outgassing (and capture of samples of both the water and the escaping gas). The expected low-particulate concentration in ESL suggests filtration capture from large volumes of water.

3.5.1. Water sampling. Water sampling has been used effectively in glacier systems and boreholes enabling characterization of geochemistry and study of hydrological and glacial processes [*Tranter et al.*, 1997; *Blake and Clarke*, 1991; *Gaidos et al.*, 2007]. Typically formed from acrylic plexiglass, such systems are not gas tight or sufficiently pressure resistant for the ESL experiment. A compact (12 cm × 85 cm) and elegant gas-tight sampler developed by *Roman and Camilli* [2007] enables collection of 8 × 20 mL samples. Unfortunately, the depth rating is only 2000 m in seawater (~20.5 MPa pressure, ESL 30.2 MPa). It also uses a single multiport commutating valve (which represents a potential single point of failure) and has not been designed to be sterilized both internally and externally. Other gas-tight samplers [*Albro et al.*, 1990; *Tabor et al.*, 1981] offer additional useful design insights, but require significant re-engineering to make them sufficiently compact for the ESL experiment. A commercially available system (AquaLab Deep Ocean Water, EnviroTech LLC) using flexible titanium bags provides water sampling and retention of gas. This system does not maintain pressure, but gas can be retained within the limits of the flexible bag. Unfortunately, it also uses a multiport valve and would require compaction for ESL. Numerous systems exist for water sampling without maintaining pressure or retaining gas. The Niskin bottle [*Niskin et al.*, 1973] rosette is frequently used in an oceanographic setting. This system consists of an array of bottles (often tubes) with sealing end caps at either end, which when released are pulled together by an elastic or spring member. Release systems are commercially available (e.g., SBE 32, Seabird Electronics Inc.) but are not sufficiently compact for the ESL sampler. More compact systems based on a multiport valve exist (Aqua Monitor, EnviroTech LLC) but do not sample gas quantitatively. Alternative designs include variants of the end seal system as used in Niskin bottles [*de Resseguier*, 2000] and syringes (to pump and retain water samples) [*Di Meo et al.*, 1999]. Some include some form of microprocessor, e.g., to enable event-driven sampling [e.g., *Ruberg et al.*, 2000].

Care must be taken in the choice of materials, which come into contact with the sample. *Doherty et al.* [2003] recommend the use of titanium and estimate the contamination caused by commercially available (weakly alloyed) grades. They conclude that for seawater retained in 5 L bottles for 4 h, only titanium (approximately 1.3k%–18k% contamination) and tin (1.4%–118% contamination) quantification would be severely affected and that iron contamination would be less than 3.1% (~1.6 pM). Polymeric materials can also be problematic for trace gas analysis [*Doherty et al.*, 2003]. Other trace metal analysis systems using polymer bottles and acid washing [*Bell et al.*, 2002] release acid into the environment. Stainless steel is problematic for trace iron analysis. Even when coated and excluded from the interior of the sample, bottle contamination may result [*Hunter et al.*, 1996; *Martin et al.*, 1990].

Alternative strategies include sampling via a pipe or tube with a processing system above the hydrostatic level. Sampling can either be by direct suction pumping [*Stukas et al.*, 1999], which depressurizes the sample and results in degassing, or using an innovative 'U' tube design, which maintains the sample at pressure [*Freifeld et al.*, 2005]. In each case, the tube represents single point of failure, and for the U-tube, extensive processing equipment is required at the surface. Another approach collects only the analyte of interest by passing sample water through an adsorption column [*Johnson et al.*, 1987], but this limits the number of analyses that can be performed.

3.5.2. Particles. Particles may be analyzed by extraction (e.g., concentration using filtration) from water samples, but this may be problematic in the dilute samples expected in ESL. In situ systems for particle sampling typically use a pump and filter [e.g., *Johnson et al.*, 1987; *Behar et al.*, 2006]. Commercially available systems exist (e.g., WTS-LV, McLane Research Laboratories Inc.) but would require adaptation to enable miniaturization and sterilization required for ESL.

3.5.3. Corer technology. This technology is used widely in oceanographic, geological, and industrial (e.g., oil and gas) settings resulting in a significant market serviced by a number of commercial products. Innovations include pressurized systems for acquisition of gas hydrates [*Abegg et al.*, 2008], hydrostatically powered hammer corers [*Kristoffersen et al.*, 2006], and line-operated percussion corers [*Chambers and Cameron*, 2001; *Neale and Walker*, 1996], freezing sampling [*Neale and Walker*, 1996] and designs for fine-grained

[*Jahnke and Knight*, 1997] or soft [*Blomqvist*, 1991] sediments. Lacustrine sediment cores have been obtained (UWITEC gravity and percussion corers) from perennially ice-covered former subglacial Lake Hodgson in a region 93.4 m beneath the ice surface [*Hodgson et al.*, 2009]. Acquisition of short sediment cores has long provenance and enables microbial, faunal [*Tita et al.*, 2000], input rate [*Guevara et al.*, 2005], organic degradation [*Sun and Wakeham*, 1994], sediment, and paleosedimentology [*Meriläinen et al.*, 2000] studies. A number of systems for acquisition of a short core are available commercially.

3.6. Implications for ESL Sampler and Corer Technologies

The requirements for microbial control, multiple samples, small size, reliability, and compatibility with a wide range of organic geochemistry, hydrochemistry, and microbiological analysis are not met by existing water and particulate sampler technologies. In addition for water samples, the requirement for gas analysis necessitates a pressure-tolerant (gas-tight) design. While gas-tight sampling has been demonstrated widely, no system exists which meets the aforementioned constraints. Custom-made water sampler systems and particulate samplers, which draw on elements of existing systems, are therefore required for ESL.

Existing sediment sampler technologies are in contrast, far simpler, robust, and amenable to microbial control. This suggests the use of commercial solutions for both the large percussion corer and the short core mounted on the probe. The probe-mounted short core will require careful selection and testing to ensure effective sampling of the sediment-water interface and upper sediment.

3.7. Microbial Control

Microbial control during subglacial lake access experiments is recommended by the Scientific Committee on Antarctic Research (SCAR) [*Alekhina et al.*, 2009] and by the U.S. National Academies [*Committee on Principles of Environmental Stewardship for the Exploration and Study of Subglacial Environments, N.R.C.*, 2007]. Both Committees express a number of principles pertaining to stewardship of these unique environments, which may be summarized as follows:

1. Possible damage and contamination is minimized to safeguard the scientific value of subglacial lakes and to conserve these pristine environments.

2. The results of sampling and microbial analysis must not be confounded by contamination.

3. Any access to the base of Antarctic ice sheets should assume underlying liquid water forming part of a drainage network requiring particular attention to upstream sites.

4. That a living ecosystem is possible necessitating prevention of any permanent alteration of the biology (including introduction of alien species) or habitat.

5. Any contamination should not be expected to change the measurable properties of the environment and should be expected to have a less than minor and/or transitory impact.

6. That minor, transitory, and undetectable impacts are acceptable in pursuit of scientific understanding and that these should be mitigated as far as possible.

These principles have not as yet been translated into standards, methods, or verification protocols. These details are in development by the subglacial access community. However, microbial control is routinely employed in medical, pharmaceutical, food, and space exploration industries.

Microbial control can be achieved by cleaning, disinfection, and/or sterilization. "Sterilization" is defined [*Allen et al.*, 1997] as a process used to render an object free from viable microorganisms, including bacterial spores and viruses. "Disinfection" is defined as a process used to reduce the number of viable microorganisms but which may not inactivate some viruses or bacterial spores. "Cleaning," an essential prerequisite for disinfection and sterilization, is defined as a process which physically removes contamination and hence reduces the microbial load but does not necessarily render microorganisms nonviable [*Allen et al.*, 1997]. Cleaning can remove grease, soil, and about 80% of microorganisms, but chemical disinfection is essential to completely neutralize most viable microorganisms [*Ayliffe et al.*, 1992]. Surfaces that are clean and dry will not support the growth of most bacteria [*Wilson*, 2001].

Sterility assurance level (SAL) is a term used in microbiology to describe the probability of a single unit being nonsterile after it has been subjected to the sterilization process, according to ISO 11139: 2006 specifications of the International Organization for Standardization [*Rutala and Weber*, 1999; *McDonnell and Moselio*, 2009; *Mosley*, 2006]. Aseptically produced products are generally considered to have a SAL of 10^{-3} or more, while physical sterilization technologies used in component preparation resulted in a SAL of at least 10^{-6} [*Agalloco*, 2004].

3.7.1. Microbial control standards.
The pharmaceutical industry has very strict regulations for the production of sterile drug products by aseptic processing [*Agalloco*, 2004]. The drug products can be either produced under aseptic processing or sterilized at a terminal stage. The U.K. National Health Service (NHS) has recently published a policy on cleaning, disinfection, and sterilization [*Holmes*, 2008]. Also, National Specifications for Cleanliness in the NHS are available [*National Patient Safety Agency*, 2007]. Bacterial contamination is closely monitored in the food

industry [*Haysom and Sharp*, 2005]. Both the U.K. *Food Standards Agency* [2004] and the European Union have published guidelines on food hygiene practices. The Food and Drug Administration utilizes the ISO 14644 standards for manufacturing facilities in the pharmaceutical industry (see Table 3).

These standards control the environment of manufacture, but require that components and materials enter the process precleaned. They are most suitable for equipment assembly and further processing. While control to ISO 14644 class 5 would achieve a high level of cleanliness, it would require extensive infrastructure, which is not easily or economically available to the Ellsworth consortium. ISO 14644 class 6 and below also require significant infrastructure but are unlikely on their own to enable sufficient control particularly for our most stringent requirement, i.e., maintaining contamination below that which will compromise our measurement of samples with potentially very low numbers of endogenous microorganisms.

The most applicable standards to our project are those used by the space exploration industry. The space science community has a long history of microbial control both for forward contamination (i.e., protection of the explored environment) and backward contamination (resulting from the return of samples). The acceptable terminal sterilization bioburden level for Viking vehicles is 30 bacterial spores per vehicle. A series of guidelines has been published by NASA for the purpose of planetary protection [*Barengoltz*, 2005; *DeVincenzi et al.*, 1996]. Ice coring in Mars employed alcohol, bleach, and flame sterilization, and very low counts using the ATP luciferin-luciferase assay were enough to validate sterility [*Steele et al.*, 2006]. Contamination control during the MARTE drilling project (intended for Mars missions) was achieved by using a suite of procedures depending on material compatibility, laminar flow assembly for hardware, and bagging during transport and testing [*Miller et al.*, 2008]. Microbial load was validated by a negative ATP luciferin-luciferase assay, which corresponds to less than four cell number equivalents per square centimeters [*Eigenbrode et al.*, 2009]. It must be noted that extrapolating cell numbers from the luciferin assay is difficult; hence, the method is semiquantitative.

3.7.2. Disinfection and sterilization. For microbial control, disinfection and sterilization can be either physical or chemical. The current approaches have been thoroughly reviewed [*Rutala and Weber*, 1999; *McDonnell and Moselio*, 2009], but a short review follows.

3.7.3. Physical methods. Physical methods include wet and dry heat, radiation, and filtration. Wet heat, or autoclaving, typically involves sample exposure to steam for 15 min at 121°C or 3 min at 134°C. Autoclaving still remains the most popular method for sterilization of health care surgical equipment [*National Patient Safety Agency*, 2007] and glass and elastomeric components used in the pharmaceutical industry [*Agalloco*, 2004]. This method is problematic for electronics, water sensitive, or temperature-intolerant materials, which are used on the ESL probe. However, autoclaving can be attractive for robust subsystems (e.g., the water sampler bottle). Dry heat is used extensively for sterilization of glassware and reduces bacterial endotoxins and spores resulting in a SAL of 10^{-6} [*Agalloco*, 2004]. Until 2005, dry heat was the approved sterilization technique of NASA for planetary protection for instruments and probes [*DeVincenzi et al.*, 1996]. It remains the most practical technique for large hardware, but NASA has replaced it for electronics parts with hydrogen peroxide vapor (HPV; see combined physical and chemical treatments which are reviewed below) [*Chung et al.*, 2008]. Dry heat is not applicable to most polymers, sensors, and electronic systems used in the ESL probe. While suitable for components

Table 3. Pharmaceutical Industry Permissible Limits for Cleanliness as Per ISO 14644 Standard[a]

Clean Area Contamination ($\geq$ 0.5 μm particles/ft^3)	ISO 14644 Designation	$\geq$ 0.5 μm particles/m^3	Microbiological Active Air Action Levels[b] (CFU m^{-3})	Microbiological Settling Plates Action Levels[c]
100	5	3,520	1[d]	1[d]
1000	6	35,200	7	3
10,000	7	352,000	10	5
100,000	8	3,520,000	100	50

[a]Reproduced from *Department of Health and Human Services* [2004].
[b]The number of colony-forming units (CFU) per cubic meter of air.
[c]Number of CFU collected on a 90 mm culturing media (settling) plate with 4 h exposure.
[d]Samples from Class 100 (ISO 5) environments should normally yield no microbiological contaminants.

(e.g., metal items prior to assembly), larger units require extensive facilities. Alcohol and flame sterilization is another form of dry heat treatment that is particularly attractive because of its simplicity [*Richardson*, 1987]. However, flame sterilization is restricted to resistant materials, and there is also a lack of data on its efficacy, making its use for ESL unattractive.

Radiation methods involve the irradiation of the samples with energy as particles or electromagnetic waves. Ionizing radiation treatment with beta or gamma rays is very effective but costly, requires an isolated site, and may affect the bulk properties of polymers by disruption of chemical bonds [*Henn et al.*, 1996]. The nonionizing alternative of UV radiation treatment is suitable for heat-sensitive materials and thus could be used for electronic components. Also, UV irradiation will irreversibly degrade DNA. The most effective wavelength for sterilizing bacterial spores from *Bacillus atrophaeus* (a robust spore forming model organism) is from 235 to 300 nm [*Halfmann et al.*, 2007]. UV is attractive for the ESL experiment in a number of applications (e.g., borehole and tether sterilization), as systems can be compact, noncontact, and fast-acting. However, it is only effective on exposed surfaces and is nonpenetrating, necessitating alternative strategies for recessed components and closed systems.

Filtration is the last alternative of physical disinfection and sterilization. Sterile filtration can be performed in a normal flow or a dead-end configuration, and microorganisms are excluded from filtered liquids or gases [*van Reis and Zydney*, 2001]. First, gases are usually filtered using depth filters, which are made of mineral, glass, or cotton wool. Second, membrane filters retain particles based on their size depending on the diameter of membrane pores. Finally, nucleation track filters are similar to membrane filters, but consist of irradiated thin polycarbonate films.

Sterilizing filters were originally manufactured with a 0.45 μm pore-size specification, but current regulatory standards have decreased the pore size to 0.22 μm to enable retention of bacterium *Brevundimonas diminuta*, formerly known as *Pseodomonas diminuta* [*Jornitz et al.*, 2003]. However, some organisms cannot be retained by 0.22 μm filters [*Wallhausser*, 1979]; thus, a 0.1 μm pore size filter has to be used if viruses, mycoplasmas, and small bacteria need to be retained. Sterilizing filtration for liquids has been critically reviewed by the Parenteral Drug Association, a leading authority of pharmaceutical science in the United States [*Antonsen et al.*, 2008]. A biological filtration system, which uses a modular cartridge to remove all bacteria and many viruses, as small as 45 nm, has been incorporated in the concept design of the hot water drill used for the exploration of Lake Ellsworth [*Siegert et al.*, 2006].

3.7.4. Chemical treatment. Chemical treatment includes use of hypochlorite (household bleach), acids, bases, alcohols, halogens, epoxides, phenols, metals, oxidizing agents, quaternary ammonium compounds, and aldehydes. The method is attractive, particularly in preparation of materials prior to encapsulation for shipping, because of the limited infrastructure required to achieve high efficacy. However, handling of these materials requires care, and they are less attractive on site in Antarctica, as risk reduction and disposal will require further on site infrastructure. Treatment with 2-glutaraldehyde for 2–3 h has been traditionally used for the inactivation of most viruses, vegetative bacteria, and mycobacteria on surgical instruments but is now less common due to its high chemical toxicity [*Manzoor et al.*, 1999]. However, toxic effects can be significantly diminished by subsequent washes of the sample with a saline solution [*Tosun et al.*, 2003]. Ethylene oxide gas (EtO) is traditionally used for sterilization of plastics and other nonheat stable components used for packing of drug products [*Agalloco*, 2004]. Despite environmental and occupational safety hazards, it is still used in the pharmaceutical industry [*Mendes et al.*, 2007]. It is suitable for heat-sensitive materials and thus could be used for electronic components. Commercial chemical disinfectants include Virkon© [*Hernandez et al.*, 2000], which contains suspected neurotoxicant sodium dodecylbenzenesulfonate [*Gasparini et al.*, 1995]. Enzymes and peptides, although not strictly "chemicals," can also be used for disinfection and sterilization because of their biocidal activity. Different types of enzyme include esterases, nucleases, and lysozyme. Lysozyme, for example, hydrolyzes cell walls and membrane components and can be an effective biocide against bacteria, fungi, protozoan, and viruses [*Benkerroum*, 2008]. Biocidal peptides include nisin and magainin. Nisin is a lactic acid bacteria metabolite, which is commonly used as a food preservative because of its antimicrobial properties [*Maqueda and Rodriguez*, 2008]. Similarly, magainin, a peptide first extracted from the African clawed frog *Xenopus laevis* in the 1980s [*Zasloff*, 1987], has since revolutionized the use of antimicrobial peptides for food industry and agriculture [*Meng and Wang*, 2010].

3.7.5. Combined physical and chemical treatments. The most recently developed techniques for disinfection and sterilization combine physical and chemical treatments and are termed as physiochemical. The most common example is the aforementioned HPV. HPV is suitable for heat-sensitive materials and, thus, could be used for electronic components. HPV acts as a biocide using the free radicals produced when hydrogen peroxide is heated beyond its liquid-vapor conversion point and forced to evaporate.

After decontamination, the HPV can be converted (either catalytically or using ventilation) into water vapor and oxygen, leaving no toxicity. HPV is currently used for planetary protection by NASA [*Chung et al.*, 2008] and has been found to completely deactivate bacterial spores of *Clostridium botulinum*, *Clostridium* spp. and *Geobacillus stearothermophilus* dried onto stainless steel surfaces [*Johnston et al.*, 2005]. Large-scale applications of this method are currently developed for the decontamination of whole buildings from *Bacillus anthracis* [*Wood and Blair Martin*, 2009]. This method is attractive for ESL in both the preparation of engineering structures and on site. The duration of ventilation required to enable complete conversion to water and oxygen is volume and temperature dependent (e.g., 2 h at 20°C for large rooms). The suitability of the method must therefore be carefully assessed for moving objects (e.g., the tether or drill hose) or during time-critical processes (e.g., loading the probe into the borehole). Other alternative physiochemical treatments employ plasmas, which are highly energized gases. Low-temperature plasma treatment (LTPT) is suitable for heat-sensitive materials, such as electronic components. LTPT exposes any microorganisms present in the sample to an electrical discharge with biocidal effects [*Moisan et al.*, 2001]. Low-pressure plasma treatment is used for surgical instruments and usually includes a UV irradiation step for genetic material destruction [*Kyltan and Rossi*, 2009]. Chlorine dioxide vapor is suitable for heat-sensitive materials and thus could be used for electronic components. Large-scale applications of this method are currently developed for the decontamination of whole buildings from *B. anthracis* [*Wood and Blair Martin*, 2009]. While effective, these methods are less suited to the ESL experiment either because of disposal, the infrastructure required, or flexibility.

3.7.6. Other techniques. Techniques which could be defined as neither physical nor chemical include anodic protection [*Nakayama et al.*, 1998] and freeze-thaw cycling [*Walker et al.*, 2006]. Also, special sample manipulation is used for sediments [*Lanoil et al.*, 2009], permafrost [*Vishnivetskaya et al.*, 2000], or ice cores [*Bulat et al.*, 2009; *Christner et al.*, 2005]. Material is removed from the innermost portion of the solid sample, while the outer layers of the core protect the sample used in the measurement. While these processes may be applicable to the treatment of samples, they are not suitable for the engineered structures used in the ESL probe systems, as they would either be ineffective, would create engineering challenges, or a better result could be obtained using alternative methods.

3.8. Microbial Assessment

Microbial assessment is undertaken once disinfection and sterilization has been affected to probe components. This is required to verify the processes which we use to control microbiology. This provides assurance that we are working in accordance with the recommendations (SCAR [*Alekhina et al.*, 2009], U.S. National Academies [*Committee on Principles of Environmental Stewardship for the Exploration and Study of Subglacial Environments, N.R.C.*, 2007]). However, it is likely that our requirement to prevent confounding of samples analyzed for microbiology is harder to achieve. Assessment methods also help us to develop techniques that enable us to reach this target.

A number of techniques are available to assess and validate the efficacy of disinfection and sterilization. While microscopy (see below) may be performed on engineered surfaces, other techniques require creation of an aqueous sample. This can be achieved using swabs or washing both with and without subsequent concentration steps (e.g., centrifugation). Such additional sample preparation steps can induce errors and sample contamination, which must be accounted for in protocol design.

The simplest assessment methods provide data on cell numbers. Quantitative information can be gained utilizing fluorescence microscopy [*Kepner and Pratt*, 1994] or flow cytometry [*Hoefel et al.*, 2003; *Lebaron et al.*, 1998; *Lemarchand et al.*, 2001]. Fluorescent cell staining dyes are used to increase contrast and include 4',6-diamidino-2-phenylindole (DAPI) [*Kepner and Pratt*, 1994] and can be used to discriminate between live and dead cells [*Boulos et al.*, 1999]. Low cell density requires careful contamination control, differential measurement versus blanks, and may be aided by a preconcentration step to raise the measured cell number above any background. However, such additional processing can also lead to contamination and loss of whole cells.

A total viable count enumerates cell density or population typically using serial dilution, culturing on appropriate growth media, and counting of colony-forming units (CFU) [*Miles et al.*, 1938]. As growth rates are culture- and species-dependent, the result is semiquantitative. Different culture media can be used to selectively grow different target microorganism, but only culturable species can be investigated. The process is also slow, particularly for psychrophilic organisms, which take weeks or months to culture. Therefore, there is limited applicability to field monitoring under Antarctic conditions.

Finally, adenosine triphosphate (ATP) and other similar biomarkers can be used as a marker of cell presence on probe components. ATP concentration can be measured using the bioluminescence luciferin-luciferase assay [*Lin and Cohen*,

1968]. The concentration of ATP found in solution in environmental samples is subject to processing and matrix effects as ATP binds efficiently to surfaces [*Webster et al.*, 1984].

More complex analyses for microbial load assessment include qualitative and quantitative nucleic acid detection. Tested surfaces can be swabbed and followed by total DNA or RNA extraction and purification. The pure nucleic acids can then be quantified by UV spectrophotometry and gel electrophoresis using intercalating chemical dyes like ethidium bromide, SYBR®, RiboGreen®, or Hoechst stains. The sensitivity of intercalating dyes is within the range of 10 ng mL^{-1} to 50 µg mL^{-1} of nucleic acids. To improve sensitivity and detection limits, quantitative molecular biology techniques can also be employed for detection and speciation of contaminant organisms. The state-of-the-art methods include quantitative polymerase chain reaction (qPCR), fluorescence in situ hybridization (FISH), denaturing gradient gel electrophoresis (DGGE), terminal-restriction fragment length polymorphisms (T-RFLP), and automated ribosomal intergenic spacer analysis (ARISA).

qPCR is a quantitative DNA amplification protocol with real-time detection of the amount of amplified DNA using a fluorescent reporter molecule. Use of generic markers and fluorescent probes for terrestrial bacteria and archaea and comparison of the results of samples of unknown concentration with a series of standards can determine accurately the amount of template DNA in samples and help to assess how DNA-free surfaces really are. The technique can be applied to difficult sample matrices such as soil [*Picard et al.*, 1992]. FISH, in optimum conditions, is a quantitative nucleic acid detection technique that employs fluorescent probes with sequences complementary to known sequences. In optimum conditions, it is a nondestructive (i.e., cells remain intact) and can therefore be used together with flow cytometry or microscopy. However, to ensure the fluorescent probes gain access to cellular contents, the cell membrane must be made permeable. This process is rarely perfect, either resulting in impermeable membranes and unstained cells, or cell lysis. Both problems result in an underestimation of populations in uncompensated analyses. As in qPCR, generic markers for terrestrial bacteria and archaea could be used to validate whether our sterilization methods are effective. FISH also targets 16S ribosomal RNA (rRNA); thus, only viable cells or recently moribund cells are detected. Most recently, FISH has been used to monitor bacterial community fouling on polyvinylidene fluoride and polyethylene filter membranes used for freshwater treatment [*Fontanos et al.*, 2010].

DGGE is another DNA-based technique, which is most often used to look at the community structure of samples rather than to quantitatively measure cell numbers. It can separate (often amplified) DNA amplicons of PCR of the same length but with different sequences and involves the construction of clone libraries [*Muyzer and Smalla*, 1998]. DGGE has been used to study bacterial populations in a variety of samples, from antique paintings [*Piñar et al.*, 2001] to processed water from waste water plants [*Boon et al.*, 2002]. The disadvantage of DGGE is that it may lack the specificity to enable separation of unknown species from known ones, and if used in the Antarctic environment where new species might inhabit, novel organisms may be overlooked.

T-RFLP is based on the analysis of bacterial 16S rRNA, which is a small subunit of nucleic acid, which is highly conserved in bacteria [*Liu et al.*, 1997]. Archaeal diversity in deep-sea sediments has been estimated using T-RFLP [*Luna et al.*, 2009]. Length Heterogeneity-PCR is very similar to T-RFLP [*Suzuki et al.*, 1998] and similarly takes advantage of naturally occurring sequence length variation. A further variant is amplified rDNA restriction analysis [*Liu et al.*, 1997]. These variations can be small, and therefore, the technique requires single base pair resolution of sequence length. The 16S rRNA genes are often targeted.

ARISA was originally developed for studying populations in soil [*Borneman and Triplett*, 1997] and then also applied to freshwater samples [*Fisher and Triplett*, 1999]. ARISA involves total community DNA isolation, PCR amplification using a fluorescent forward primer and ARISA-PCR fragments analyzed using automated gel electrophoresis. Nucleic acid analysis in Antarctic samples is a challenge, as gene sequences for novel organisms are not known a priori, which is a requirement for methods such as qPCR and FISH.

As an alternative to microbial enumeration and speciation, chemical tracers can be used to estimate transfer of contaminant materials after disinfection and sterilization. This technique is particularly useful, if the limit of detection of the chemical tracer is lower than that of the contaminant. Multiple tracers could also be used to verify the source of contamination [*Smith et al.*, 2004]. For example, the drill fluid could be loaded with a tracer, and the surfaces of engineering structures loaded with another. Tracers can also be used in solid samples, such as ice or sediment cores, in order to evaluate penetration of contamination [*Christner et al.*, 2005]. Perfluorocarbon and fluorescent-microspheres-based tracers have been used widely in the International Ocean Drilling Programme [*D'Hondt et al.*, 2004; *Lever et al.*, 2006]. The use of tracers for subglacial exploration would have to be carefully controlled to prevent the tracer from becoming a chemical contaminant in the environment or in the samples.

*3.9. Implications for Microbial Control for the
ESL Experiment*

The protection of subglacial lake environments and assurance of microbiological sample validity through microbial control remains the most challenging aspect of the development of probe and support system technologies for the ESL experiment. While progress has been made in the development of universally accepted standards for subglacial environmental protection, further work is required to translate these to agreed targets, methods, and verification protocols. In contrast, the science of microbial control and assessment is advanced, particularly in analogous disciplines such as space science and in health care. While microbial control and assessment techniques will need to be modified and assessed when applied to ESL technologies, widely used techniques are promising in this application. The development of appropriate methods of microbial control and assessment should be undertaken in parallel with the design and construction of technologies for the ESL experiment.

3.10. Targeted Reliability

The short duration during which the borehole into ESL will remain open necessitates a high degree of reliability of the engineered systems used in the experiment. There are a number of techniques that may be applied to engineering systems to enable estimation of risk enabling a reliability assessment [*O'Connor*, 2002]. Inclusion of this assessment within the design cycle and feedback into this process enables design to a reliability or availability target. Availability is the probability that a sequence of successful phases of the deployment will take place, at the time required. A similar analysis may be used with existing technology to assess availability during operational phases of a deployment [*Brito and Griffiths*, 2011]. Targeted reliability has been applied to the design and operation of AUVs [e.g., *Brito and Griffiths*, 2011] including under-ice missions. Methods of assessment include the use of expert judgment [*Brito and Griffiths*, 2009], Markov chain models [*Brito and Griffiths*, 2011] (which can be combined with Bayesian theory, Monte Carlo methods, and event trees [*Furukawa et al.*, 2009; *Chu and Sun*, 2008]), fault tree analysis (FTA) and mean time between failure (MTBF) analysis [*O'Connor*, 2002].

3.11. Implications for Reliability of ESL Technologies

The use of targeted reliability in engineering design is gaining acceptance in the development of platforms and systems for environmental science. The existing tools and techniques can be adapted for use in the development of Antarctic subglacial lake probe technologies, and this should be undertaken for the ESL experiment.

4. THE DEVELOPMENT OF ESL PROBE TECHNOLOGIES

In addition to risk management through documentation, peer review, training, and testing, we are currently using availability analysis employing a Markov chain model of the deployment process to determine reliability targets for systems used in different phases of the deployment. The reliability of systems and subsystems is estimated using FTA. The system FTA model combines MTBF (determined from datasheets, testing, or expert judgment) of components to calculate the overall system reliability.

At time of writing, we have developed an initial design concept and are undertaking detailed design of components, subsystems, and systems. Within this concept design, we have attempted to minimize risk and cost by keeping the number and complexity of elements to a minimum and by using proven commercial off-the-shelf (COTS) technology wherever possible. This limits the risk and cost of bespoke system development.

Scientific return is maximized by the combined use of instrumentation returning real-time data and acquisition of water and sediment samples for postretrieval analyses. This provides redundancy and enables informed deployment of the sampler systems. An overview of the concept design is given below.

A key engineering challenge is to enable microbial control while maintaining minimal cost and reliability in an extreme yet delicate environment. The standards that we propose to attain and the methods selected to enable microbial control and verification are discussed below prior to discussion of the engineering structures. This illustrates that microbial control is both central and dominant in our design process.

4.1. Microbial Control

In anticipation of the development of detailed standards and protocols approved by the subglacial community, we have proposed standards and identified a number of microbial control and assessment techniques that could be applied to the experiment. These pragmatic approaches have been developed in response to the principles recommended by SCAR [*Alekhina et al.*, 2009] and U.S. National Academies [*Committee on Principles of Environmental Stewardship for the Exploration and Study of Subglacial Environments, N.R.C.*, 2007]. In addition, they also prevent confounding of microbial analysis of samples.

4.1.1. Standards. We propose the following standards:

1. That exogenous microflora populations are reduced to prevent confounding of sample analysis. This will be achieved using population reduction techniques (described below) to reduce the exogenous background below the detection limit of the analysis techniques used (see Table 2) wherever possible. Where this is not possible (and we do not foresee this), the exogenous microorganism population must be sufficiently low and repeatable to allow the use of correction using an appropriate control and differential measurement, for example, parallel analysis of the water that was stored in one of the sample bottles while in transit to the experiment site.

2. That the final assessment of all engineered structures should verify an exogenous population at or below the detection limit of the analysis used.

3. That following final assessment, a method that is proven to reduce population further is applied. The efficiency of this final population reduction step will be quantified using a positive control contaminant on representative models.

4.1.2. Verification and assessment. To enable assessment of the efficacy of each method, we will use a positive control bacterial species to contaminate engineered surfaces and components. The level of contamination will be assessed (see below), a microbial reduction protocol applied, and a repeated measurement of population used to calculate the reduction achieved. We propose to use both adherent (*Pseudomonas fluorescens*) and nonadherent (*Escherichia coli*) species. This method allows accurate efficacy assessment while minimizing error by raising the number of cells well above the limit of detection.

4.1.3. Analytical methods

4.1.3.1. Qualitative. We propose visualization of cells with fluorescence microscopy poststaining with DAPI, 5-cyano-2,3-di(*p*-tolyl)tetrazolium chloride (CTC), Light Green SF Yellowish (Acid Green) or LIVE/DEAD® BacLight™ Bacterial Viability Kit (Invitrogen). Fluorescence microscopy is advantageous, as it has a low limit of detection (one cell per field of view) and allows discrimination of live/dead cells. Wherever possible, this will be performed directly on engineered surfaces. This is preferred to analysis of samples collected by eluting or swabbing, as these additional processing steps can introduce errors and further contamination.

4.1.3.2. Quantitative. We propose bacterial enumeration of liquid samples, eluents, and eluted swab samples using flow cytometry post staining with SYBR Green II. Cytometry is advantageous, as it is quantitative and enables rapid enumeration of large numbers of cells. However, it is difficult to achieve contamination-free analysis and to discriminate target cells from the background if very low cell numbers are present. We propose to use this technique, where microscopy and qPCR are not possible, and we will use positive controls (see above), as then, the cell numbers will be well above the limit of detection. To reduce the limit of detection and to investigate use for low cell numbers, we are developing appropriate controls and the use of preconcentration (filtration and centrifugation).

4.1.3.3. Semiquantitative. We propose qPCR enumeration using domain-specific primers (Archaea, Bacteria, and Eukaryota). This will be performed on swab/eluted samples prior to cleaning to estimate the total population in each domain. This will be repeated postcleaning to assess efficacy. The bacteria primers will also be used in the evaluation of population reduction techniques using positive control species (see above). qPCR is a robust (if experimental contamination is controlled) and quick analytical technique with very low detection limits [*Burns and Valdivia*, 2008]. The technique could also be employed in the field with modest infrastructure. Also, ATP luciferin/luciferase assays will be undertaken.

Each of these techniques will be used in the preparation of engineered systems. Microscopy facilities will also be available on site at a minimum. In all cases, the effect of sample matrix variation on the result will be quantified and resuspension/elution with standard buffers used if matrix effects become dominant.

4.1.4. Population reduction methods. We propose the following methods:

1. Hydrogen peroxide vapor will be used in both construction and at the field site to reduce exogenous microorganism populations. Despite requiring a dedicated machine, heat, and ventilation for a significant duration (~2 h), this technique is attractive because it enables treatment of engineered structures with complex topography and small recesses, it can be used on a wide range of polymers and all electronic components, it has high and proven efficacy, and does not result in a toxic end product requiring disposal.

2. Autoclaving will be used in the construction and preparation of the probe (and the water sampler in particular). This method offers a proven and convenient method of treating resistant structures and is effective for closed volumes (e.g., water retained within a sample bottle). On-site facilities will be available as a backup, as little infrastructure is required.

3. UV radiation (254 nm, 30 W) will be used at the field site for treatment of the probe, the drill hose, and wellhead structures (including the air-filled section of the borehole). A minimum dose of 30 mJ cm^{-2} will be applied to moving surfaces and higher doses used in static applications. UV has high efficacy, is portable, requires modest infrastructure, and is fast acting on surfaces with limited topography.

4. Chemical wash (70% ethanol and also hypochlorite) will be used in preparation of equipment where persistent microorganisms are encountered. We will only use 70% ethanol on-site to reduce the complexity of environmental protection and site clean up.

4.1.5. Control through design. To improve the efficacy and extent of application of these methods, there are a number of steps that can be taken during engineering design. These are primarily the following:

1. Materials selection. In general, hard materials (e.g., titanium) are easier to clean than those with thick oxides (aluminum) or soft materials (elastomers and rubbers which may be porous). Titanium will be used extensively on the probe. This eases microbial control but also enables trace Fe analysis (see Table 2) and reduction of the thickness of load-bearing structures giving more room for ancillary equipment. Samples of all candidate materials will be exposed to cleaning and the population control measures to identify any material degradation and the efficacy of these treatments.

2. Minimization of recesses. Recesses and intricate surface topography has been shown to promote microbial growth [*Ploux et al.*, 2009]. Autoclaving is the only procedure that can reliably kill organisms in blind recesses but cannot be used for all materials, components, and subsystems. It is therefore desirable to limit the number and extent of recesses, which we will do through design. For example, seals are traditionally placed in recessed grooves; we will explore alternative geometries and alternative sealing designs. Where a recess cannot be avoided, we will use sterile liquids (for pressure communication and compensation) and elastomeric capping (potting) to provide a recess-free and cleanable surface. All fluidic systems (e.g., the valve and pump system for the water sampler) are designed to enable flushing to enable cleaning with HPV and/or chemical wash.

3. Limited handling. The design should facilitate operation while requiring minimal handling. This requires simplicity, durability, and reliability. For example the probe is designed to operate without being touched after final assembly, cleaning, and bagging. Targeted reliability design has and will be used to ensure the sterile bagging is not opened to affect repairs or adjustments.

4. Containment. Once the engineered systems are assembled and cleaned, they must be protected against recontamination. All the systems are designed to be placed within protective environments such as sterile bags, which protect them against unavoidable handling.

4.1.6. Methodological control. In each phase of the ESL experiment, we will ensure efficient and effective microbial control. In the "construction phase," we will use a combination of postmanufacture cleaning and population reduction to ensure components are clean. The population reduction methods selected (from the shortlist above) will depend on the material and design of the component. Assessment and verification will be undertaken at a process level in all cases and at a component level where required. The components will be assembled where necessary (e.g., where an inaccessible void is created such as in a gas-filled pressure case) in a clean room environment to ISO 14644 (cleanliness for equipment used in clean rooms) working to Class 100,000 (ISO 8) of this standard (Pharmaceuticals industry permissible limits for cleanliness of equipment in clean rooms as per ISO 14644). Terminal cleaning (i.e., at the end of the assembly) will be used in all cases and may be sufficient for simple structures and subsystems. Subsequent to final terminal cleaning, all equipment will be placed in a protective environment (e.g., heat-sealed bagging or hard case). For transport, the bagged probe will be placed inside a lightweight 20' ISO container together with the winch, tether, and an HPV generator. This sealed container will be shipped to site without breach of access. All other ancillary equipment will be shipped in sterile bags and protected from mechanical damage. Microbial control procedures used immediately prior and during deployment are described in the section below describing the probe support systems.

4.2. The Probe

The probe is heavily negatively buoyant, is tethered to the surface, and has only simple maneuverability (depth control via tether and limited rotation). A melting probe design has not been used because of the difficulties using this technology (e.g., sediment accumulation at the drill tip) in deep ice and because it would be difficult to rapidly retrieve such a probe. A borehole will be created prior to deployment using a hot water drill giving a 36 cm hole that will refreeze at approximately 6 mm h^{-1}.

To facilitate microbial control, we have developed a design with minimal components, reduced dead volumes, few exposed threads, minimized recesses, and using materials suited to microbial control protocols. The probe will be cleaned during manufacture and assembly and delivered to the drill site in a protective bag. Prior to insertion into the borehole, the exterior of this bag will be cleaned inside a sealed wellhead structure. The tether will be cleaned prior to entering the borehole.

The probe (see Figure 1) is ~3.5 m in length, 20 cm in diameter and consists of two gas-filled pressure cases separated by three carousels of water samplers (see Figure 3), all attached to a central core that is attached to the tether. The bottom pressure case houses the majority of the instrument package and is tipped with a short gravity core sediment sampler (increasing total length to ~4 m). A schematic of the probe systems and their interconnections is depicted in Figure 2.

The upper pressure case contains the power and communications link to the tether. We are completing a trade study of the use of an onboard microprocessor and data logger to enable continued operation (with reduced functionality, e.g., sampling at predetermined intervals) and archiving of instrument data in case of communications failure. The trade study evaluates the impact of this design decision on the reliability of the system (i.e., the likelihood of data and sample return) using formal methods (see above) to assess any advantage over a simpler (and hence less prone to fault) design that controls the probe and logs all data at the ice surface via the tether. Probe-to-surface communications (two-way) will be via an optical link and backup wire modem. The data link is provided by two (one is redundant for robustness) pairs (one of each pair is at the surface end of the tether) of optical multiplexers (Focal 907-HDM). Each pair will provide $1\times$ High Definition Video channel, $1\times$ RS422 and $2\times$ RS232. An additional daughterboard with each pair enables an additional $16\times$ RS232 channels. Power will be supplied through the tether as high voltage DC and down converted within the upper pressure case (e.g., using DC-DC converters (Lambda)). Duplicate power supplies will be used for the 5 and 12 V systems for redundancy. A bespoke multiplexor and power control board will monitor the status of the multiplexors and power supplies and will select between systems in the event of an error. We are also evaluating the use of on board batteries that are sufficient to complete the mission but with limited video footage. The upper pressure case also includes an upward pointing camera and lights to enable imaging of the underside of the overlying ice.

4.2.1. Probe instrumentation.

Including instrumentation within the probe enables acquisition of data on the properties of the lake. The approximate chemical and physical properties of the lake have been estimated from geophysical survey, consideration of glaciological history of the lake, and estimates of ice melt and accretion rates [*Siegert et al.*, 2006]. This enables estimation of the required performance and measurement range for each of the instrument systems and testing in simulated lake conditions in the lab. The communications link with the surface allows this data to be recorded and available in real time at the surface. This enables the operations team to plan and execute deviations from the deployment and sampling plan in response to environmental conditions as detailed in the requirements specification. COTS technology will be used to obtain all in situ data. This enables purchase at an early stage in the project and facilitates extended testing prior to deployment. This testing, coupled with quality control measures implemented by the suppliers reduces the risk of instrumentation failure during the deployment. The instruments selected are all available with titanium casings (or will be converted reducing chemical and biological sample confounding) and have designs amenable to microbial control. The extent of microbial control possible is being determined experimentally in our laboratories. The instrumentation package is based around the 320 plus CTD instrument (Idronaut, Italy). This has been selected, as it has a proven pedigree, is available in a variant with conductivity range suited to subglacial deployment, and has the capability to include proven electrochemical sensors for Eh, pH, and O_2. In addition, the manufacturer has facilities to test and develop these sensors in simulated lake conditions and at the low temperatures expected in transport. A Midas SVX2 (Valeport, United Kingdom) is also included; this provides duplicate CTD data which enhances robustness through redundancy. In addition, it provides measurement of sound velocity with an accuracy of ±0.03 m s^{-1}. A dissolved oxygen optode (e.g., model 3830 AADI, Norway) will be included to provide duplicate oxygen data and to mitigate the risk of electrochemical (Idronaut) sensor damage due to electrolyte freezing during transport. A camera (Iconix High Definition Colour Video, CA, USA) and light system are included to provide images of the ice borehole, the lake, and sediment. Our preferred light design includes LEDs for flood lighting and a tungsten filament lamp with retroreflector to create a long-range (>15 m) spot with power over a wide spectrum, maximizing color intensity. This gives a high-performance imaging system, while minimizing the surface area occupied on the tip of the probe. Sonar ranging systems will be employed for measuring the distance from the probe to both the lake floor and ice ceiling. These measurements will be augmented with laser spot altimetry imaged using the onboard cameras. The measurement of altitude from the lake floor is particularly important prior to and during acquisition of the short core.

Instrumentation systems for nutrient, dissolved gas, and other biogeochemical parameters are commercially available but have not been included in the Lake Ellsworth instrumentation suite. This decision is motivated by the difficulty in attaining microbial control, by poorly matched performance to the conditions expected in ESL, by physical space on the probe, and by a desire to increase robustness through simplicity.

4.2.2. Sampler systems

4.2.2.1. The water sampler. The water sampler (see design concept depicted in Figure 3) consists of three carousels each containing eight pressure-tolerant bottles (tubes) capped at each end with pressure-tolerant valves. This design enables preservation of samples at in situ conditions and quantitative measurement of gas content. The inlet valves are each connected to the lake via short inlet tubes, while the outlet valves are connected to a common pump, which pulls sample through the tubes when the appropriate valves are open. This design limits the interaction of the sample with the engineered system and limits the flushing volume required to obtain a discrete sample. Valves at each end of the bottle are actuated simultaneously using a gear and rod driven via a magnetic coupling attached to an electric motor inside a pressure case. The simple exposed design facilitates cleaning and robustness. The pressure case also contains electronics that interface to the multiplexer unit (via RS 232) and control the valve and pump motors. We are comparing the reliability of this design with one that includes a battery and a more advanced water sampler control unit to enable continued operation (i.e., sampling at predetermined depths) should the tether power or communications link fail. Commercial valves and sample bottles (Swagelock) are available with a pressure rating of

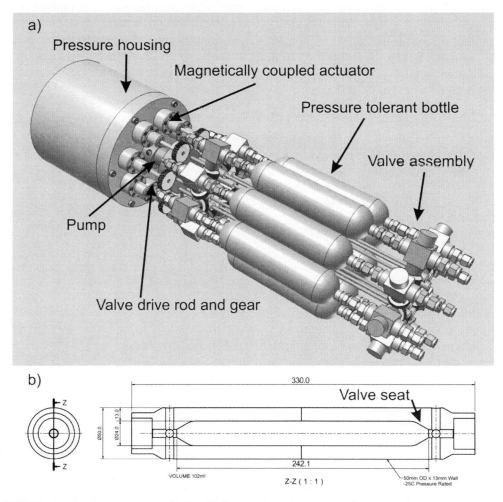

Figure 3. Water sampler design concept: (a) 3D CAD rendering of water sampler carousel concept using commercial valves and (b) engineering drawing of titanium bottle design to withstand sample freezing to −25 °C. The valve seat (cup) for bespoke cone and cup valves (not shown) is annotated.

34 MPa (bottles) and 46 MPa (valves). While this is sufficient to return liquid samples from the lake bottom to the surface at or above approximately −2°C, this is not sufficient to resist the additional pressure should the sample be cooled further. In addition, these components are not available in Titanium as standard, but are manufactured in stainless steel, which is problematic for iron metrology. To address these problems, we have developed a bottle and valve design that can be manufactured in titanium and resists greater pressures (see Figure 3b). These are formed by welding two separately machined halves together with a bespoke internal cone and cup valve included before welding. This design facilitates flushing and cleaning of the bottle and valve and is designed to withstand an internal pressure of 226 MPa.

4.2.2.2. The particle sampler.

The particle sampler is integrated with the water sampler, and there are duplicate systems in each carousel. The sampler is based on filter membranes and a gear pump (GJ-N23, IDEX Health & Science Group) that pulls a measured volume through the filter. The pump is controlled by the electronics unit within the water sampler carousel pressure case. This unit enables measurement of the number of pump rotations (and hence volume passed through the filters). The power consumption and rotation rate can also be used to estimate back pressure and can therefore be used to estimate filter clogging. The particle samplers will not be run simultaneously to further mitigate the risk of rapid clogging before a sufficient depth range of the lake is sampled. We are currently evaluating variants of the design using MicropreSure, 0.45 μm mixed ester of cellulose membrane, filters (Millipore, UK). In the first design, three filters are connected in parallel to the pump to reduce the back pressure and to enable the flow rate required (>3.3 L min^{-1}). The filters are contained within the sterile plastic housing in which they are supplied. This has the advantage of low-cost, while facilitating microbial control. In the second design, the same filter material is used but is repackaged into a cone or blind-ended tube, which is contained within a retainer within a pressure-tolerant water sample bottle. This design also enables microbial control (using the same methods applied to the water sampler) and maintains the pressure of the sample until analysis. This method is extremely robust and allows the filter samples to be treated in the same way as the water samples on site.

4.2.2.3. The short corer.

The short corer (UWITEC, Austria) will be tested with a range of sediment types including with indistinct sediment water interfaces. We propose to use a double "orange peel" core catcher (UWITEC) to enable retention of loose sediment but are developing bespoke core catcher solutions (both flap and multiple

offset elastomeric iris designs) to provide risk mitigation. Prior to acquisition of the short core, the distance between the corer tip and the sediment water interface will be measured using sonar.

4.3. The Tether

The tether provides a flexible mechanical link between the probe, the top sheave (of the gantry), and the winch system provided by the probe support system (see below). The tether is also used to control the large corer. A tether that meets the preliminary specification of both the probe and corer has been identified (for details see Table 4). Detailed design of the probe, corer, and support systems, together with assessment of microbial control procedures, is underway to confirm the suitability of this choice. Optical and electrical links are provided by conductors within the cable, while strength members (aramid/aromatic polyester fiber) take the mechanical load. An encapsulating sheath (polyurethane) provides mechanical protection and prevents microbial egress from the tether interior. Preliminary tests evaluating microbial control on the outer surface of the encapsulation have been conducted. These incubated sections were cleaned with Teknon Biocleanse™ biocidal cleaner (Fisher, United Kingdom) and uncleaned sections with synthetic (simulated) lake water for 1 week. Samples of the unexposed synthetic lake water and of both water incubated with the cleaned and uncleaned sections were

Table 4. Specification of Tether Used to Provide Mechanical, Power, and Communication Links to the Probe

Tether Property	Specification
Optical conductors	Six single-mode communications fibers inside a protective stainless steel tube placed in the center of the tether layup. Water blocked
Electrical power conductors	4×2.5 mm^2 (12 AWG) copper conductors (operating voltage >340 DC). Water blocked
Electrical communication conductors	2×20 AWG twisted pairs. Water blocked
Binder	Mylar
Strength	Kevlar (Aramid) or Vectran (aromatic polyester) to provide breaking strain >6000 kg
Weight in air	332 kg km^{-1}
Weight in sea water	132 kg km^{-1}
Weight in pure water (0°C)	137 kg km^{-1}
Estimated volume on reel (4 km tether)	3 m^3

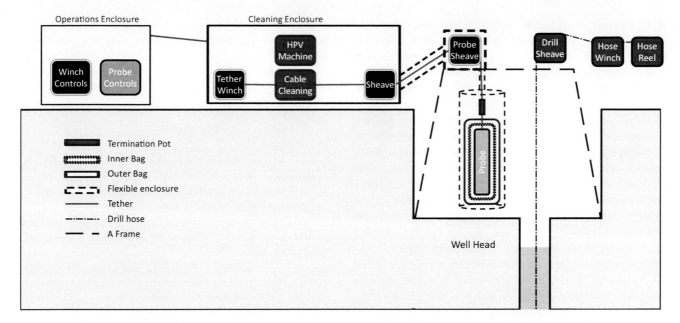

Figure 4. Schematic of probe support systems illustrating key components and zones for microbial control. The three main enclosures are (1) the operations enclosure which houses the probe (or corer) operator, winch operator, and scientist; (2) the cleaning enclosure, which contains the tether winch, equipment for microbial control (depicted here using HPV (that permeates the enclosure) and a separate cable cleaning system); and (3) the wellhead that provides an enclosed space for operations above the borehole. The winches and reels for the hot water drill (HWD) is depicted outside the wellhead and cleaning enclosure; however, we are investigating the use of a cleaning unit immediately above the wellhead, which would act on the hose and tether.

analyzed with flow cytometry SYBR Green II dye. This enumerated bacterial populations and confirmed that the polyurethane sheath could be cleaned and did not harbor significant microflora uncleaned. Further quantitative tests are underway to confirm this initial result.

4.4. Probe Support Systems

Probe support systems present at the ice surface are illustrated in Figure 4 together with representations of the probe and hot water drill. The three main enclosures are (1) the operations enclosure, which houses the probe (or corer) operator, winch operator, and scientist; (2) the cleaning enclosure, which contains the tether winch, equipment for microbial control; and a separate cable cleaning system, and (3) the wellhead that provides an enclosed space for operations above the borehole.

The operations enclosure includes the probe command and control unit, video screens for visualization of images and data from the probe, and facilities of data logging and recording. The system is based on two (for redundancy) ruggedized PCs with data acquisition cards to interface to other systems. The data link from the command and control unit passes through one (of the two) multiplexor units where an optical link is generated. Power systems for the probe (also used for the large corer) are also housed in this unit. The optical link and power supply connects the operations enclosure to the cleaning enclosure.

The winch is housed inside the cleaning enclosure and consists of a lightweight reel, electric rotation control with indexing (cable out measurement), a render (a torque limiting clutch) rotating optical connections, and slip rings for electrical power connection. The HPV machine is used, together with heat and ventilation, to further clean the tether winch and the outside of the probe protective bagging prior to deployment.

As illustrated in Figure 4, the probe top sheave is connected to the cleaning enclosure via a flexible link. This enables the probe and tether to pass over the sheave in a protective environment and to be positioned in the wellhead ready for deployment.

The presence of gas hydrates and high gas concentrations in the lake are being investigated by the Lake Ellsworth Consortium. We are applying formal risk estimation techniques to evaluate the blowout risk. If required, blowout protection will be developed to mitigate the risk that the water in the borehole is ejected by expanding gas. In our current design, the borehole itself is capped with a gland that is securely fixed to the wellhead structure, which includes a large plate weighted with snow. This gland includes a shut-off valve and lining through the porous fern ice to enable blowout protection. Our current calculations suggest that pressurization of the air-filled headspace (nominally 270 m at hydrostatic equilibrium) to only 5 bar would sufficiently compress any outgassing enough to prevent blowout. The details of the blowout protection system are in development, but it is likely that an additional high-pressure water input (using sterile and stored hot water drill water) will connect to the gland beneath the shut-off valve to enable large quantities of water to be added to the borehole under pressure. This would enable any clathrate or supersaturated water to be pushed back into the lake and would provide hydrostatic head. This would rapidly reestablish equilibrium, and any remaining gas pressure at the gland could be vented while replacing this volume with water. This will be done if outgassing in the lake or borehole forces water up the borehole depressurizing the lake. The addition of large quantities of high-pressure water compresses any gas and re-pressurizes the borehole and the lake preventing a runaway situation.

The gland, lining, and any exposed fern ice at the top of the borehole will be irradiated with UV prior to deployment of the probe. Together with the heat and filter sterilization used in the HWD operation, this ensures that a microbial-controlled environment extends from the ice surface to the lake. The flexible enclosure link is then connected to the gland (and further irradiated with UV). The probe and tether is then lowered through the gland to be deployed.

The probe then descends to the lake to conduct the experiment as per the requirements specified above. On retrieval, the probe is guided from the lake into the ice borehole with the aid of conic guides on the top pressure vessel and images from the upward-looking instrumentation. On assent, the probe exterior is flushed with borehole water and passes back through the gland into the controlled environment of the sealed wellhead. This enables two-way contamination control at the drill site. All probe surfaces, samples, and data will be preserved for appropriate analysis to maximize the impact of this pioneering project.

5. OUTLOOK

The development of probe technologies and methodologies for clean access for the direct measurement and sampling of ESL will provide significant equipment and knowledge for future exploration of pristine and extreme environments. This will include the novel probe and probe management systems specifically developed to enable measurement of pristine environments. All equipment developed for the project is designed for multiple use, and key items will be NERC capital assets made available for future subglacial studies.

Acknowledgment. This research was undertaken by the Lake Ellsworth Consortium and funded by the UK Natural Environment Research Council.

REFERENCES

Aamot, H. W. C. (1968), Self-contained thermal probes for remote measurements within an ice sheet, paper presented at International Symposium on Antarctic Glaciological Exploration (ISAGE), Int. Assoc. of Sci. Hydrol., Hanover, N. H.

Aamot, H. W. C. (1970), Development of a vertically stabilized thermal probe for studies in and below ice sheets, *J. Eng. Ind.*, *92*(2), 263–268.

Abegg, F., H.-J. Hohnberg, T. Pape, G. Bohrmann, and J. Freitag (2008), Development and application of pressure-core-sampling systems for the investigation of gas- and gas-hydrate-bearing sediments, *Deep Sea Res., Part I*, *55*(11), 1590–1599.

Agalloco, J. (2004), Aseptic processing: A review of current industry practice, *Pharm. Technol.*, *28*(10), 126–150.

Albro, C., et al. (1990), *An Innovative Water Sampling Device for Use in Hydrographic Profiling Systems - The WOCE Water Sampler, Marine Instrumentation 90*, 82–86, West Star Productions, Spring Valley, Calif.

Alekhina, I., P. Doran, T. Naganuma, G. di Prisco, B. Storey, W. Vincent, J. Wadham, and D. Walton (2009), Code of conduct for the exploration and research of subglacial aquatic environments, Sci. Comm. on Antarct. Res.

Allen, B., T. Austin, N. Forrester, R. Goldsborough, A. Kukulya, G. Packard, M. Purcell, and R. Stokey (2006), Autonomous docking demonstrations with enhanced REMUS technology, paper presented at Oceans 2006, IEEE, Boston, Mass.

Allen, K. W., H. Humphreys, and R. F. Sims-Williams (1997), Sterilization of instruments in general practice: What does it entail?, *Public Health*, *111*(2), 115–117.

Antonsen, H. R. K., et al. (2008), Sterilizing filtration of liquids. Technical report no. 26 (revised 2008), *PDA J. Pharm. Sci. Technol.*, *62*(5)Suppl. TR26, 2–60.

Ayliffe, G. A. J., J. R. Babb, and C. R. Bradley (1992), Sterilization of arthroscopes and laparoscopes, *J. Hosp. Infect.*, *22*(4), 265–269.

Baehr, M. M., and M. D. DeGrandpre (2002), Under-ice CO_2 and O_2 variability in a freshwater lake, *Biogeochemistry*, *61*(1), 95–113.

Bagshaw, E. A., M. Tranter, J. L. Wadham, A. G. Fountain, and M. Mowlem (2011), High resolution monitoring reveals dissolved oxygen dynamics in an Antarctic cryoconite hole, *Hydrol. Processes*, in press.

Barengoltz, J. (2005), *A Review of the Approach of NASA Projects to Planetary Protection Compliance*, NASA, Washington, D. C.

Behar, A., F. Carsey, A. Lane, and H. Engelhardt (2001), The Antarctic Ice Borehole Probe, paper presented at Aerospace Conference, 2001, IEEE Proceedings, Big Sky, Mont. 10–17 March 2001.

Behar, A., J. Matthews, K. Venkateswaran, J. Bruckner, and J. Jonsson (2006), A deep sea hydrothermal vent bio-sampler for large volume in situ filtration of hydrothermal vent fluids, *Cah. Biol. Mar.*, *47*(4), 443–447.

Bell, J., J. Betts, and E. Boyle (2002), MITESS: A moored in situ trace element serial sampler for deep-sea moorings, *Deep Sea Res., Part I*, *49*(11), 2103–2118.

Bell, R. J., R. T. Short, F. H. W. van Amerom, and R. H. Byrne (2007), Calibration of an in situ membrane inlet mass spectrometer for measurements of dissolved gases and volatile organics in seawater, *Environ. Sci. Technol.*, *41*(23), 8123–8128.

Benkerroum, N. (2008), Antimicrobial activity of lysozyme with special relevance to milk, *Afr. J. Biotechnol.*, *7*(25), 4856–4867.

Bentley, C. R., and B. R. Koci (2007), Drilling to the beds of the Greenland and Antarctic ice sheets: A review, *Ann. Glaciol.*, *47*(1), 1–9.

Bentley, M. J., P. Christoffersen, D. A. Hodgson, A. M. Smith, S. Tulaczyk, and A. M. Le Brocq (2011), Subglacial lake sediments and sedimentary processes: Potential archives of ice sheet evolution, past environmental change, and the presence of life, in *Antarctic Subglacial Aquatic Environments, Geophys. Monogr. Ser.*, doi: 10.1029/2010GM000940, this volume.

Biele, J., S. Ulamec, J. Garry, S. Sheridan, A. D. Morse, S. Barber, I. Wright, H. Tüg, and T. Mock (2002), Melting probes at Lake Vostok and Europa, *Proceedings of the Second European Workshop on Exo-Astrobiology*, edited by H. Lacoste, pp. 253–260, ESA Publ. Div., Noordwijk, The Netherlands.

Bindschadler, R. A., M. A. King, R. B. Alley, S. Anandakrishnan, and L. Padman (2003), Tidally controlled stick-slip discharge of a West Antarctic ice stream, *Science*, *301*(5636), 1087–1089.

Blake, E. W., and G. K. C. Clarke (1991), Subglacial water and sediment samplers, *J. Glaciol.*, *37*(125), 188–190.

Blomqvist, S. (1991), Quantitative sampling of soft-bottom sediments — problems and solutions, *Mar. Ecol. Prog. Ser.*, *72*(3), 295–304.

Boon, N., W. Windt, W. Verstraete, and E. M. Top (2002), Evaluation of nested PCR-DGGE (denaturing gradient gel electrophoresis) with group specific 16S rRNA primers for the analysis of bacterial communities from different wastewater treatment plants, *FEMS Microbiol. Ecol.*, *39*(2), 101–112.

Borneman, J., and E. Triplett (1997), Molecular microbial diversity in soils from eastern Amazonia: Evidence for unusual microorganisms and microbial population shifts associated with deforestation, *Appl. Environ. Microbiol.*, *63*(7), 2647–2653.

Boulos, L., M. Prévost, B. Barbeau, J. Coallier, and R. Desjardins (1999), LIVE/DEAD® BacLight™: Application of a new rapid staining method for direct enumeration of viable and total bacteria in drinking water, *J. Microbiol. Methods*, *37*(1), 77–86.

Brito, M. P., and G. Griffiths (2009), Results of expert judgments on the faults and risks with Autosub3 and an analysis of its campaign to Pine Island Bay, Antarctica, 2009, paper presented at International Symposium on Unmanned Untethered Submersible Technology (UUST 2009), Autonomous Undersea Syst. Inst. (AUSI), Durham, N. H., 23–26 August 2009.

Brito, M. P., and G. Griffiths (2011), A Markov chain state transition approach to establishing critical phases for AUV reliability, *IEEE J. Oceanic Eng.*, in press.

Bruchhausen, P. M., J. A. Raymond, S. S. Jacobs, A. L. DeVries, E. M. Thorndike, and H. H. DeWitt (1979), Fish, crustaceans, and the sea-floor under the Ross Ice Shelf, *Science*, *203*(4379), 449–451.

Bruhn, F. C., F. D. Carsey, J. Kohler, M. Mowlem, C. R. German, and A. E. Behar (2005), MEMS enablement and analysis of the miniature autonomous submersible explorer, *IEEE J. Oceanic Eng.*, *30*(1), 165–178.

Bulat, S. A., I. A. Alekhina, V. Y. Lipenkov, V. V. Lukin, D. Marie, and J. R. Petit (2009), Cell concentrations of microorganisms in glacial and lake ice of the Vostok ice core, East Antarctica, *Microbiology*, *78*(6), 808–810.

Burns, M., and H. Valdivia (2008), Modelling the limit of detection in real-time quantitative PCR, *Eur. Food Res. Technol.*, *226*(6), 1513–1524.

Caimi, F. M., D. M. Kocak, F. Dalgleish, and J. Watson (2008), Underwater imaging and optics: Recent advances, paper presented at OCEANS 2008, IEEE, Quebec City, Que., Canada.

Camilli, R., and A. Duryea (2007), Characterizing marine hydrocarbons with in situ mass spectrometry, paper presented at OCEANS 2007, IEEE, Vancouver, B. C., Canada.

Carr, M. H., et al. (1998), Evidence for a subsurface ocean on Europa, *Nature*, *391*, 363–365.

Carsey, F., A. Behar, A. L. Lane, V. Realmuto, and H. Engelhardt (2002), A borehole camera system for imaging the deep interior of ice sheets, *J. Glaciol.*, *48*(163), 622–628.

Chambers, J. W., and N. G. Cameron (2001), A rod-less piston corer for lake sediments: An improved, rope-operated percussion corer, *J. Paleolimnol.*, *25*(1), 117–122.

Christner, B. C., J. A. Mikucki, C. M. Foreman, J. Denson, and J. C. Priscu (2005), Glacial ice cores: A model system for developing extraterrestrial decontamination protocols, *Icarus*, *174*(2), 572–584.

Chu, G. Q., and J. H. Sun (2008), Quantitative assessment of building fire risk to life safety, *Risk Anal.*, *28*(3), 615–626.

Chung, S., R. Kern, R. Konkol, J. Barengoltz, and H. Cash (2008), Vapor hydrogen peroxide as alternative to dry heat microbial reduction, *Adv. Space Res.*, *42*(6), 1150–1160.

Commitee on Principles of Environmental Stewardship for the Exploration and Study of Subglacial Environments N.R.C. (2007), *Exploration of Antarctic Subglacial Aquatic Environments: Environmental and Scientific Stewardship*, Natl. Acad. Press, Washington, D. C.

Craven, M., I. Allison, R. Brand, A. Elcheikh, J. Hunter, M. Hemer, and S. Donoghue (2004), Initial borehole results from the Amery Ice Shelf hot-water drilling project, *Ann. Glaciol.*, *39*, 531–539.

Craven, M., F. Carsey, A. Behar, J. Matthews, R. Brand, A. Elcheikh, S. Hall, and A. Treverrow (2005), Borehole imagery of meteoric and marine ice layers in the Amery Ice Shelf, East Antarctica, *J. Glaciol.*, *51*(172), 75–84.

Daly, K. L., and R. H. Byrne (2004), Chemical and biological sensors for time-series research: Current status and new directions, *Mar. Technol. Soc. J.*, *38*(2), 121–143.

Dawber, M., and R. D. Powell (1995), Epifaunal distributions at Antarctic Marine-ending Glaciers: Influences of ice dynamics and sedimentation in The Antarctic Region: Geological evolution and processes, paper presented at VII International Symposium on Antarctic Earth Sciences, Terra Antarctica, Siena, Italy.

Degrandpre, M. D. (1993), Measurement of seawater pCO_2 using a renewable-reagent fiber optic sensor with colorimetric detection, *Anal. Chem.*, *65*(4), 331–337.

Department of Health and Human Services (2004), Guidance for industry: Sterile drug products produced by aseptic processing— Current good manufacturing practice, report, U.S. Food and Drug Admin., Silver Spring, Md.

de Resseguier, A. (2000), A new type of horizontal in situ water and fluid mud sampler, *Mar. Geol.*, *163*(1–4), 409–411.

DeVincenzi, D. L., P. Stabekis, and J. Barengoltz (1996), Refinement of planetary protection policy for Mars missions, *Adv. Space Res.*, *18*(1–2), 311–316.

D'Hondt, S., et al. (2004), Distributions of microbial activities in deep subseafloor sediments, *Science*, *306*(5705), 2216–2221.

Di Meo, C. A., J. R. Wakefield, and S. C. Cary (1999), A new device for sampling small volumes of water from marine microenvironments, *Deep Sea Res., Part I*, *46*(7), 1279–1287.

Doble, M. J., A. L. Forrest, P. Wadhams, and B. E. Laval (2009), Through-ice AUV deployment: Operational and technical experience from two seasons of Arctic fieldwork, *Cold Reg. Sci. Technol.*, *56*(2–3), 90–97.

Doherty, K. W., C. D. Taylor, and O. C. Zafiriou (2003), Design of a multi-purpose titanium bottle for uncontaminated sampling of carbon monoxide and potentially of other analytes, *Deep Sea Res., Part I*, *50*(3), 449–455.

Eigenbrode, J., L. G. Benning, J. Maule, N. Wainwright, A. Steele, and H. E. F. Amundsen (2009), A field-based cleaning protocol for sampling devices used in life-detection studies, *Astrobiology*, *9*(5), 455–465.

Engelhardt, H. F., W. D. Harrison, and B. Kamb (1978), Basal sliding and conditions at the glacier bed as revealed by bore-hole photography, *J. Glaciol.*, *20*(84), 469–508.

Engelhardt, H. F., N. Humphrey, B. Kamb, and M. Fahnestock (1990), Physical conditions at the base of a fast moving Antarctic ice stream, *Science*, *248*(4951), 57–59.

Falkner, K. K., et al. (2005), Dissolved oxygen extrema in the Arctic Ocean halocline from the North Pole to the Lincoln Sea, *Deep Sea Res., Part I*, *52*(7), 1138–1154.

Falkowski, P., and D. A. Kiefer (1985), Chlorophyll a fluorescence in phytoplankton: Relationship to photosynthesis and biomass, *J. Plankton Res.*, *7*(5), 715–731.

Fisher, M. M., and E. W. Triplett (1999), Automated approach for ribosomal intergenic spacer analysis of microbial diversity and its application to freshwater bacterial communities, *Appl. Environ. Microbiol.*, *65*(10), 4630–4636.

Fontanos, P. M., K. Yamamoto, F. Nakajima, and K. Fukushi (2010), Identification and quantification of the bacterial community on the surface of polymeric membranes at various stages of biofouling using fluorescence in situ hybridization, *Sep. Sci. Technol.*, *45*(7), 904–910.

Food Standards Agency (2004), *FSA Guidance on the Requirements of Food Hygiene Legislation*, Northern Ireland, U. K.

Freifeld, B. M., R. C. Trautz, Y. K. Kharaka, T. J. Phelps, L. R. Myer, S. D. Hovorka, and D. J. Collins (2005), The U-tube: A novel system for acquiring borehole fluid samples from a deep geologic CO_2 sequestration experiment, *J. Geophys. Res.*, *110*, B10203, doi:10.1029/2005JB003735.

French, L. C., F. Carsey, W. Zimmerman, G. French, F. S. Anderson, P. Shakkotti, J. Feldman, R. Bonitz, J. Hall, and C. Harmon (2001), Cryobots: An answer to subsurface mobility on planetary icy environments, paper presented at 6th International Symposium on Artificial Intelligence, Robotics, and Automation in Space, EMS Technol., Montreal, Que., Canada.

Fricker, H. A., and T. Scambos (2009), Connected subglacial lake activity on lower Mercer and Whillans ice streams, West Antarctica, 2003–2008, *J. Glaciol.*, *55*(190), 303–315.

Fukasawa, T., S. Hozumi, M. Morita, T. Oketani, and M. Masson (2006), Dissolved methane sensor for methane leakage monitoring in methane hydrate production, paper presented at OCEANS 2006, IEEE, Boston, Mass.

Furukawa, K., J. B. Cologne, Y. Shimizu, and N. P. Ross (2009), Predicting future excess events in risk assessment, *Risk Anal.*, *29*(6), 885–899.

Gaidos, E., et al. (2007), A simple sampler for subglacial water bodies, *J. Glaciol.*, *53*(180), 157–158.

Gary, M., N. Fairfield, W. C. Stone, D. Wettergreen, G. A. Kantor, and J. M. Sharp Jr. (2008), 3D mapping and characterization of Sistema Zacatón from DEPTHX (DEep Phreatic THermal eXplorer, paper presented at Karst Conference 2008: 11th Multidisciplinary Conference on Sinkholes and the Engineering and Environmental Impacts of Karst: Integrating Science and Engineering to Solve Karst Problems, ASCE, Tallahassee, Fla., September 22–26, 2008.

Gasparini, R., T. Pozzi, R. Magnelli, D. Fatighenti, E. Giotti, G. Poliseno, M. Pratelli, R. Severini, P. Bonanni, and L. De Feo (1995), Evaluation of in vitro efficacy of the disinfectant Virkon, *Eur. J. Epidemiol.*, *11*(2), 193–197.

Gordon, S., M. Sharp, B. Hubbard, C. Smart, B. Ketterling, and I. Willis (1998), Seasonal reorganization of subglacial drainage inferred from measurements in boreholes, *Hydrol. Processes*, *12*(1), 105–133.

Guevara, R., A. Rizzo, R. Sánchez, and M. Arribére (2005), Heavy metal inputs in Northern Patagonia lakes from short sediment core analysis, *J. Radioanal. Nucl. Chem.*, *265*(3), 481–493.

Halfmann, H., B. Denis, N. Bibinov, J. Wunderlich, and P. Awakowicz (2007), Identification of the most efficient VUV/UV radiation for plasma based inactivation of Bacillus atrophaeus spores, *J. Phys. D Appl. Phys.*, *40*(19), 5907.

Hansen, B. L., and L. Kersten (1984), An in situ sampling thermal probe, *Ice Drilling Technology*, edited by G. Holdsworth et al., *USA CRREL Spec. Rep. 84-34*, 119–122, Cold Reg. Res. and Eng. Lab., Hanover, N. H.

Hanson, A. K. (2005), Cabled and autonomous submersible chemical analyzers for coastal observations, *Abstr. Pap. Am. Chem. Soc.*, *229*, U152–U153.

Harper, J. T., N. F. Humphrey, and M. C. Greenwood (2002), Basal conditions and glacier motion during the winter/spring transition, Worthington Glacier, Alaska, U.S.A, *J. Glaciol.*, *48*, 42–50.

Harrison, W. D., and B. Kamb (1973), Glacier bore-hole photography, *J. Glaciol.*, *12*(64), 129–137.

Harrison, W. D., M. Truffer, K. A. Echelmeyer, D. A. Pomraning, K. A. Abnett, and R. H. Ruhkick (2004), Probing the till beneath Black Rapids Glacier, Alaska, USA, *J. Glaciol.*, *50*, 608–614.

Hart, J. K., J. C. Rose, K. Martinez, and R. Ong (2009), Subglacial clast behaviour and its implication for till fabric development: New results derived from wireless subglacial probe experiments, *Quat. Sci. Rev.*, *28*, 597–607.

Haysom, I. W., and A. K. Sharp (2005), Bacterial contamination of domestic kitchens over a 24-hour period, *Br. Food J.*, *107*(7), 453–466.

Henn, G. G., C. Birkinshaw, and M. Buggy (1996), A comparison of the effects of γ-irradiation and ethylene oxide sterilization on the properties of compression moulded poly-D,L-lactide, *J. Mater. Sci.*, *7*(10), 591–595.

Hernandez, A., E. Martro, L. Matas, M. Martin, and V. Ausina (2000), Assessment of in-vitro efficacy of 1% Virkon® against bacteria, fungi, viruses and spores by means of AFNOR guidelines, *J. Hosp. Infect.*, *46*(3), 203–209.

Hodgson, D. A., S. J. Roberts, M. J. Bentley, E. L. Carmichael, J. A. Smith, E. Verleyen, W. Vyverman, P. Geissler, M. J. Lengd, and D. C. W. Sanderson (2009), Exploring former subglacial Hodgson Lake, Antarctica. Paper II: Palaeolimnology, *Quat. Sci. Rev.*, *28*(23–24), 2310–2325.

Hoefel, D., W. L. Grooby, P. T. Monis, S. Andrews, and C. P. Saint (2003), Enumeration of water-borne bacteria using viability assays and flow cytometry: A comparison to culture-based techniques, *J. Microbiol. Methods*, *55*(3), 585–597.

Holmes, C. (2008), Cleaning, disinfection and sterilisation policy, *Issue, CP24 0908*, West, East and North Hertfordshire Primary Care Trusts, Natl. Health Serv., London, U. K.

Hunter, C. N., R. M. Gordon, S. E. Fitzwater, and K. H. Coale (1996), A Rosette System for the collection of trace metal clean seawater, *Limnol. Oceanogr.*, *41*(6), 1367–1372.

Jaffe, J. S. (2005), Performance bounds on synchronous laser line scan systems, *Opt. Express*, *13*(3), 738–748.

Jaffe, J. S. (2010), Enhanced extended range underwater imaging via structured illumination, *Opt. Express*, *18*(12), 12,328–12,340.

Jahnke, R. A., and L. H. Knight (1997), A gravity-driven, hydraulically-damped multiple piston corer for sampling fine-grained sediments, *Deep Sea Res., Part I*, *44*(4), 713–718.

Jenkins, A., P. Dutrieux, S. S. Jacobs, S. D. McPhail, J. R. Perrett, A. T. Webb, and D. White (2010), Observations beneath Pine Island Glacier in West Antarctica and implications for its retreat, *Nat. Geosci.*, *3*, 468–472.

Johnson, B. D., P. J. Wangersky, and X. L. Zhou (1987), An in situ pump sampler for trace materials in seawater, *Mar. Chem.*, *22*(2–4), 353–361.

Johnson, K. S., and L. J. Coletti (2002), In situ ultraviolet spectrophotometry for high resolution and long-term monitoring of nitrate, bromide and bisulfide in the ocean, *Deep Sea Res., Part I*, *49*(7), 1291–1305.

Johnston, M. D., S. Lawson, and J. A. Otter (2005), Evaluation of hydrogen peroxide vapour as a method for the decontamination of surfaces contaminated with Clostridium botulinum spores, *J. Microbiol. Methods*, *60*(3), 403–411.

Jornitz, M. W., J. E. Akers, J. P. Agalloco, R. E. Madsen, and T. H. Meltzer (2003), Considerations in sterile filtration. Part II: The sterilizing filter and its organism challenge: A critique of regulatory standards, *PDA J. Pharm. Sci. Technol.*, *57*(2), 88–96.

Kasser, P. (1960), Ein leichter thermischer Eisbohrer als Hilfsgerät zur Installation von Ablationsstangen auf Gletschern, *Pure Appl. Geophys.*, *45*(1), 97–114.

Kaufmann, E., G. Kargl, N. I. Kömle, M. Steller, J. Hasiba, F. Tatschl, S. Ulamec, J. Biele, M. Engelhardt, and J. Romstedt (2009), Melting and sublimation of planetary ices under low pressure conditions: Laboratory experiments with a melting probe prototype, *Earth Moon Planets*, *105*(1), 11–29.

Kelty, J. R. (1995), An in situ sampling thermal probe for studying global ice sheets, Ph.D. dissertation, p. 189, Univ. of Nebr., Lincoln.

Kepner, R. L., and J. R. Pratt (1994), Use of fluorochromes for direct enumeration of total bacteria in environmental-samples — past and present, *Microbiol. Rev.*, *58*(4), 603–615.

Kocak, D. M., and F. M. Caimi (2005), The current art of underwater imaging - With a glimpse of the past and vision of the future, *Mar. Technol. Soc. J.*, *39*(3), 5–26.

Kocak, D. M., F. R. Dalgleish, F. M. Caimi, and Y. Y. Schechner (2008), A focus on recent developments and trends in underwater imaging, *Mar. Technol. Soc. J.*, *42*(1), 52–67.

Kristoffersen, Y., E. Lien, K. Festervoll, S. Ree, K. Ðardahl, and Ò. Hosøy (2006), The hydrostatic corer Selcore—a tool for sediment sampling and geophysical site characterization, *Mar. Geol.*, *229*(1–2), 101–112.

Kühl, M., R. N. Glud, J. Borum, R. Roberts, and S. Rysgaard (2001), Photosynthetic performance of surface-associated algae below sea ice as measured with a pulse-amplitude-modulated (PAM) fluorometer and O_2 microsensors, *Mar. Ecol.*, *223*, 1–14.

Kylian, O., and F. Rossi (2009), Sterilization and decontamination of medical instruments by low-pressure plasma discharges: Application of $Ar/O_2/N_2$ ternary mixture, *J. Phys. D Appl. Phys.*, *42*(8), 85,207–85,213.

Lanoil, B., M. Skidmore, J. C. Priscu, S. Han, W. Foo, S. W. Vogel, S. Tulaczyk, and H. Engelhardt (2009), Bacteria beneath the West Antarctic Ice Sheet, *Environ. Microbiol.*, *11*(3), 609–615.

Lebaron, P., N. Parthuisot, and P. Catala (1998), Comparison of blue nucleic acid dyes for flow cytometric enumeration of bacteria in aquatic systems, *Appl. Environ. Microbiol.*, *64*(5), 1725–1730.

Lemarchand, K., N. Parthuisot, P. Catala, and P. Lebaron (2001), Comparative assessment of epifluorescence microscopy, flow cytometry and solid-phase cytometry used in the enumeration of specific bacteria in water, *Aquat. Microb. Ecol.*, *25*(3), 301–309.

Lever, M. A., et al. (2006), Trends in basalt and sediment core contamination during IODP Expedition 301, *Geomicrobiol. J.*, *23*(7), 517–530.

Lin, S., and H. P. Cohen (1968), Measurement of adenosine triphosphate content of crayfish stretch receptor cell preparations, *Anal. Biochem.*, *24*(3), 531–540.

Lipps, J. H., T. E. Ronan, and T. E. Delaca (1979), Life below the Ross Ice Shelf, Antarctica, *Science*, *203*(4379), 447–449.

Liu, W., T. L. Marsh, H. Cheng, and L. J. Forney (1997), Characterization of microbial diversity by determining terminal restriction fragment length polymorphisms of genes encoding 16S rRNA, *Appl. Environ. Microbiol.*, *63*(11), 4516–4522.

Luna, G., K. Stumm, A. Pusceddu, and R. Danovaro (2009), Archaeal diversity in deep-sea sediments estimated by means of different terminal-restriction fragment length polymorphisms (T-RFLP) protocols, *Curr. Microbiol.*, *59*(3), 356–361.

Makinson, K. (1993), The Bas hot-water drill: Development and current design, *Cold Reg. Sci. Technol.*, *22*(1), 121–132.

Manzoor, S. E., P. A. Lambert, P. A. Griffiths, M. J. Gill, and A. P. Fraise (1999), Reduced glutaraldehyde susceptibility in Mycobacterium chelonae associated with altered cell wall polysaccharides, *J. Antimicrob. Chemother.*, *43*(6), 759–765.

Maqueda, M., and J. M. Rodriguez (2008), Antimicrobial lactic acid bacteria metabolites, in *Molecular Aspects of Lactic Acid Bacteria for Traditional and New Applications*, edited by B. Mayo, P. López, and G. Pérez-Martínez, pp. 167–208, Research Signpost, Trivandrum, Kerala, India.

Martin, J. H., R. M. Gordon, and S. E. Fitzwater (1990), Iron in Antarctic waters, *Nature*, *345*(6271), 156–158.

McDonnell, G., and S. Moselio (2009), Sterilization and disinfection, in *Encyclopedia of Microbiology*, edited by M. Schaechter, pp. 529–548, Academic Press, Oxford, U. K.

McMinn, A., and C. Ashworth (1998), The use of oxygen microelectrodes to determine the net production by an Antarctic sea ice algal community, *Antarct. Sci.*, *10*(1), 39–44.

McPhail, S. D., et al. (2009), Exploring beneath the PIG Ice Shelf with the Autosub3 AUV, paper presented at Oceans 2009-Europe, 2009. Oceans '09, IEEE, Bremen, Germany.

Mendes, G. C. C., T. R. S. Brandâo, and C. L. M. Silva (2007), Ethylene oxide sterilization of medical devices: A review, *Am. J. Infect. Control*, *35*(9), 574–581.

Meng, S., H. Xu, and F. Wang (2010), Research advances of antimicrobial peptides and applications in food industry and agriculture, *Curr. Protein Pept. Sci.*, *11*(4), 264–273.

Meriläinen, J. J., J. Hynynen, A. Palomäki, P. Reinikainen, A. Teppo, and K. Granberg (2000), Importance of diffuse nutrient loading and lake level changes to the eutrophication of an originally oligotrophic boreal lake: A palaeolimnological diatom and chironomid analysis, *J. Paleolimnol.*, *24*(3), 251–270.

Miles, A. A., S. S. Misra, and J. O. Irwin (1938), The estimation of the bactericidal power of the blood, *J. Hyg.*, *38*(6), 732–749.

Miller, D. P., R. Bonaccorsi, and K. Davis (2008), Design and practices for use of automated drilling and sample handling in MARTE while minimizing terrestrial and cross contamination, *Astrobiology*, *8*(5), 947–965.

Mock, T., G. S. Dieckmann, C. Haas, A. Krell, J.-L. Tison, A. L. Belem, S. Papadimitriou, and D. N. Thomas (2002), Micro-optodes in sea ice: A new approach to investigate oxygen dynamics during sea ice formation, *Aquat. Microb. Ecol.*, *29*(3), 297–306.

Mock, T., M. Kruse, and G. S. Dieckmann (2003), A new microcosm to investigate oxygen dynamics at the sea ice water interface, *Aquat. Microb. Ecol.*, *30*(2), 197–205.

Moisan, M., J. Barbeau, S. Moreau, J. Pelletier, M. Tabrizian, and L. H. Yahia (2001), Low-temperature sterilization using gas plasmas: A review of the experiments and an analysis of the inactivation mechanisms, *Int. J. Pharm.*, *226*(1–2), 1–21.

Mosley, G. A. (2006), Sterility Assurance Level: The term and its definition continues to cause confusion in the industry, *Pharm. Microbiol. Forum Newslett.*, *14*(5), 2–14.

Mueller, R., R. S. Brown, H. Hop, and L. Moulton (2006), Video and acoustic camera techniques for studying fish under ice: A review and comparison, *Rev. Fish Biol. Fish.*, *16*(2), 213–226.

Muyzer, G., and K. Smalla (1998), Application of denaturing gradient gel electrophoresis (DGGE) and temperature gradient gel electrophoresis (TGGE) in microbial ecology, *Antonie van Leeuwenhoek*, *73*(1), 127–141.

Nakayama, T., H. Wake, K. Ozawa, N. Nakamura, and T. Matsunaga (1998), Electrochemical prevention of marine biofouling on a novel titanium-nitride-coated plate formed by radio-frequency arc spraying, *Appl. Microbiol. Biotechnol.*, *50*(4), 502–508.

National Patient Safety Agency, U. K. (2007), *National Specifications for Cleanliness in the NHS*, London, U. K.

Neale, J. L., and D. Walker (1996), Sampling sediment under warm deep water, *Quat. Sci. Rev.*, *15*(5–6), 581–590.

Nicholls, K. W., et al. (2006), Measurements beneath an Antarctic ice shelf using an autonomous underwater vehicle, *Geophys. Res. Lett.*, *33*, L08612, doi:10.1029/2006GL025998.

Niskin, S. J., D. A. Segar, and P. R. Betzer (1973), New Niskin sampling bottles without internal closures and their use for collecting near bottom samples for trace metal analysis, *Eos Trans. AGU*, *54*(11), 1110.

O'Connor, P. (2002), *Practical Reliability Engineering*, 4th ed., John Wiley, Chichester, U. K.

Picard, C., C. Ponsonnet, E. Paget, X. Nesme, and P. Simonet (1992), Detection and enumeration of bacteria in soil by direct DNA extraction and polymerase chain-reaction, *Appl. Environ. Microbiol.*, *58*(9), 2717–2722.

Piñar, G., C. Saiz-Jimenez, C. Schabereiter-Gurtner, M. T. Blanco-Varela, W. Lubitz, and S. Rölleke (2001), Archaeal communities in two disparate deteriorated ancient wall paintings: Detection, identification and temporal monitoring by denaturing gradient gel electrophoresis, *FEMS Microbiol. Ecol.*, *37*(1), 45–54.

Ploux, L., K. Anselme, A. Dirani, A. Ponche, O. Soppera, and V. Roucoules (2009), Opposite responses of cells and bacteria to micro/nanopatterned surfaces prepared by pulsed plasma polymerization and UV-irradiation, *Langmuir*, *25*(14), 8161–8169.

Plueddemann, A. J., G. Packard, J. Lord, and S. Whelan (2008), Observing arctic coastal hydrography using the REMUS AUV, paper presented at 2008 IEEE/OES Autonomous Underwater Vehicles (AUV), IEEE, Woods Hole, Mass.

Powell, R. D., M. Dawber, J. N. McInnes, and A. R. Pyne (1995), Observations of the grounding-line area at a floating glacier terminus, paper presented at Int. Symp. on Glacial Erosion and Sedimentation, Reykjavík, Iceland.

Prien, R. D. (2007), The future of chemical in situ sensors, *Mar. Chem.*, *107*(3), 422–432.

Richardson, P. S. (1987), Flame sterilisation, *Int. J. Food Sci. Technol.*, *22*(1), 3–14.

Roman, C., and R. Camilli (2007), Design of a gas tight water sampler for AUV operations, paper presented at OCEANS 2007 - Europe, IEEE, Aberdeen, U. K.

Ross, N., et al. (2011), Ellsworth Subglacial Lake, West Antarctica: Its history, recent field campaigns, and plans for its exploration, in *Antarctic Subglacial Aquatic Environments*, *Geophys. Monogr. Ser.*, doi: 10.1029/2010GM000936, this volume.

Ruberg, S. A., and B. J. Eadie (2000), Remotely deployable water sampler, in OCEANS 2000 MTS/IEEE Conference and Exhibition, vol. 1, pp. 113–117, IEEE, New York.

Rutala, W. A., and D. J. Weber (1999), Infection control: The role of disinfection and sterilization, *J. Hosp. Infect.*, *43*(Suppl. 1), S43–S55.

Seidel, M. P., M. D. DeGrandpre, and A. G. Dickson (2008), A sensor for in situ indicator-based measurements of seawater pH, *Mar. Chem.*, *109*(1–2), 18–28.

Short, R. T., D. P. Fries, S. K. Toler, C. E. Lembke, and R. H. Byrne (1999), Development of an underwater mass-spectrometry system for in situ chemical analysis, *Meas. Sci. Technol.*, *10*, 1195–1201.

Short, R. T., S. K. Toler, G. P. G. Kibelka, D. T. Rueda Roa, R. J. Bell, and R. H. Byrne (2006), Detection and quantification of chemical plumes using a portable underwater membrane introduction mass spectrometer, *TrAC, Trends Anal. Chem.*, *25*(7), 637–646.

Siegert, M. J., et al. (2006), Exploration of Ellsworth Subglacial Lake: A concept paper on the development, organisation and execution of an experiment to explore, measure and sample the environment of a West Antarctic subglacial lake, *Rev. Environ. Sci. Biotechnol.*, *6*, 161–179.

Smith, A. W., D. E. Skilling, J. D. Castello, and S. O. Rogers (2004), Ice as a reservoir for pathogenic human viruses: Specifically, caliciviruses, influenza viruses, and enteroviruses, *Med. Hypotheses*, *63*(4), 560–566.

Steele, A., M. K. Schweizer, V. Starke, M. D. Fries, J. Eigenbrode, L. Benning, and H. E. F. Amundsen (2006), In situ contamination monitoring of sterile sampling and coring procedures in the arctic, *Astrobiology*, *6*(1), 209.

Stokey, R., M. Purcell, N. Forrester, T. Austin, R. Goldsborough, B. Allen, and C. von Alt (1997), A docking system for REMUS, an autonomous underwater vehicle, paper presented at OCEANS '97 MTS/IEEE Conference Proceedings, IEEE, Halifax, Nova Scotia, Canada, 06–09 Oct. 1997.

Stone, D. B., and G. K. C. Clarke (1996), In situ measurements of basal water quality and pressure as an indicator of the character of subglacial drainage systems, *Hydrol. Processes*, *10*(4), 615–628.

Stone, D. B., G. K. C. Clarke, and E. W. Blake (1993), Subglacial measurement of turbidity and electrical-conductivity, *J. Glaciol.*, *39*(132), 415–420.

Stone, W., et al. (2010), ENDURANCE engineering: Two years under the ice in Antarctica, paper presented at Astrobiology Science Conference 2010, NASA Astrobiol. Program, League City, Tex., April 26–29, 2010.

Stukas, V., C. S. Wong, and W. K. Johnson (1999), Sub-part per trillion levels of lead and isotopic profiles in a fjord, using an ultra-clean pumping system, *Mar. Chem.*, *68*(1–2), 133–143.

Suggett, D. J., et al. (2009), Interpretation of fast repetition rate (FRR) fluorescence: Signatures of phytoplankton community structure versus physiological state, *Mar. Ecol. Prog. Ser.*, *376*, 1–19.

Sun, M. Y., and S. G. Wakeham (1994), Molecular evidence for degradation and preservation of organic-matter in the anoxic Black Sea Basin, *Geochim. Cosmochim. Acta*, *58*(16), 3395–3406.

Suzuki, M., M. S. Rappe, and S. J. Giovannoni (1998), Kinetic bias in estimates of coastal picoplankton community structure obtained by measurements of small-subunit rRNA gene PCR amplicon length heterogeneity, *Appl. Environ. Microbiol.*, *64*(11), 4522–4529.

Tabor, P. S., J. W. Deming, K. Ohwada, H. Davis, M. Waxman, and R. R. Colwell (1981), A pressure-retaining deep ocean sampler and transfer system for measurement of microbial activity in the deep-sea, *Microb. Ecol.*, *7*(1), 51–65.

Tengberg, A., et al. (2006), Evaluation of a lifetime-based optode to measure oxygen in aquatic systems, *Limnol. Oceanogr. Methods*, *4*, 7–17.

Tita, G. S., G. Desrosiers, and M. Vincx (2000), New type of hand-held corer for meiofaunal sampling and vertical profile investigation: a comparative study, *J. Mar. Biol. Assoc. U. K.*, *80*(1), 171–172.

Tosun, N., F. Akrinar, and A. Dogan (2003), Effects of glutaraldehyde on synovial tissue, *J. Int. Med. Res.*, *31*(5), 422–427.

Tranter, M., M. J. Sharp, G. H. Brown, I. C. Willis, B. P. Hubbard, M. K. Nielsen, C. C. Smart, S. Gordon, M. Tulley, and H. R. Lamb (1997), Variability in the chemical composition of in situ subglacial meltwaters, *Hydrol. Processes*, *11*(1), 59–77.

Trenerry, L. J., A. McMinn, and K. G. Ryan (2002), In situ oxygen microelectrode measurements of bottom-ice algal production in McMurdo Sound, Antarctica, *Polar Biol.*, *25*(1), 72–80.

Tüg, H. (2003), *Rechnergesteuerte Schmelzsonde zur Ermittlung unterschiedlicher Messparameter im Eisbereich*, D. P. Patentschrift, Germany.

Ulamec, S., J. Biele, O. Funke, and M. Engelhardt (2007), Access to glacial and subglacial environments in the Solar System by melting probe technology, *Rev. Environ. Sci. Biotechnol.*, *6*(1), 71–94.

van Reis, R., and A. Zydney (2001), Membrane separations in biotechnology, *Curr. Opin. Biotechnol.*, *12*(2), 208–211.

Vishnivetskaya, T., S. Kathariou, J. McGrath, D. Gilichinsky, and J. M. Tiedje (2000), Low-temperature recovery strategies for the isolation of bacteria from ancient permafrost sediments, *Extremophiles*, *4*(3), 165–173.

Vogel, S. W., et al. (2008), Subglacial Environment exploration-concept and technological challenges for the development and operation of a Sub-Ice ROV'er (SIR) and advanced Sub-Ice instrumentation for short and long-term observations, paper presented at 2008 IEEE/OES Autonomous Underwater Vehicles (AUV), IEEE, Woods Hole, Mass.

Vogel, S. W., R. D. Powell, I. Griffith, K. Anderson, T. Lawson, S. A. Schiraga, and T. Lawson (2007), Ice-Borehole ROV- A new tool for subglacial research, paper presented at Workshop on AUV Science in Extreme Environments, Scott Polar Research Institute, Cambridge, U. K., 11–13 April 2007.

Wadhams, P., J. P. Wilkinson, and S. D. McPhail (2006), A new view of the underside of Arctic sea ice, *Geophys. Res. Lett.*, *33*, L04501, doi:10.1029/2005GL025131.

Walker, V. K., G. R. Palmer, and G. Voordouw (2006), Freeze-thaw tolerance and clues to the winter survival of a soil community, *Appl. Environ. Microbiol.*, *72*(3), 1784–1792.

Wallhausser, K. H. (1979), Is the removal of microorganisms by filtration really a sterilization method?, *J. Parenter. Drug Assoc.*, *33*(3), 156–170.

Webster, J. J., G. J. Hampton, and F. R. Leach (1984), ATP in soil: A new extractant and extraction procedure, *Soil Biol. Biochem.*, *16*(4), 335–342.

Wiens, D. A., S. Anandakrishnan, J. P. Winberry, and M. A. King (2008), Simultaneous teleseismic and geodetic observations of the stick-slip motion of an Antarctic ice stream, *Nature*, *453*, 770–774.

Wilson, J. (2001), *Infection Control in Clinical Practice*, 2nd ed., Bailliere Tindall, Kidlington, U. K.

Wood, J. P., and G. Blair Martin (2009), Development and field testing of a mobile chlorine dioxide generation system for the decontamination of buildings contaminated with Bacillus anthracis, *J. Hazard. Mater.*, *164*(2–3), 1460–1467.

Zasloff, M. (1987), Magainins, a class of antimicrobial peptides from Xenopus skin: Isolation, characterization of two active forms, and partial cDNA sequence of a precursor, *Proc. Natl. Acad. Sci. U. S. A.*, *84*(15), 5449–5453.

Zimmerman, W., F. Scott Anderson, F. Carsey, P. Conrad, H. Englehardt, L. French, and M. Hecht (2002), The Mars '07 North Polar Cap deep penetration cryo-scout mission, in 2002 IEEE Aerospace Conference Proceedings, vol 1, pp. 305– 315, IEEE, New York.

Zimmerman, W., R. Bonitz, and J. Feldman (2001), Cryobot: An ice penetrating robotic vehicle for Mars and Europa, in 2001 IEEE Aerospace Conference Proceedings, vol. 1, pp. 311– 323, IEEE, New York.

A. D. Beaton, W. Harris, M. Tranter, and J. Wadham, School of Geographical Sciences, University of Bristol, Bristol BS8 1SS, UK.

M. Bentley, Department of Geography, University of Durham, Durham DH1 3LE, UK.

D. Blake, L. Capper, R. Clarke, H. Corr, C. Hill, R. Hindmarsh, D. A. Hodgson, E. King, K. Makinson, D. Pearce, A. Smith, and A. Tait, British Antarctic Survey, Madingley Road, Cambridge CB3 0ET, UK.

M. P. Brito, R. Brown, C. F. A. Floquet, L. Fowler, G. Griffiths, M. C. Mowlem, K. Saw, M.-N. Tsaloglou, E. M. Waugh, and J. B. Wyatt, National Oceanography Centre, European Way, Southampton SO14 3ZH, UK. (mcm@noc.soton.ac.uk)

C. Cockell, Planetary and Space Sciences Research Institute, Open University, Milton Keynes MK7 6AA, UK.

H. Lamb, Institute of Geography and Earth Sciences, Aberystwyth University, Aberystwyth SY23 2AX, UK.

B. Maher, Lancaster Environment Centre, Lancaster University, Lancaster LA1 4YB, UK.

J. Parnell, School of Geosciences, University of Aberdeen, Aberdeen AB24 3UF, UK.

J. Priscu, Department of Land Resources and Environmental Sciences, Montana State University, Bozeman, MT 59717, USA.

W. Rivera, Centro de Estudios Científicos, Arturo Prat 514, Valdivia, Chile.

N. Ross and M. J. Siegert, School of GeoSciences, University of Edinburgh, Edinburgh EH8 9XP, UK.

B. Whalley, School of Geography, Archaeology, and Palaeoecology, Queen's University Belfast, Belfast BT7 1NN, UK.

J. Woodward, School of Built and Natural Environment, Northumbria University, Newcastle upon Tyne NE1 8ST, UK.

Vostok Subglacial Lake: Details of Russian Plans/Activities for Drilling and Sampling

Valery Lukin

Russian Antarctic Expedition, Arctic and Antarctic Research Institute, St. Petersburg, Russia

Sergey Bulat

Petersburg Nuclear Physics Institute, Russian Academy of Sciences, Gatchina, Russia

The Russian Federation has developed a national project involving the drilling and sampling of Vostok Subglacial Lake, East Antarctica. The objective is to explore this extreme icy environment, using a variety of techniques to identify the forms and levels of life that exist there. The project is funded by the Russian Federal Service ROSHYDROMET. In the 2009/2010 season, drilling operations were restarted at a depth of 3559 m via new borehole 5G-2, successfully reaching a new depth of approximately 3650 m. New accretion ice, including the inclusion-rich "thermophile-containing" horizon (around 3608 m) was again recovered and will be studied to assess the previous scenario and findings. In 2010/2011, the drill will carefully continue to deepen the borehole leaving a 10- to 15-m ice cork and will in season 2011/2012 enter the lake, allowing water to rise up dozens of meters within borehole 5G-2 and subsequently freeze. During the same or following season (2012/2013), borehole 5G-2 will be redrilled to acquire rapidly frozen lake water for complex investigations. In the following season, 2013/2014, a special set of strictly decontaminated biophysical instruments, developed at the Petersburg Nuclear Physics Institute, will be lowered into the water body, with a battery of ocean observatory sensors, cameras, fluorimeters-spectrometers, and special water samplers on board several submersible titan modules. Such activities are in line with environmental stewardship in the exploration of unique aquatic environments under the Antarctic ice sheet.

1. INTRODUCTION

By the end of the 20th century, subglacial lakes were found to be a typical feature of bedrock landscapes under the Antarctic ice sheet. More than 150 subglacial water bodies have been discovered using airborne radio echo sounding and satellite altimetry surveys. The Vostok Subglacial Lake (also referred to herein as Lake Vostok), situated in East Antarctica under the Russian Antarctic station of the same name, is the largest and most studied among them.

The first official information regarding the discovery of this large subglacial water body was presented by the corresponding member of the Russian Academy of Science, Professor A.P. Kapitsa, at the session of the Scientific Committee on Antarctic Research (SCAR) in August 1994 in Rome. The following year, the first publication of this unique natural phenomenon appeared in the journal *Nature*, prepared by a group of Russian and British investigators [*Kapitsa et al.*,

Antarctic Subglacial Aquatic Environments
Geophysical Monograph Series 192
Copyright 2011 by the American Geophysical Union
10.1029/2010GM000951

1996]. This paper summarized, for the first time, the outcomes of Soviet seismic and U.S.-British airborne radio echo sounding studies carried out in the 1960s–1970s in the vicinity of the Russian Antarctic Vostok Station. It also presented an analysis of satellite altimetry data collected in the central areas of East Antarctica during the 1990s, using Russian theoretical calculations of possible glacial melting at its base due to friction forces and vertical pressure gradients contributing to the occurrence of ice-melting temperatures. From the 1995/1996 season, the Russian Antarctic Expedition (RAE) began systematic and planned studies of this natural water body, which by then had been named by its discoverers after the Russian (Soviet) Antarctic Vostok Station, located over the southern part of this lake. These studies included seismic sounding of the ice sheet via the reflected wave method and aimed to obtain values of its total thickness, lake water strata thickness, and, if possible, bottom sediment thickness [*Masolov et al.*, 2006]. These studies were later supplemented by ground-based radio echo sounding to determine the lake's shoreline, configuration, and overlying glacier body thickness [*Masolov et al.*, 1999].

2. GEOPHYSICAL STUDIES AND LAKE SETTING

During the period from 1995 to 2008, a total of 318 seismic soundings were conducted by the reflected wave method, and 5190 km of ground-based radar profiling were performed. It was determined that Vostok Subglacial Lake has an area of about 16,000 km^2. The west shore of the lake is significantly dissected by bays and inlets, whereas the east shore is relatively even (Plate 1). Lake length is about 280 km in the submeridional direction, and the width is 30–70 km. Ice sheet thickness changes northward over the lake, increasing from 3500 to 4200 m. The average water layer thickness of Vostok Subglacial Lake is equal to about 300 m; total water body volume is around 6343 km^3. In the general plan, the lake is subdivided into two unequal parts. The first (southern) is the deeper of the two but smaller in size, with an area of approximately 70 × 30 km and a predominant water layer thickness of about 800 m. The second (northern) part is relatively shallow and occupies a territory of about 180 × 60 km.

The lake water body is situated in a steep-sided trough, with slopes exceeding 15° and their significant height in some places greater than 1500 m. The near-bottom region of the Vostok trough is, in general, a hillocky plain with an average absolute altitude of about 900 m. Relative relief appears to be quite insignificant, not more than 100 m with maximum slopes of up to 4°. The hillocky plain occupies an area of about 5800 km^2, which comprises more than one third of the entire territory.

The southern and southwestern parts of the Vostok trough have two pronounced basins. The former is the deepest and largest in size (around 60 × 30 km), with a depth equal to about 400 m and an average slope steepness of around 8°. The near-bottom regions of the valley are flatter and are located at an absolute altitude of approximately 1500 m. The second basin measures 45 × 15 km, while its relative relief and slopes are also insignificant. The near bottom of the basin is situated at an absolute height of about 1150 m.

3. DRILLING TOWARD THE LAKE AND ENVIRONMENTAL ISSUES

Along with investigation of subglacial lake characteristics by remote sensing methods, deep drilling of the ice sheet that began in 1970 at Vostok Station was continued. Originally, the main aim of the project was to perform palaeoclimatic reconstructions in Central Antarctica. By measuring the concentration of carbon dioxide, methane, and deuterium in the ice core, it is possible to determine climatic fluctuations that have been taking place for the last 420 ka. Four full climatic cycles of air temperature changes (glaciation and deglaciation) have been determined in this area of East Antarctica, comprising, on average, about 100 ka each [*Petit et al.*, 1999]. In the 1980s, this project received international status with participation of researchers from the USSR (Russia), France, and the United States. Drilling was carried out by means of technological and engineering facilities developed by specialists at the Drilling Division of the St. Petersburg Mining Institute [*Kudryashov et al.*, 2002]. A mixture of kerosene and foranes was used as a drilling fluid, making it possible to create a nonfreezing fluid with the density inside the borehole equal to that of ambient ice (0.91 g cm^{-3}). This counteracted the development of the mountain pressure effect, leading to narrowing of the bottom part of the borehole's diameter. It is worth noting that similar drilling fluid was used by European specialists during ice drilling in Greenland and Antarctica within the framework of the European EPICA Project at both the French-Italian Dome C station and the German Kohnen Base. In 1999, when the depth of ice borehole 5G-1 at Vostok Station had reached 3623 m, drilling was stopped. This was under the recommendation of SCAR experts, who advised Russian specialists to first develop measures ensuring safety from incidental penetration of the borehole's kerosene-based mixture into the relict waters of the lake and, thereby, protect it from environmental contamination. At the same time, the depth of the borehole bottom was approximately 130 m from the upper boundary of the lake water layer, creating necessary economic preconditions for using this borehole for entering Lake Vostok. The thickness of the glacier (around 3750 m) was determined by two independent methods (seismic logging and radio echo sounding).

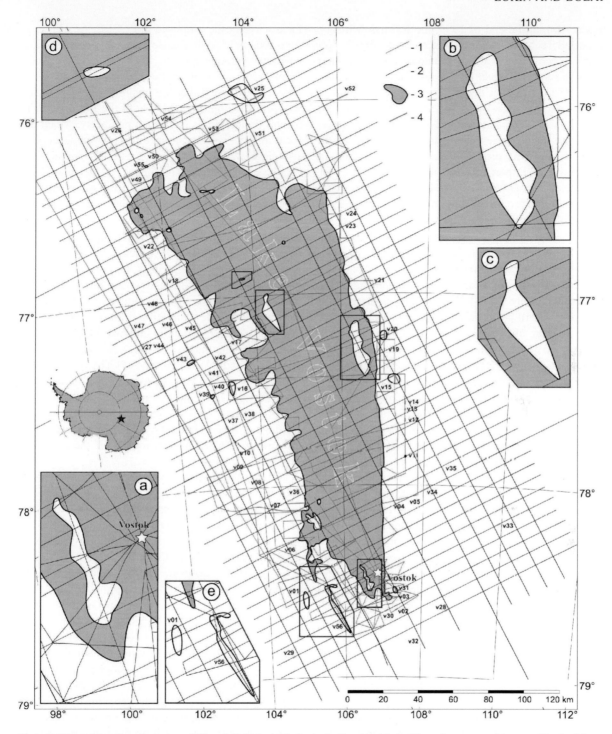

Plate 1. Map of the shoreline area of Vostok Subglacial Lake including islands inside and water cavities outside the lake [*Popov et al.*, 2010]. Numbers indicate the following: 1, Russian RES profiles; 2, U.S. RES profiles; 3, water table of Lake Vostok and subglacial water caves; and 4, fragments of subglacial water caves. Water caves are numbered in red.

For this reason, at the end of 1999, the Russian Ministry for Science and Technology announced an open competition to develop an ecologically clean technology for sampling water from the surface layer of Vostok Subglacial Lake, through the existing deep ice borehole at Vostok Station. The winners proved to be specialists at the St. Petersburg Mining Institute and the Arctic and Antarctic Research Institute. Creation of the technologies was completed in late 2000, and in March 2001, the project was approved by Russian State Environmental Expert Examination. The main idea for ecologically clean penetration of lake surface water was the use of a thermal drill for drilling the lower near-water part of the glacier. At the moment of drill contact with lake water, an undercompensation effect is artificially created due to decreasing drilling fluid level in the upper part of the borehole. Simultaneously an organic silicon fluid is injected into the lower part of the borehole near its bottom, separating the kerosene-Freon mixture from lake water. After the contact with water, which will be recorded (possibly along with the foot of the glacier above the lake) by special cameras mounted on a drill head, the drill will be raised to the borehole surface, while water from the lake's upper layer will rise upward in the borehole, due to the artificially created pressure difference, to a height equal to the decreased upper level of the drilling fluid in the borehole. Being less dense than lake water (0.91 versus 1.00 g cm^{-3}), the drilling fluid cannot, under any physical conditions, rush downward and spread into the lake water body. The region of water/drilling fluid contact will be limited to the borehole area (136.78 cm^2 at a borehole diameter of 132 mm). As a hydrophobic liquid, kerosene cannot mix with water, and thus, a rather clear interface is preserved, which will also prevent further water contamination. As it rises upward in the borehole and comes into thermal contact with the lateral sides of the ice borehole, lake water will begin to cool and eventually freeze. Reconstructed in a natural way, the ice cork can be gradually drilled in the future to obtain fresh fast-frozen water samples from the surface layer of Lake Vostok for subsequent hydrochemical and microbiological analyses.

The results of the aforementioned expert examination were presented by the Russian Delegation at Antarctic Treaty Consultative Meeting (ATCM) XXIV in July 2001 in St. Petersburg. The following year in Warsaw at ATCM XXV, the Russian Federation circulated the draft Comprehensive Environmental Evaluation (CEE) for penetration of Lake Vostok water through deep borehole 5G-1, aiming to sample the surface layer of lake water. In spite of the fact that the document was presented with some procedural breaches, members of the Committee for Environmental Protection (CEP) showed great interest in this Russian presentation and organized a special Inter-commission Contact Group (ICG)

for discussing the proposed draft CEE. At ATCM XXVI in June 2003 in Madrid, Russia introduced a revised draft CEE, where the comments of the ICG were positively considered or taken into account. In the final report of ATCM XXVI, it was acknowledged that the document developed by Russia was in full compliance with both the Protocol on Environmental Protection of the Antarctic Treaty, which came into force in 1998, and the procedural requirements of the CEP, established within the structure of the Antarctic Treaty System the same year. The comments made by some CEP parties were as follows: (1) While the Committee recognizes the importance of the long-term science goals for subglacial lake exploration, the Russian draft CEE provides insufficient consideration to reduce the potential environmental risks posed by the activity. (2) Insufficient information is provided regarding the special organic silicon drilling fluid to support the conclusion that it is "ecologically clean." (3) The treatment of alternatives to the proposed activity is inadequate and should include alternative solutions. (4) The draft CEE does not adequately identify and discuss gaps in knowledge, particularly relating to the question of ice/water interface conditions and lake chemistry. (5) The draft CEE does not adequately address the risk of accidental release of drilling fluid into the lake and the potential consequences of this spill. (6) Consistent with Annex 1, Article 3, paragraph 2(g), contingency plans should be developed to deal promptly and effectively with unforeseen impacts if the activities do not proceed as predicted.

In light of these concerns, the CEP recommended that the Russian Federation be urged to consider carefully this advice, making any revisions to the final CEE as necessary to address the above insufficiencies and produce a final CEE fully consistent with the requirements of Annex 1 of the Protocol [*ATCM XXVI*, 2003].

Since 2003, European colleagues of Russian scientists carried out unplanned full-scale testing of the Russian technology under natural conditions in Greenland (Denmark) and Queen Maud Land in Antarctica (Germany). In conducting drilling operations in deep glacial layers of these areas, the drills unexpectedly penetrated the water layers of local subglacial "aquifers," with no geophysical data regarding their existence available before the start of drilling. The Danish and German specialists were able to remove the drills to the surface without allowing them to freeze. Simultaneously, an increase in drilling fluid (a mixture of kerosene and foranes) level was registered in the borehole due to the effect of water rising under pressure. A year after the incident in Greenland, drilling of the quickly frozen water, which had risen in the borehole, was conducted. The samples obtained were then subject to international chemical and microbiological expert examination. Investigation revealed that kerosene and microbe contamination of the new ice core was extremely severe in its upper part (bag 4, 3039 m), but 6 m further down (bag 15, 3045 m), neither kerosene nor cells were revealed [*Bulat et al.*, 2005]. Unfortunately, drilling operations at the German Kohnen Base were suspended, making it impossible to investigate fast-frozen water in this borehole.

Study of the gaseous and isotopic composition of the Vostok ice core showed that below a depth of 3539 m, ice is formed of frozen lake water and has no relation to the overlying ice sheet layer, which is of atmospheric origin [*Jouzel et al.*, 1999]. Lake ice is subdivided into two types by the presence of mineral inclusions which are observed from 3539 to 3608 m (accretion ice type 1). Below 3608 m, the lake ice does not contain mineral inclusions and is referred to as accretion ice type 2. Its structure is composed of large ice crystals up to 1.5 m in size but without the prominent orientation of crystalline axes.

4. BIOLOGICAL FINDINGS

Initial microscopy observations were conducted on strictly decontaminated ice samples (i.e., meeting standards for both ice geochemistry and sterile conditions). Two samples were selected: one from accretion ice 1 (3651 m) and another from glacier ice (2054 m) used as a reference. Different microscopy methods were utilized, including specially implemented techniques (e.g., fluorescence, laser confocal, and scanning electron microscopy). All results were negative, with no microbial cells from the ice identified with confidence. Several attempts were made to culture on various media, but these were also unsuccessful.

Flow fluorocytometry was applied for cell counting in strictly decontaminated ice samples. Repeated measurements indicated values not exceeding 19–24 cells mL^{-1} in both accretion and glacial ice [*Bulat et al.*, 2009] (Table 1). This result and those from the microscopic observations and culturing attempts are rather different from data published by other investigators, with observed cell concentrations here two or three orders of magnitude lower than data published previously [*Abyzov et al.*, 2001; *Christner et al.*, 2001, 2006; *Karl et al.*, 1999; *Priscu et al.*, 1999]. The difference between the results may be a result of incomplete decontamination of the ice from the drilling fluid, since cells have been observed in a core coating [*Petit et al.*, 2005] and several phylotypes identified in drilling fluid [*Alekhina et al.*, 2007].

It is worth noticing that the deepest (3400 and 3600 m) and comparatively warm (−10°C and −6°C, respectively) borehole horizons were dominated by two phylotypes of *Sphingomonas* (a well-known degrader of polyaromatic hydrocarbons). These results indicate the persistence of bacteria in extremely cold, hydrocarbon-rich environments

Table 1. Microbial Cell Concentrations and Bacterial Phylotypes Detected at High Confidence in Vostok Ice Core

Ice Type	Sample Depth (m)	Cells (per mL)	Phylotype[a]
Snow	4.0–4.3	0–0.02	
Glacier	122	1.9	
	2005	2.4	
	2054	3–24	
	3489	0	
Accretion I	3561	4–9	*Hydrogenophilus thermoluteolus* (99%) (DQ422863)
	3607	ND[b]	*Hydrogenophilus thermoluteolus* (100%) (AF532060) Uncultured bacterium[c] (91%) (AF532061)
	3608	0–19	
Accretion II	3613	3	
	3621	2	
	3622	0.6	
	3635	4.7	
	3650	3.1	
	3650	4777[d]	
	3659	12	

[a]Percent similarity with closest sequence in GenBank, GenBank acc. no.
[b]ND, not determined.
[c]OP11 candidate division.
[d]Untreated ice surface.

within the borehole and show the potential for contamination of ice and subglacial water samples during lake penetration. However, these bacteria are unlikely to thrive in the lake water due to its specific extreme conditions [*Petit et al.*, 2005]. Indeed, as *de Wit and Bouvier* [2006] state, "Everything is everywhere, but, the environment selects."

The multiple trial DNA study showed that accretion ice is essentially microbial DNA-free. Up to now, only two bacterial phylotypes recovered from accretion ice 1 passed the contaminant controls and criteria. Among the findings, DNA signatures of a (facultative) chemolithoautotroph *Hydrogenophilus thermoluteolus* proved to be the true thermophile (optimum temperature of +50°C–52°C) [*Bulat et al.*, 2004]. This result at 3607m depth was then confirmed for another sample at 3561m depth [*Lavire et al.*, 2006]. As lake water temperature is at freezing point (~−2°C, due to pressure), the appropriate niche for thermophiles should be found outside the lake, likely at depth within the deep bedrock faults encircling the lake. There, warm anoxic conditions and sediments rich in CO_2, along with probable hydrogen emission by water radiolysis are likely to be conducive to thermophile survival. Sporadic seismotectonic activity (faults remain active today [*Studinger et al.*, 2003]) may have flushed out some material from these veins into the lake and accretion ice 1 [*Bulat et al.*, 2004; *Petit et al.*, 2005]. A second phylotype recovered from the same ice horizon (3607 m) was assigned as an uncultured bacterium of the OP11 Candidate division, reserved for only uncultured representatives, and proven to be fully unclassified (less than 91% similarity with closest entries in GenBank). As a genuine candidate from lake ice (or water), the metabolic profile of this phylotype remains uncertain. Thus, our results leave open the question of whether life (and which forms) exists within the water of Vostok Subglacial Lake.

5. RECENT DRILLING PROGRESS

During the 2003/2004 and 2004/2005 seasons, RAE examined the conditions of borehole 5G-1 in order to make a decision regarding the continuation of drilling. In the 2005/2006 season, RAE resumed operations in deep ice borehole 5G-1 at Vostok Station, drilling 27 m of ice and consequently reaching a depth of 3650 m. This work continued during 2006/2007, but on 13 January 2007, the drill became stuck at a depth of 3659 m. Attempts to recover it resulted in the power cable being broken at the place of its connection with the drill. Emergency measures to rescue the damaged drill from the borehole were performed, and by the end of February 2007, the drill was finally recovered. A number of actions were subsequently undertaken in order to widen the borehole diameter and clean its base of the debris left behind by the drill recovery maneuvers. In June 2007, drilling operations continued once more.

However, at the end of October 2007, there was another accident, this time at a depth of 3667.8 m, which again resulted in breakage of the carrying cable. This accident was caused by the loss of control of drilling fluid level and density. In the 2007/2008 season, work was carried out to eliminate these problems (involving elevating drill fluid level and increasing its density at the borehole bottom) and to prevent borehole narrowing (collapsing) near its base. During the 2008/2009 season, numerous efforts were undertaken to recover the drill, but they were all unsuccessful. As the limited period of seasonal activity at Vostok Station was coming to an end, it was decided to drill a new borehole, deviating from 5G-1. It was decided to start the deviation at a depth of 3580 m, in order to once more drill through the 3600- to 3609-m horizon in which the maximum concentration of mineral particles was previously observed and to avoid the lost damaged drill in borehole 5G-1 by 1.5 m. The new borehole was named 5G-2. By the end of January 2009, drilling operations in this new borehole reached a depth of 3599 m, starting from a so-called moon-shaped ice core and finishing with a full-

diameter ice core, clearly demonstrating a successful bypass of the earlier accident in borehole 5G-1. In the 2009/2010 season, drilling operations in borehole 5G-2 were successfully continued, reaching a depth of ~3650 m (Plate 2).

6. PLANS FOR THE FUTURE

Studies of Vostok Subglacial Lake were recognized as a research priority by the Russian Federation Government, with the issuing of the Order of the RF Government No. 730 R of 2 June 2005, "On activity of the Russian Antarctic Expedition during the period 2006–2010." Investigation of Lake Vostok was also placed on the list of main projects in the third stage of the subprogram "Study and Research of the Antarctic," under the Federal Target Program "Global Ocean" for 2008–2012.

It is obvious that the overall aim of any investigation of a natural water body, including the subglacial lakes of Antarctica, is to study its geochemistry, hydrophysical characteristics, and, most importantly, biological content and biodiversity in the water column and bottom sediments. It should be kept in mind that no highly precise remote sensing techniques will, under any conditions, replace contact methods of measurement and laboratory analyses of the specimens from Mars, the Moon, and other planets and satellites. This also applies to Vostok Subglacial Lake. At present, using geophysical sounding methods and various laboratory analyses of accretion ice from the deep borehole at Vostok Station, a series of characteristic features of the water body was obtained. They include the following: (1) area and configuration of the water table; (2) spatial distribution of thickness of the water layer, overlying glacier and lake bottom relief; (3) crystalline structure, electrical conductivity, and gaseous composition of the glacier ice core samples (atmospheric origin) taken at the deep drilling point; (4) thickness of the ice composed of frozen lake water (accretion ice), vertical distribution of the concentration of mineral inclusions and gaseous content, crystalline structure of this ice at the drilling point of the deep ice borehole; (5) concentration of microbial cells and microbial diversity within the thick glacier above the lake at the drilling point of the deep ice borehole, including surface snow (0–3539 m), accretion ice type 1 (3539–3608 m) and type 2 (below 3608 m); (6) chemical, physical, and microbiological characteristics of the drilling fluid used in the deep ice borehole at Vostok Station; and (7) spatial distribution of absolute heights and drift vectors of the day surface of the glacier above the lake area.

In addition, a number of features of Vostok Subglacial Lake can be inferred from the hypothetical and mathematical modeling of glacier "behavior" above the lake, as well as its water strata and the lake itself. These proposed features are not based on measurements and hence represent only assumed characteristics. However, in scientific publications devoted to studies of Lake Vostok, they are often published together with the results of remote sensing and laboratory measurements and are therefore distinguished only by specialists. The inferred features include the following: (1) origin of lake water strata due to processes of melting at the glacier base, (2) processes of ice melting at the foot of the glacier in the northern part of the lake and ice accretion in its southern part, (3) nature of lake ice accretion and its growth with respect to glacier motion along the flow line from Dome B, (4) age of lake ice (it is not possible to determine the age of accretion ice by isotopic methods due to the lack of material), (5) chemical characteristics of the water strata (fresh or salt), (6) pattern of horizontal circulation of lake water strata and processes of vertical convective water mixing, and (7) thickness of the bottom sediment layer.

In order to prove or verify much of the above, it is necessary to enter the lake, perform sampling of the liquid water, and arrange new deep boreholes in the ice on the northern, western, and eastern lake peripheries, as well as behind its shore limits. Unfortunately, the latter requires large additional logistical/financial expenses, which are unachievable for RAE in the near future. As a result, at the present time, the main direction of RAE activity in the study of Lake Vostok is the penetration of the water layer through deep ice borehole 5G-2, the exploration of lake water strata by means of sounding contact systems, and collecting and analyzing,

under laboratory conditions, water and bottom sediment samples.

So, after completing remote sensing and laboratory analyses of ice cores from the deep borehole at Vostok Station, the main priority with respect to studying Lake Vostok is to penetrate the lake water strata and assess their characteristics by contact methods. Entering the lake using the aforementioned technology should be realized around the 2011/2012 season. By this time, RAE will hopefully have been granted a permit for using this technology, in compliance with the procedure adopted in the Russian Federation. RAE also has to submit the Final CEE of this technology used for lake penetration and surface water sampling, to the Committee for Environmental Protection of the Antarctic Treaty System. This document should provide answers to all comments recorded in the Final Report of ATCM XXVI in 2003 in Madrid. Such a system of consideration and approval complies with the adopted procedure within the Protocol of Environmental Protection of the Antarctic Treaty, for consideration of CEEs of different types of activity in Antarctica.

During the 2009/2010 season, drilling operations in borehole 5G-2 were continued. Penetration of the lake water column is intended by the end of the 2011/2012 season. The following season after the rising and freeze of lake water within borehole 5G-2, sampling of a new ice core from the fast-frozen water will be conducted. Before entering the lake water strata, an ice cork 10–15 m thick will be left in the lower part of the borehole. The final decision regarding the thickness of this cork will be made based on a study of

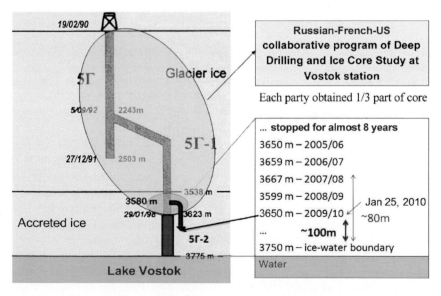

Plate 2. Diagram of Vostok 5G-1[2] borehole.

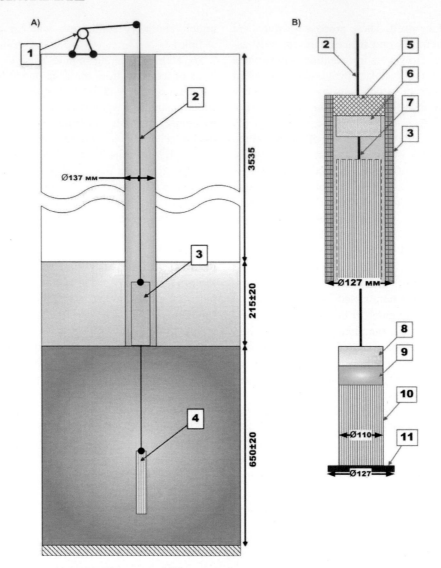

Plate 3. (a) Diagram of technology used for exploring the water column of Vostok Subglacial Lake within borehole 5G-2, (b) with principal schemes of transportation and measurement-exploration modules. Numbers indicate the following: 1, drill winch; 2, load-carrying electrical cable; 3, transportation module; 4, exploration module with respective sensors or water sampler; 5, power supply engine; 6, winch; 7, steel cable; 8, 12-V battery set; 9, microprocessor; 10, exploration unit; 11, hermetically sealing lid. The borehole is shown filled with drilling fluid (yellow-green shading).

the crystalline structure of the newly drilled ice core. The investigation of the water strata of Lake Vostok by contact methods will commence in the 2013/2014 season.

For this purpose, the Arctic and Antarctic Research Institute was commissioned by the Federal Service for Hydrometeorology and Environmental Monitoring (the lead body of the Federal Executive of Russia, responsible for governance and control of RAE activities) to declare an open competition for the design and production of the experimental measurement systems and auxiliary engineering equipment, which would directly investigate the water

strata of Lake Vostok. The winner was the Petersburg Nuclear Physics Institute (PNPI RAS), a conglomeration of a few institutes and companies specializing in different areas. In addition to its significant scientific potential in the area of physics, this institute has a comprehensive Design Bureau and industrial experimental facilities capable of developing and producing a variety of measuring and engineering systems able to operate under extreme environmental conditions.

Until now, there have been no analogs of such equipment, as the devices developed at the research centers of the U.S.

National Aeronautic and Space Administration were never tested under conditions similar to those found at Vostok Subglacial Lake. However, similar devices are to be constructed in the future within the United Kingdom-funded Subglacial Lake Ellsworth Program. In the case of Lake Vostok, measuring and sampling systems will be constructed on a modular basis, within which will be installed ready-to-use units, various cameras, and sensor assemblies that perform well under the extreme conditions of the Global Ocean and outer space. The measurement systems will be divided into separate main blocks (hydrophysical and hydrobiochemical fluorimeter-spectrometers and water sampler) in order to increase their reliability. Combining these complex systems within a single block will significantly complicate its design and increase the risk of failure of sensors and working units. Another important task in the design of these systems is tackling the problem of sterilizing the blocks, as they pass through an ice borehole filled with a kerosene-Freon mixture before entering the lake water strata. This problem will be solved by creating a special hermetically sealed transportation unit, which will contain the preliminarily decontaminated block systems (Plate 3). In terms of its size, shape, and control system, the transportation unit is similar to an electrical mechanical drill, connected to the control panel of the surface drilling complex by means of the electrical cable of the drilling winch. This approach enables delivery of sterile measuring systems to the place of drilling fluid (silicone oil) and lake water body contact, as well as raising the measurement systems upward on the return journey without making contact with the drilling fluid in the borehole.

The transportation unit is comprised of a pipe, the length of which varies depending on the task from 3 to 13 m. The pipe consists of separate sections with a total length of 3000 mm and a diameter of 127 mm, which can be hermetically connected with each other. In the upper part of the unit, there is a connection block containing the current-carrying wire used presently for borehole drilling at Vostok Station. The transportation unit is equipped with its own heating system to provide the internal temperature required by the measurement systems during ascent and descent. In the hermetical block in the upper part of the pipe, an electric motor is installed, which operates the lowering and lifting system for the measurement instruments. Under the electric motor block, there is a system which enables the automatic ascent and descent of the sampling and sounding devices, with a cable length of 750 m and a lifting capacity of up to 120 kg. Each measurement-sampling block, together with the video cameras and thermal imager attached to the lower part of the cable located inside the transportation unit, will be alternately launched toward the lake water strata to perform the observation program and register parameters. The

transportation unit and measuring modules will be sterilized by gamma-radiation at PNPI RAS, with subsequent decontamination by ozone treatment at Vostok Station.

The measurement or exploration modules are self-contained, with their own control block and electrical power system not connected with the surface in terms of energy or information transfer.

In order to sample from different horizons of the Lake Vostok water column, the water sampling device has its technical basis in samplers used in studies of deep oil wells. These have a reinforced hull and an extended set of sensors. The volume of the collected sample is more than 550 cm^3, while the sampler mass is not more than 6.2 kg. To control sampler operation, a single-board microprocessor is used, which will operate via a specially programmed software package.

The water sample will be poured from the sampler chamber to a special container by means of specially developed devices, ensuring transfer of the sample without contact with the environment under conditions equivalent to a class 1000 clean laboratory. Such a laboratory will be assembled inside the Vostok Station drilling complex prior to beginning investigations of lake water strata.

As a basis for creating the hydrophysical module instruments investigating Lake Vostok water, a standard oceanographic device with the maximum number of recorded water medium parameters has been chosen. The most suitable for this purpose is a complex designed by "Seabird electronics"; the SBE 19 plus V2 SEACAT PROFILER with a strengthened hull and extended set of sensors. This standard oceanographic device has been adjusted to fit inside the transportation unit in order to be delivered to the lake water strata. Its external diameter is 108 mm. To accommodate it within the transportation unit, the layout of sensors was rearranged, and a unified single-board computer was included in the final assembly to control its operation and record information via an autonomous power supply system.

The sounding hydrobiochemical complex used to investigate Lake Vostok strata is a two-module structure arrayed in the titanium pipe with an external diameter of 110 mm. It includes three video cameras (frontal, side, and rear views) with necessary illumination. The instrument was developed taking into account its operational capacity under low temperature (up to −70°C) and high pressure (450 dbar) conditions.

One of the modules contains a power supply unit, while the second contains a fluorimeter, which carries out fluorescent spectroscopy of complex organic compounds after their excitation by a laser emission impulse. To register mostly tryptophan and tyrosine along with archaeal cofactor F$_{420}$ and possibly flavones, two lasers are used as the excitation light source. One of them is a solid-state impulse laser with an excitation length of 266 nm and an average power of 20 MW, while the second is a diode impulse laser with an emission wavelength of 405 nm and an average power of 50 MW. Both lasers form a stationary fluorescence exciting block and allow us to ensure excitation of the aforementioned organic compounds in the water.

As a control computer, a small single-board processor with low power consumption and an extended temperature range was developed to operate the sampler, hydrophysical and hydrobiochemical devices.

Construction and assembly of test models of the transportation unit and measuring systems was completed in 2009. Field testing and full-scale trials at the fairly deep (up to 230 m) Lake Ladoga (Leningrad region and Karelia Republic) are planned for late 2010.

Finally, we would like to emphasize that despite "Subglacial Lake Vostok Entry and Water/Sediment Exploration" being a Russian national project, we are open to international collaboration. This is both in terms of logistics, as well as the replication of some highly contamination-prone analyses, dealing with the search and authentication of life in lake water/sediments, which may include some quite unusual life forms. In addition, we would like to assure the international community that activities dealing with entry and further exploration of this unique subglacial Antarctic aquatic environment, via the use of submersible modules full of on-board instruments, will be performed strictly in line with appropriate rules of environmental stewardship.

Acknowledgment. This work is a contribution to Project 2 of the Subprogram 'Antarctica' of the Russian Federal Targeted Program 'World Ocean.'

REFERENCES

Abyzov, S., I. Mitskevich, M. Poglazova, N. Barkov, V. Lipenkov, N. Bobin, B. Koudryashov, V. Pashkevich, and M. Ivanov (2001), Microflora in the basal strata at Antarctic ice core above the Vostok lake, *Adv. Space Res.*, *28*, 701–706.

Alekhina, I., D. Marie, J. R. Petit, V. Lukin, V. Zubkov, and S. Bulat (2007), Molecular analysis of bacterial diversity in kerosene-based drilling fluid from the deep ice borehole at Vostok, East Antarctica, *FEMS Microbiol. Ecol.*, *59*, 289–299.

ATCM XXVI (2003), Final Report of the XXVI Antarctic Treaty Consultative Meeting, report, Agencia Esp. de Coop. Int., Madrid, Spain, 9/20 June.

Bulat, S., et al. (2004), DNA signature of thermophilic bacteria from the aged accretion ice of Lake Vostok, Antarctica: Implications for searching for life in extreme icy environments, *Int. J. Astrobiol.*, *3*, 1–12.

Bulat, S., I. Alekhina, J. R. Petit, J. P. Steffensen, and D. Dahl-Jensen (2005), Bacteria and archaea under Greenland Ice Sheet:

NGRIP 'red' ice issue. EGU, General Assembly, Vienna, Austria, 24–29 April 2005, *Geophys. Res. Abstr.*, *7*, 05298.

Bulat, S., I. Alekhina, V. Lipenkov, V. Lukin, D. Marie, and J. R. Petit (2009), Cell concentrations of microorganisms in glacial and lake ice of the Vostok ice core, East Antarctica, *Microbiology*, *78*, 808–810.

Christner, B., E. Mosley-Thompson, L. Thompson, and J. Reeve (2001), Isolation of bacteria and 16S rDNAs from Lake Vostok accretion ice, *Environ. Microbiol.*, *3*, 570–577.

Christner, B., G. Royston-Bishop, C. Foreman, B. Arnold, M. Tranter, K. Welch, B. Lyons, A. Tsapin, M. Studinger, and J. Priscu (2006), Limnological conditions in subglacial Lake Vostok, Antarctica, *Limnol. Oceanogr.*, *51*, 2485–2501.

de Wit, R., and T. Bouvier (2006), 'Everything is everywhere, but, the environment selects'; what did Baas Becking and Beijerinck really say?, *Environ. Microbiol.*, *8*, 755–758.

Jouzel, J., J. R. Petit, R. Souchez, N. I. Barkov, V. Y. Lipenkov, D. Raynaud, M. Stievenard, N. I. Vassiliev, V. Verbeke, and F. Vimeux (1999), More than 200 meters of lake ice above subglacial Lake Vostok, Antarctica, *Science*, *286*, 2138–2141.

Kapitsa, A. P., J. K. Ridley, G. D. Robin, M. J. Siegert, and I. A. Zotikov (1996), A large deep freshwater lake beneath the ice of central East Antarctica, *Nature*, *381*, 684–686.

Karl, D., D. Bird, K. Bjorkman, T. Houlihan, R. Shackelford, and L. Tupas (1999), Microorganisms in the accreted ice of Lake Vostok, Antarctica, *Science*, *286*, 2144–2147.

Kudryashov, B. B., N. I. Vasiliev, R. N. Vostretsov, A. N. Dmitriev, V. M. Zubkov, A. V. Krasilev, P. G. Talalay, N. I. Barkov, V. Y. Lipenkov, and J. R. Petit (2002), Deep ice coring at Vostok Station (East Antarctica) by an electromechanical drill, *Mem. Natl. Inst. Polar Res.*, *56*, 91–102.

Lavire, C., P. Normand, I. Alekhina, S. Bulat, D. Prieur, J. L. Birrlen, P. Fournler, C. Hanni, and J. R. Petit (2006), Presence of *Hydrogenophilus thermoluteolus* DNA in accretion ice in the subglacial Lake Vostok, Antarctica, assessed using rrs, cbb and hox, *Environ. Microbiol.*, *8*, 2106–2114.

Masolov, V., S. Popov, V. Lukin, A. Sheremetyev, and A. Popkov (2006), Russian geophysical studies of Lake Vostok, Central East Antarctica, in *Antarctica. Contributions to Global Earth Sciences*, edited by D. K. Fütterer et al., pp. 135–140, Springer, Berlin.

Masolov, V. N., et al. (1999), Earth science studies in the Lake Vostok region: Existing data and proposals for future research, in *Subglacial Lake Exploration–Workshop Report and Recommendations, Addendum*, pp. 1–18, Cambridge Univ., Cambridge, U. K.

Petit, J. R., et al. (1999), Climate and atmospheric history of the past 420,000 years from the Vostok ice core, Antarctica, *Nature*, *399*, 429–436.

Petit, J. R., I. Alekhina, and S. Bulat (2005), Lake Vostok, Antarctica: Exploring a subglacial lake and searching for life in an extreme environment, in Lectures in Astrobiology, vol. 1, in *Advances in Astrobiology and Biogeophysics*, edited by M. Gargaud et al., pp. 227–288, Springer, Berlin.

Popov, S. V., et al. (2010), Shore line of Vostok Subglacial Lake, East Antarctica, paper presented at IPY Oslo Science Conference, Det Norske Veritas, Oslo, 8–12 June.

Priscu, J. C., et al. (1999), Geomicrobiology of subglacial ice above Lake Vostok, Antarctica, *Science*, *286*, 2141–2144.

Studinger, M., et al. (2003), Ice cover, landscape setting and geological framework of Lake Vostok, East Antarctica, *Earth Planet. Sci. Lett.*, *205*, 195–210.

S. Bulat, Petersburg Nuclear Physics Institute, Russian Academy of Sciences, Leningrad region, Gatchina 188300, Russia. (bulat@omrb. pnpi.spb.ru; sergey.bulat@ujf-grenoble.fr)

V. Lukin, Russian Antarctic Expedition, Arctic and Antarctic Research Institute, 38 Bering str., St. Petersburg 199397, Russia. (lukin@aari.nw.ru)

Siple Coast Subglacial Aquatic Environments: The Whillans Ice Stream Subglacial Access Research Drilling Project

Helen Amanda Fricker,[1] Ross Powell,[2] John Priscu,[3] Slawek Tulaczyk,[4] Sridhar Anandakrishnan,[5,6] Brent Christner,[7] Andrew T. Fisher,[4] David Holland,[8] Huw Horgan,[5] Robert Jacobel,[9] Jill Mikucki,[10] Andrew Mitchell,[11] Reed Scherer,[2] and Jeff Severinghaus[1]

The Whillans Ice Stream Subglacial Access Research Drilling (WISSARD) project is a 6-year (2009–2015) integrative study of ice sheet stability and subglacial geobiology in West Antarctica, funded by the Antarctic Integrated System Science Program of National Science Foundation's Office of Polar Programs, Antarctic Division. The overarching scientific objective of WISSARD is to assess the role of water beneath a West Antarctic Ice Stream in interlinked glaciological, geological, microbiological, geochemical, hydrological, and oceanographic systems. The WISSARD's important science questions relate to (1) the role that subglacial and ice shelf cavity waters and wet sediments play in ice stream dynamics and mass balance, with an eye on the possible future of the West Antarctic Ice Sheet and (2) the microbial metabolic and phylogenetic diversity in these subglacial environments. The study area is the downstream part of the Whillans Ice Stream on the Siple Coast, specifically Subglacial Lake Whillans and the part of the grounding zone across which it drains. In this chapter, we provide background on the motivation for the WISSARD project, detail the key scientific goals, and describe the new measurement tools and strategies under development that will provide the framework for conducting an unprecedented range of scientific observations.

1. INTRODUCTION

The latest report of the Intergovernmental Panel on Climate Change recognized that one of the greatest uncertainties in assessing future global sea level change is ice sheet stability in a changing climate. This stems from a poor understanding of ice sheet dynamics and ice sheet vulnerability to oceanic and atmospheric warming [Lemke et al., 2007].

[1]Scripps Institution of Oceanography, La Jolla, California, USA.

[2]Department of Geological and Environmental Sciences, Northern Illinois University, DeKalb, Illinois, USA.

[3]Department of Land Resources and Environmental Sciences, Montana State University, Bozeman, Montana, USA.

[4]Department of Earth and Planetary Sciences, University of California, Santa Cruz, California, USA.

[5]Department of Geosciences, Pennsylvania State University, University Park, Pennsylvania, USA.

[6]Earth and Environmental Systems Institute, Pennsylvania State University, University Park, Pennsylvania, USA.

[7]Department of Biological Sciences, Louisiana State University, Baton Rouge, Louisiana, USA.

[8]Courant Institute of Mathematical Sciences, New York University, New York, New York, USA.

[9]Department of Physics, St. Olaf College, Northfield, Minnesota, USA.

[10]Earth Science Department and Environmental Studies Program, Dartmouth College, Hanover, New Hampshire, USA.

[11]Center for Biofilm Engineering, Montana State University, Bozeman, Montana, USA.

Antarctic Subglacial Aquatic Environments
Geophysical Monograph Series 192
Copyright 2011 by the American Geophysical Union
10.1029/2010GM000932

Disintegration of the West Antarctic Ice Sheet (WAIS) alone would contribute a ~3.5 m rise in global sea level [Bamber et al., 2009]. Furthermore, WAIS has recently been shown to be warming faster than previously thought [Steig et al., 2009]; this has raised scientific concern that the WAIS is potentially susceptible to internal or ocean-driven instability. A number of glaciological studies in West Antarctica have highlighted the importance of understanding ice sheet interactions with water, including studies at the basal boundary where ice streams come in contact with active subglacial hydrologic systems [Gray et al., 2005; Fricker et al., 2007; Bell, 2008] and at the seaward margin where the ice sheet is exposed to oceanic forcing [Rignot and Jacobs, 2002; Anandakrishnan et al., 2007; Alley et al., 2007; Pollard and DeConto, 2009]. Water and wet sediments play an important role in determining the rate of ice stream flow and in triggering changes in flow rates on short timescales [Hulbe and Fahnestock, 2007; Peters et al., 2007; Vaughan and Arthern, 2007]. Evidence for a direct link between ice stream dynamics and subglacial lakes is emerging: e.g., on Byrd Glacier drainage of a subglacial lake has been linked to an increase in ice velocity of 10% sustained for 14 months [Stearns et al., 2008]. At grounding zones, thermal ocean forcing may represent a key mechanism for destabilizing the WAIS, thereby increasing its contribution to global sea level rise [Oppenheimer, 1998; Rignot and Jacobs, 2002; Shepherd et al., 2004; Walker et al., 2008; Pollard and DeConto, 2009] and potentially altering ocean and atmospheric circulation. Scherer et al. [2008] demonstrated that the Ross Ice Shelf (RIS) and the WAIS responded to warming during an early Pleistocene interglacial (MIS-31) and that, at least during that interglacial, retreat of the WAIS preceded warming-induced retreat of the Greenland Ice Sheet. Similar scenarios are known to have occurred repeatedly throughout Pliocene times when atmospheric pCO_2 and global temperatures were similar to those predicted for the coming centuries [Naish et al., 2009; Pollard and DeConto, 2009].

In addition to their potentially important role in ice sheet dynamics, subglacial environments in Antarctica are an unexplored component of the biosphere. The low temperatures (~0°C), complete darkness, and direct isolation from the atmosphere for millions of years make them one of the most extreme environments on our planet. Despite their perceived inhospitable nature, the available data imply that Antarctic subglacial environments may support a diversity of microorganisms. The metabolic diversity discovered range from chemolithoautotrophs that use energy from redox chemistry to fix CO_2 [Priscu and Christner, 2004; Mikucki and Priscu, 2007; Priscu et al., 2008, Lanoil et al., 2009] to heterotrophic bacteria that utilize the reduced carbon fixed by the chemolithoautotrophs. As such, the subglacial ecosystem may function in much the same way as deep-sea vent ecosystems. Priscu and Christner [2004] and Priscu et al. [2008] estimated that prokaryotic organic carbon within the subglacial Antarctic ecosystem exceeds that of all surface freshwater lakes and rivers on our planet, which are considered to be biogeochemically important systems. Mineral weathering of basement rock and sediments is extremely rapid [Mitchell et al., 2006; Mitchell and Brown, 2008] and appears to be enhanced by microbial activity [Wadham et al., 2010]. Recent estimates suggest that the amount of weathering beneath the Antarctic ice sheet rivals that of the Amazon River in terms of geochemical input to the oceans and that much of this weathering is microbially mediated [Wadham et al., 2010]. Despite the potential importance of subglacial ecosystems to biogeochemical cycles, these environments have yet to be sampled in a comprehensive, integrated fashion.

2. BACKGROUND

2.1. Subglacial Hydrology (and Lakes)

Subglacial lakes and grounding zones of ice streams have been identified as high-priority targets for scientific investigations by U.S. and international research communities [National Research Council (NRC), 2007]. The presence of water beneath the ice sheet has long been known; however, the subglacial environment is one of the most inaccessible places in the world and is consequently poorly studied. Thus, our knowledge of distribution and flux of subglacial water and the processes that drive subglacial hydrology remains limited. There have been isolated field campaigns to certain regions of the ice sheet, notably including the upstream portion of the Whillans and Kamb ice streams [Engelhardt et al., 1990; Kamb, 2001; Engelhardt, 2005]. Water or wet sediment was encountered at the base of many of the boreholes, including a shallow water cavity beneath the Kamb Ice Stream. Subsequent ground-based radar surveys [Jacobel et al., 2009] provided new data on the areas of basal melt. Improved ice sheet models are helping to predict where water may flow on a regional basis [LeBrocq et al., 2009].

Despite the lack of field data, our knowledge of Antarctica's subglacial environment has dramatically changed over the last 5 years. A major breakthrough was the realization that subglacial lake activity (both filling and draining) causes changes in surface elevation of the ice sheet that can be detected by satellite instruments. A subglacial lake inventory tallied 145 lakes in 2005 [Siegert et al., 2005], but such lakes were thought to be inactive. Since 2005, satellite results have shown that the lakes can be "active": Lake volumes can fluctuate, and rapid fluxes of water can occur periodically between them via subglacial floods [*Gray et al.*, 2005; *Wingham et al.*, 2006; *Fricker et al.*, 2007]. Many subglacial lakes (124) have now been designated as active [*Smith et al.*, 2009].

A complex and dynamic subglacial hydrology has the potential to rapidly change the basal conditions of an ice sheet. On Alpine glaciers and the Greenland Ice Sheet, there is considerable evidence of seasonal acceleration and in response to meltwater penetration to the bed and likely subglacial flooding/lubrication [e.g., *Joughin et al.*, 2008; *van de Wal et al.*, 2008]. These events lead to transient speedup as well as transient slowdown of the overlying ice, although the long-term effect of meltwater penetration is still poorly known. These events are analogous to the Antarctic subglacial lake drainage phenomena, although the trigger for subglacial lake drainage is not linked to environmental forcing [*Bell*, 2008]. The effect of subglacial lake activity on ice dynamics remains largely unquantified, however, and there has only been one observation of glacier speedup coinciding with a flooding event (on Byrd Glacier [*Stearns et al.*, 2008]). The reason there have been so few observations of such speedups in Antarctica is mainly a lack of velocity data that must be acquired simultaneously with elevation data.

Since subglacial lakes exist along a hydrologic continuum, the transfer of water from one lake to the next lake downstream ultimately releases subglacial water to the ocean across the grounding line. To date, there are no observations of how injection of water into the sub-ice-shelf cavity might alter water properties, or circulation in the cavity, although a recent modeling study shows that subglacial outflows increase basal melt rates near the grounding line (A. Jenkins, personal communication, 2009).

2.2. Grounding Zones

A better understanding of grounding zones is essential because of their influence on the stability of ice sheets and their consequent role in sea level rise under future global warming scenarios. Although there have been several successful efforts to drill through ice shelves to access sub-ice-shelf ocean cavities [e.g., *Craven et al.*, 2009; *Clough and Hansen*, 1979], direct study and exploration of ice sheet grounding zones have been limited to a few relatively accessible locations. Only two grounding lines have actually been observed and sampled in Antarctica: those of the tidewater cliff at Blue Glacier, McMurdo Sound, and at Mackay Glacier's floating glacier tongue [*Powell et al.*, 1996; *Dawber and Powell*, 1998]. Grounding zones are currently being mapped using satellite imagery to better understand their structure and dynamics [e.g., *Fricker et al.*, 2009]; however, no studies to date have combined remote sensing and fieldwork in a targeted grounding zone investi-

gation. Importantly, this combined approach has not been applied to the study of the sensitive ice streams draining WAIS. Whillans Ice Stream provides a dynamic system for investigating the control of subglacial hydrological and sedimentary processes on the rate of ice discharge to the ocean. Observations and measurements are also needed to constrain sedimentary conceptual [e.g., *Powell and Alley*, 1997] and quantitative models [*Alley et al.*, 2007] that link flux and accumulation of sediment beneath ice streams to grounding zone instability. Direct and indirect measurements from the grounding zone will verify and improve hypothesized models for the migration of modern grounding zones of WAIS and their sedimentary footprints, by improving our understanding of ice and sediment fluxes and their recent history. Integration of these data is needed to assess the future vulnerability of the WAIS.

One subproject of the Whillans Ice Stream Subglacial Access Research Drilling (WISSARD) plans to collect data from the Whillans Ice Stream grounding zone to allow estimation of basal melt rates in the sub-ice-shelf cavity near the grounding zone and to help constrain models used to assess the future behavior of the WAIS. Such data are needed to elucidate the microscale and mesoscale processes controlling basal melting rates in grounding zone environments to improve parameterizations of ice-shelf mass balance and quantitative models of sub-ice-shelf cavity circulation [e.g., *Holland and Jenkins*, 2001; *Jenkins and Holland*, 2002; *Makinson*, 2002; *Holland et al.*, 2003]. Despite substantial recent progress in the numerical modeling of ice shelf-ocean interactions, field investigations of processes occurring beneath ice shelves, necessary to establish boundary conditions, have been scarce. This is especially true for grounding zones [*Steffen et al.*, 2009].

Further driving this need for in situ data is the idea put forward by *Holland et al.* [2008] that warming of surface waters need not be necessary to enhance sub-ice-shelf warming because the warmer waters that penetrate ice-shelf cavities upwell from depth and thus regional atmospheric warming may not be required on short timescales. Changes in oceanic circulation caused by atmospheric/oceanic dynamics may alter the flow of existing deeper warm water masses onto the continental shelf [e.g., *Jacobs*, 2006]. *Holland et al.* [2008] also show that if a steady warming of offshore waters were to take place, then melting of the ice-shelf base would increase at an accelerating rate. They further emphasize that each ice shelf has a nonlinear melt temperature curve such that melt rate sensitivity varies with both topography and temperature. Hence, each region needs to be assessed independently based on local conditions.

Other studies substantiate that in situ data are also required from the upstream side of a grounding line because subglacial bed conditions are an important interactive and feedback component of the ice stream-ice shelf system. The pattern of basal melting and freezing beneath an ice shelf may shift as the shelf evolves [*Walker and Holland*, 2007], and this distribution can strongly affect ice stream response [*Walker et al.*, 2008]. By explicitly modeling the sub-ice-shelf ocean, *Walker et al.* [2008] were able to apply climatic forcing in a way to avoid arbitrary specifications of ice-shelf basal melting to try and isolate the effect of subglacial bed rheology. In these experiments, they found that the applied oceanic warming exerts the greatest influence over the evolution of the ice shelf-ice stream, with the seaward-sloping bed limiting the resulting grounding line retreat, even when a significant portion of the ice shelf is melted. However, for the same oceanic warming, there is a noticeable increase in thinning and acceleration of the ice stream, leading to greater flux across the grounding line as the basal rheology is changed from linear-viscous toward plastic. *Walker et al.* [2008] concluded that these results should be tested using a combination of field, laboratory, and remote sensing data.

WISSARD data sets are wide ranging and include remote-sensing data (seismic and radar surveys and GPS and satellite monitoring) together with direct measurements. Direct measurements include multibeam sonar and sub-bottom profiling, hydrological and oceanographic measurements (conductivity, temperature, pressure, current strength, and direction), ice velocity and deformation data, sediment cores, and in situ sediment strength measurements. Therefore, this study will greatly increase those sparse data available on sub-ice-shelf oceanic circulation processes and ice dynamics of the RIS and will allow testing of the newly established ideas and models of ice sheet-ice shelf interactions and potential future behavior of WAIS.

2.3. Subglacial Biology and Biogeochemistry

Recent biologic investigations of Antarctic subglacial environments support the hypothesis that they provide a suitable habitat for life [*Priscu et al.*, 1999, 2008; *Christner et al.*, 2006; *Mikucki et al.*, 2004]. Microbiological studies indicate relatively high bacterial densities (~10^6 cell g^{-1} sediment) in sediments from beneath the nearby Kamb Ice Stream [*Lanoil et al.*, 2009]. If these abundance estimates are accurate, subglacial water and wet sediments may constitute a significant and as of yet unrecognized pool of organic carbon on Earth [*Priscu et al.*, 2008]. Just as streaming ice flow is dependent on availability and dynamics of subglacial water and wet sediments, subglacial microbial ecosystems rely on these two physical fac-

tors for a supply of water, nutrients, and energy sources [*Tranter et al.*, 2005]. Subglacial microbial ecosystems also enhance biogeochemical weathering, mobilizing elements from long-term geological storage [*Mitchell et al.*, 2006; *Mikucki et al.*, 2004, 2009]. Integrating genomic and biogeochemical measurements with glaciological and geological studies takes an ecosystem approach to the study of subglacial ecology and will allow for an assessment of structure-function relationships in this previously unexplored system.

Ice sheet drilling technologies, to date, have been designed largely for the retrieval of ice cores for paleoclimate research or for direct access to the bed for glaciological observations [*NRC*, 2007]. With the inclusion of biology in glaciological and subglacial studies, new concerns arise regarding the introduction of chemical and biological contamination during drilling operations. A recent report by the *NRC* [2007] recommended that "the numbers of microbial cells contained in or on the volume of any material or instruments added to or placed in these environments should not exceed that of the basal ice being passed through." The WISSARD project will develop, demonstrate, and execute a clean sampling strategy that meets the needs of the science questions being asked and the recommendations for environmental stewardship in Antarctica; WISSARD represents one of the largest subglacial exploration projects yet proposed, and our challenge is to recover data and samples that represent the autochthonous ecological conditions without altering the subglacial ecosystem irreversibly.

3. WISSARD SUBPROJECTS

WISSARD is interdisciplinary in nature and consists of three interrelated subprojects, each with a different focus:

1. Lake and Ice Stream Subglacial Access Research Drilling (LISSARD) focuses on investigating the role of active subglacial lakes in controlling temporal variability of ice stream dynamics and mass balance dynamics.

2. Robotic Access to Grounding-zones for Exploration and Science (RAGES) concentrates on stability of ice stream grounding zones which may be perturbed by increased thermal ocean forcing, internal ice stream dynamics, subglacial sediment flux to the grounding zone, and/or filling/draining cycles of subglacial lakes upstream from grounding zones.

3. GeomicroBiology of Antarctic Subglacial Environments (GBASE) addresses microbial metabolic and phylogenetic diversity and associated biogeochemical rock weathering and elemental transformations in subglacial lake and grounding zone environments.

The subprojects are connected through a common interest in coupled fluxes of ice, subglacial sediments, nutrients, and water, as well as by the common need to characterize and quantify physical, chemical, and biological processes operating subglacially (Plate 1). Direct observations and real-time in situ data collected during WISSARD will address fundamental scientific questions pertaining to (1) past and future marine ice sheet stability, (2) subglacial hydrologic and sedimentary dynamics, (3) subglacial metabolic and phylogenetic biodiversity, and (4) the biogeochemical transformation of major nutrients within a selected subglacial environment.

4. REGIONAL SETTING: WHILLANS ICE STREAM SUBGLACIAL LAKE AND GROUNDING ZONE SYSTEM

4.1. Subglacial Lake

Subglacial Lake Whillans is an active subglacial lake on lower Whillans Ice Stream. The lake is part of an extensive hydrological system under the Mercer and Whillans ice streams that was discovered through analysis of repeat tracks of ICESat laser altimetry data. The ICESat data detected deformation of the ice surface in response to subglacial water activity [*Fricker et al.*, 2007; *Fricker and Scambos*, 2009]. ICESat has monitored the activity of Subglacial Lake Whillans intermittently between October 2003 and October 2009. During this time, there have been two complete fill/drain cycles

(Plate 2). It is not known how long the lake had been quiescent prior to the start of the ICESat time series. However, these data do allow for a periodic drainage cycle with a residence time on the order of 3 years. In the 2007–2008 field season, a continuous GPS station was established on the lake; however, at the time of writing, the data from that GPS have not been downloaded since the 2008–2009 season; therefore, exact timing of the 2009 drainage event is not yet known. During the 2007–2008 field season, ice-penetrating radar surveys showed that ice thickness above Subglacial Lake Whillans is only 700 m and suggest that the depth of water in the lakes is >8 m even after lake drainage events [*Tulaczyk et al.*, 2008].

Subglacial Lake Whillans was selected for drilling owing to its location beneath a major West Antarctic Ice Stream that is known to have highly variable surface velocity [*Joughin et al.*, 2002; *Bindschadler et al.*, 2003; *Joughin et al.*, 2005]. Other selection factors included safety (no visible surface crevassing; verified through geophysical reconnaissance in 2007–2008) and its proximity to the grounding line (~80 km). Subglacial Lake Whillans is also accessible from McMurdo Station and has a relatively thin cover of glacial ice (700 m). For comparison, ice thicknesses over Ellsworth Subglacial Lake and Vostok Subglacial Lake, two other targets for subglacial access drilling, are ~3 and ~4 km, respectively [*Woodward et al.*, 2010; *Siegert et al.*, 2004; *Studinger et al.*, 2003]. Concern over inadvertent biological contamination of Subglacial Lake Whillans is allayed by the fact that it is a small lake at the seaward end of the hydrologic catchment.

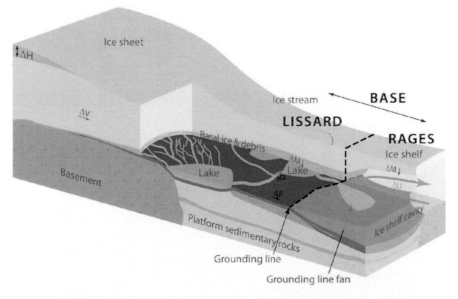

Plate 1. Schematic showing the glaciological, geological, and hydrological setting of the sector of the West Antarctic Ice Sheet to be studied by Whillans Ice Stream Subglacial Access Research Drilling subprojects [Lake and Ice Stream Subglacial Access Research Drilling (LISSARD), Robotic Access to Grounding-zones for Exploration and Science (RAGES), and GeomicroBiology of Antarctic Subglacial Environments]. Drawn by S. Vogel.

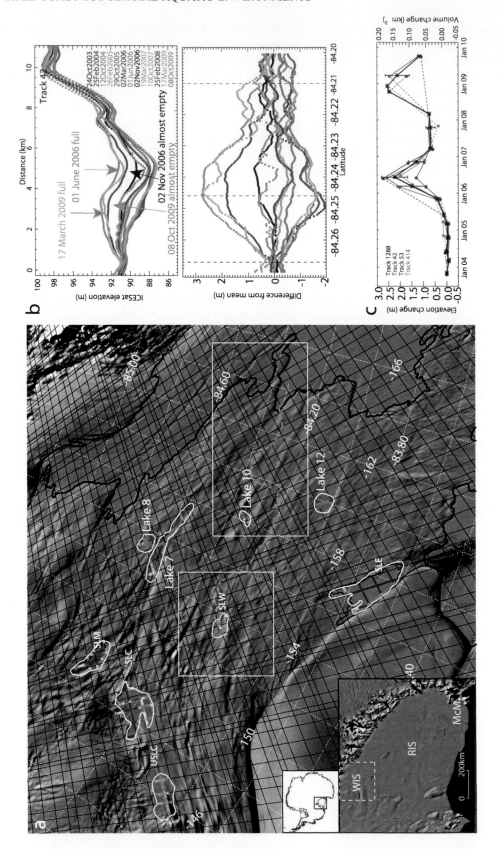

Plate 2. (a) Map of Whillans Ice Stream lake system [from *Fricker and Scambos*, 2009] with annotation of the estimated flow path from Subglacial Lake Whillans to the grounding zone. White boxes outline the extents of Plates 4a and 4b. Inset maps at left show the location of Whillans Ice Stream with respect to Antarctica (upper inset) and McMurdo Station (McM, lower inset). (b) Surface elevations (top plot) and elevation anomalies (bottom plot) for ICESat Track 42 across SLW. (c) Averaged time series of estimated surface elevation and volume changes since October 2003 for Subglacial Lake Whillans. Updated from *Fricker and Scambos* [2009], reprinted from the *Journal of Glaciology* with permission of the International Glaciological Society.

Any potential environmental impact is therefore confined to a limited area close to the ocean with a relatively short hydrologic residence time, as inferred from ICESat observations [*Fricker and Scambos*, 2009; Figure 2b].

4.2. Grounding Zone

The most likely flow path from Subglacial Lake Whillans to the grounding line has been estimated from the hydrostatic hydropotential derived from a surface digital elevation model and bedrock digital elevation model and suggests that the outflow enters the RIS cavity in a near embayment (84.35°S, 163.06°W) (Plate 4b). The final location of the grounding zone site will depend on our geophysical site surveys, which will include collecting extensive high-resolution radar and seismic data. Surveying this part of the grounding zone will allow us to determine the effect of lakes on grounding zone processes and their stability, and vice versa, in addition to their effect on subglacial sedimentary processes and ocean-induced basal melting.

One area of the grounding zone of Whillans Ice Stream has recently been the target of remote sensing and modeling studies [*Anandakrishnan et al.*, 2007; *Alley et al.*, 2007]. These studies have indicated that a sedimentary grounding zone wedge is accumulating with a consequent thickening of the Whillans Ice Stream such that ice thickness at the grounding line is greater than that of adjacent floating ice of the RIS (which is in hydrostatic equilibrium). Grounding zone wedges may stabilize the position of the grounding line so that it will tend to remain in the same place until sea level rise of at least several meters overcomes the excess ice thickness [*Anandakrishnan et al.*, 2007]. If the condition of a dominant subglacial sediment flux occurs around Antarctic grounding lines, then recent Antarctic ice volume changes cannot be attributed to sea level rise [*Alley et al.*, 2007]. Sediment-forced stability may be overcome through fast ice stream thinning across the grounding line either by melting (from ocean warmth or geothermal heat) or by subglacial lake water lubricating the bed. Understanding these interactions, rates, and magnitudes will assist in predictions of future grounding line dynamics. Furthermore, these grounding zone studies will also provide prime data for testing assertions that synchronous behavior of ice sheets on millennial timescales implies ice sheet teleconnections via either sea level or climatic forcing [cf. *Alley et al.*, 2007; *Naish et al.*, 2009; *Pollard and DeConto*, 2009]. Certainly, large sea level rises, such as the ~100 m rise at the end of the last ice age, may overwhelm the stabilizing feedback from sedimentation, but smaller sea level changes may not have synchronized the behavior of ice sheets in the past [cf. *Alley et al.*, 2007].

5. SCIENTIFIC GOALS OF WISSARD

WISSARD involves a spectrum of scientific approaches to studying sub-ice-sheet processes, including satellite remote sensing, surface-based geophysics, borehole observations and measurements, and basal and subglacial sampling (Plate 2a and Table 2). This broad research strategy permits investigation of the dynamics of subglacial environments and their potential impact on ice stream and ice sheet stability over a range of complementary spatial and temporal scales. Exact selection of drill sites will be determined before drilling occurs based on safety considerations and accessibility criteria (e.g., crevasses) derived using radar and seismic data collected by the surface geophysics teams and data from satellite images. The surface geophysical surveys will also determine the regional sedimentary and hydrologic structure of the bed, constrain long-term history of ice flow across the region, and will provide spatial context for interpretation of borehole findings.

The boreholes will be used to (1) collect samples of subglacial water, sediments, and basal ice for biological, geochemical, glaciological, sedimentological, and micropaleontological analyses; (2) measure subglacial and ice shelf cavity physical and chemical conditions and determine their spatial variability; and (3) investigate sedimentary features and components, subglacial water discharge, oceanography, and basal ice at the seaward side of the grounding zone and within the nearby sub-ice-shelf cavity using sub-ice ROVer (SIR) (Plate 3), a multisensor remotely operated vehicle (ROV); sediment cores; thermal and geotechnical probes; and oceanographic moorings, some erected as long-term observatories.

Direct sampling of Subglacial Lake Whillans and the grounding zone downstream of it is expected to yield seminal information on the glaciological, geological, and microbial dynamics of these environments and test the overarching hypothesis that active hydrological systems connect various subglacial environments and exert major control on ice sheet dynamics, geochemistry, metabolic and phylogenetic diversity, and biogeochemical transformations of major nutrients within glacial environments. Each subproject has its own set of scientific goals, which are discussed below.

5.1. LISSARD

The primary goal for LISSARD is to obtain measurements that will be used to improve treatments of subglacial hydrological and mechanical processes in models of ice sheet mass balance and stability. We expect to synthesize data and concepts developed during LISSARD to determine whether subglacial lakes play an important role in (de)stabilizing

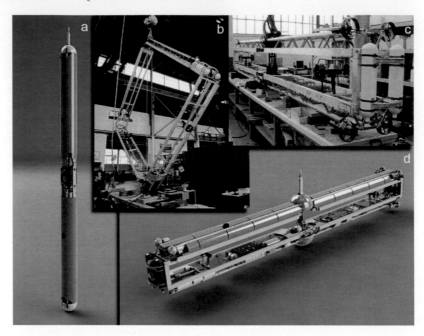

Plate 3. The sub-ice ROVer (SIR) developed for RAGES showing from left and moving anticlockwise: (a) a computer-aided design (CAD) drawing of SIR in its ice-borehole deployment mode with its fiberglass "skins" covering sensitive internal mechanisms and instrumentation, (b) SIR during its build phase unfolding from its ice-borehole deployment mode into its operational mode, (c) another of SIR during its build phase but in its operational mode showing its four rear thrusters and ballast bottles, and (d) another CAD showing the SIR open in its operational mode as it will be in the sub-ice-shelf cavity, with yellow blocks representing foam floatation on top and the instrumentation bay on the bottom.

WAIS and to recommend how their impact on subglacial hydrology and ice flow may be incorporated into models of ice sheet evolution and mass balance. In particular, we will assess whether active subglacial lakes have an important modulating effect on ice stream dynamics. Essentially, there are two scenarios that we are trying to distinguish between (1) subglacial lakes are part of the distributed subglacial hydrological system enabling ice stream lubrication or (2) subglacial lakes connect to a separate network of localized conduits draining subglacial water without significant impact on ice flow. In order to achieve this, we will establish the fundamental characteristics of subglacial hydrological systems associated with Subglacial Lake Whillans, including where water originates and whether subglacial volcanic and/or geothermal activity play a significant role in determining lake location, mass balance, and evolution.

A second goal of LISSARD is to derive information on lake history using sediment samples collected from and below the floor of Subglacial Lake Whillans. Owing to the short length of satellite records (2003–2008) compared to the activity cycles of the lakes [*Fricker and Scambos*, 2009], we have no information about lake activity before these observations began. The lake sediments are predicted to contain microfossils and cosmogenic isotopes that will help con-

strain when the most recent period of grounding zone retreat beyond these lakes occurred. Such observations may also tell us whether subglacial drainage patterns and lake connectivity have changed over timescales ranging from years to millennia. We will analyze archives of past basal water and ice flow variability contained in subglacial stratigraphy, sediment pore water, lake water, and basal accreted ice and use this information to reconstruct the history of WAIS and ice stream stability.

The final goal of LISSARD is to provide background understanding of subglacial lake environments to benefit the other subprojects (RAGES and GBASE). LISSARD's investigation of subglacial water and sediment fluxes will create a dynamic link between subglacial lakes and grounding zones that together will elucidate any change in grounding line position. LISSARD will provide data on the subglacial microbial habitat by determining what physical and chemical characteristics of subglacial hydrological and sedimentary systems impact geomicrobiological conditions.

5.2. RAGES

RAGES focuses on the processes of, and interactions among ice, subglacial water and sediment, sub-ice-shelf cavity ocean water, and sea floor sediment. RAGES field sampling integrates surface geophysical surveys with a wide variety of borehole and subglacial sediment and water cavity sampling and measurements. RAGES will primarily assess the stability of WAIS relative to the magnitudes of sediment delivery to grounding zone wedge sediment, geotechnical properties of that sediment, ice flux across the Whillans Ice Stream-RIS grounding line, melting of subglacial ice by ocean waters in the sub-ice-shelf cavity, and subglacial meltwater discharge. Of particular concern are the degrees to which the grounding zone wedge in the Whillans Ice Stream-RIS region as reported by *Anandakrishnan et al.* [2007] resists rapid ice flow to stabilize the grounding zone and the rate at which sediment is delivered to the wedge to counteract the effects of ocean-induced basal melting.

Ice flow in the transition from grounded to floating ice across the grounding zone is complex and may well be associated with sediment wedge geometry, form, and physical properties. Thus, sediment geotechnical measurements will be made, as will spatial patterns of strain and stresses throughout the ice column. Measurements of the rate of basal melting from ocean heat delivered to the grounding line will be used to constrain the scale of basal melt contribution to ice stream thinning as it goes afloat. RAGES aims to determine whether the basal melt rate has been relatively stable in the recent past using basal thermal profiles. Geothermal heat flux influences ice thinning, and evidence of geothermal activity in proximity to the grounding zone will be assessed. It is very unlikely that Subglacial Lake Whillans will be draining at the exact time of our investigations, but that record will be evaluated from glacimarine sediment cores in front of the grounding line and through deployment of long-term oceanographic observatories in the ice-shelf cavity.

RAGES will further assess the degree to which grounding zone sedimentary systems preserve important records of past WAIS dynamics, on short and relatively long time-scales. Grounding-zone sedimentary depositional systems are known from continental shelves but are notoriously difficult to sample. More precise understanding of processes and rates of grounding zone formation in modern environments will improve our interpretations of older deposits, establish constraints on past ice sheet dynamics and inform future response to external forcings, and improve ice sheet modeling. Sediment cores from the grounding zone will aid our assessment of whether the most recent period of relative grounding line stability, as documented by satellite data over about the last few decades, is part of long-term stability or an intermittent stage in a continued WAIS retreat. The cores will also provide insight into the short-term variability in grounding line position during the current period of grounding zone stability. As has been found in older conti-

nental shelf deposits [*Sjunneskog and Scherer*, 2005], these grounding zone wedge accumulations are likely to contain microfossils eroded from older sediments, and thus, the wedge may preserve evidence of Holocene and Plio-Pleistocene history of WAIS dynamics in response to past climate forcing. These fossils also provide physical constraints on subglacial shearing [*Scherer et al.*, 2004].

Lastly, RAGES assesses the role of subglacial microbial activity on the flux of weathered materials to the WAIS grounding zone, the RIS cavity, and the Southern Ocean. The project will assess biological and biogeochemical diversity and whether there are significant amounts of subglacially mobilized carbon and nutrients released at the grounding line and what their impacts might be on oceanic biogeochemical cycling. This goal provides an important link with the GBASE studies.

5.3. GBASE

GBASE will examine both the subglacial lake site (ice, bulk water, and sediments) and the grounding-zone site to provide genomic and biogeochemical data across the hydrologically linked subglacial system, thus providing an assessment of biodiversity and structure-function dynamics across these distinct subglacial environments, into the ocean. Since the biology of subglacial environments is largely unexplored, GBASE will be driven by both discovery and hypothesis-driven science. The subglacial environment associated with Whillans Ice Stream is dynamic, which is likely to impact the microbial ecosystem and its biogeochemical processes. Within this context, GBASE will assess the metacommunity structure beneath the Whillans Ice Stream and determine the metabolic and physiological properties of its constituents. International scientific committees have identified subglacial microbial ecology as one of the most important topics in contemporary glaciology [e.g., *NRC*, 2003, *Priscu et al.*, 2005; *NRC*, 2007], and the GBASE subproject will be the first to address this subject in a multidisciplinary and interdisciplinary fashion. Specifically, our study will provide new information on the biogeochemical and phylogenetic diversity of Antarctic subglacial microorganisms and their role in mobilizing nutrients to the Southern Ocean.

Molecular-based surveys have revealed that strong phylogenetic relationships exist between bacteria inhabiting subglacial sediments from around the world [*Priscu et al.*, 2008; *Nemergut et al.*, 2007; *Lanoil et al.*, 2009]. Betaproteobacteria related to the heterotrophic genus *Comamonas* and the chemolithoautotrophic genera *Gallionella* and *Thiobacillus* appear to be common in these communities. Similarly, studies of samples from Antarctic subglacial lakes and marine-influenced subglacial systems have detected phylotypes related to bacteria with metabolisms dedicated to iron and sulfur respiration and oxidation [*Priscu et al.*, 1999; *Christner et al.*, 2006; *Mikucki and*

Priscu, 2007; *Mikucki et al.*, 2009]. Thus, we predict a similar bacterially dominated community in the sediments and lake water from the Whillans Ice Stream. Processes responsible for the incorporation of debris into basal ice remain enigmatic [e.g., *Waller*, 2001]; however, that entrainment will have occurred farther up-glacier, and glacier-lake interactions are unlikely to be the source of the debris-rich ice. Microbial phylotypes present in the basal ice are therefore expected to be similar to those in upstream sediments and distinct from inhabitants of the lake water. As the sub-ice-shelf cavity is connected to marine open waters, the microbial community in these waters and sediments is hypothesized to be similar to that observed for other deep Antarctic marine water and sediment environments, respectively [e.g., *Bowman and McCuaig*, 2003]. Thus, grounding zone wedge sediments are likely to be a mixture of marine-derived and subglacially derived microbiota as that region receives input from both environs. The water and sediments will be analyzed using new DNA sequencing technologies that allow the Bacteria, Archaea, and Eukarya present to be comprehensively inventoried.

Little information exists on biological elemental cycling in subglacial environments, despite evidence that bacterial transformations of key elements in these environments may have global significance [e.g., *Sharp et al.*, 1999; *Christner et al.*, 2006]. The complete absence of light within our subglacial study sites implies that energy requirements for new organic carbon production depend on the activity of chemolithoautotrophs, with electron transport to and from inorganic material. Potential redox couples include dissolved gases and ions and solid phase minerals [e.g., *Neal et al.*, 2003]. These compounds can be derived from subglacial ice melt or from the underlying substrate via microbially mediated mineral dissolution [cf. *Hodson et al.*, 2008].

6. IMPLEMENTATION OF WISSARD

6.1. Logistics and Management

WISSARD is a complex, logistically challenging project that has required many years of planning with the National Science Foundation (NSF), the U.S. Ice Coring and Drilling Services (ICDS), and Raytheon Polar Services Corporation, the U.S. Antarctic Program's logistical subcontractor. This planning process is presently under way, and the remaining timelines for WISSARD are presented in Table 1.

6.2. Surface Geophysics

WISSARD's surface geophysical investigations (reflection seismic imaging, radar, GPS) will be conducted at both Subglacial Lake Whillans and the grounding zone sites. Surface geophysical surveys will be conducted in standard grid formats, and the continuous GPS monitoring will be established in networks to determine surface elevation and ice dynamics in the two regions of interest.

Surface geophysics will be used for site selection for both the LISSARD and RAGES drilling and will also provide critical glaciological observations, tying satellite and aero-geophysical data to borehole measurements. Dense grid surveys will provide high-resolution depictions of geological and dielectric conditions at and below the basal interface, optimizing drill site placement over Subglacial Lake Whillans and at the grounding zone sites. Geophysical data also provide a link between the high-resolution, but spatially limited, borehole observations and the low-resolution, but spatially extensive, satellite-based observations. Geophysical investigations dovetail with the borehole-based science by providing constraints on spatial variability in subglacial conditions and spatiotemporal variability in ice motion.

Site selection includes radio echo sounding and active-source seismology, along with kinematic and static GPS surveying. Radio echo sounding provides information on ice thickness, englacial stratigraphy, and bed properties. Low-frequency, high-power systems may also penetrate some distance into the bed providing information on subglacial bed properties. Active-source seismology complements the ice-thickness data acquired by radio echo sounding. Furthermore, seismic observations provide information on water depth both within the lake and in the sub-ice-shelf cavity at the grounding line. Perhaps most significantly, seismic observation establishes a seismic-stratigraphic framework for both the subglacial lake and grounding line sedimentary systems. Seismic stratigraphy is used extensively in a range of sedimentary systems to delineate stratigraphic architecture and to infer its history and genesis [e.g., *Vail et al.*, 1977, 1991]. The modern systems under study in WISSARD have never been investigated, and it is likely that new facies frameworks will be needed. To do this, we build on previous glacial and glacimarine models of system architectures, internal geometries, and facies motifs [e.g., *Powell and Alley*, 1997; *Powell and Cooper*, 2002; *Dunbar et al.*, 2008; *Dowdeswell et al.*, 2008; *McKay et al.*, 2009].

Additional surface geophysics components will acquire snapshots of the ice surface, through kinematic GPS measurements, and continuous time series of any variability in surface elevations or ice displacement, through continuous GPS. Also, passive seismic observations provide additional information on styles of deformation at the bed and possibly record the movement of subglacial water.

WISSARD affords the opportunity to link data on a number of spatial scales. At the finest resolution, the in situ borehole observations will provide direct samples of water

Table 1. Timetable for WISSARD[a]

Activity	Subproject		
	LISSARD	RAGES	GBASE
September 2009 to August 2010			
Preparation and testing of equipment for drilling, clean sampling, and measurements	√	√	√
Integration of SIR, borehole instrumentation, and drilling technologies	√	√	√
September 2010 to August 2011			
Preparation and testing of equipment for drilling, clean sampling, and measurements	√	√	√
Surface geophysics at SLW and GPS/seismic installation in SLW region	√		
GPS installation in the GZ		√	
Clean technology testing	√	√	√
Oceanographic instrumentation field test at Lake Tahoe	√	√	√
September 2011 to August 2012			
Servicing of SLW GPS/seismic array; SLW data analyses, interpretation, publication	√		
Surface geophysics at GZ and GPS/seismic installation in GZ region		√	
Drill, clean technology and instrumentation field test at McMurdo—"WISSARD shakedown"	√	√	√
September 2010 to August 2013			
Traverse to SLW	√		√
Drill and sample SLW (8 weeks)	√		√
Additional surface geophysics and servicing	√	√	
September 2013 to August 2014			
Traverse to GZ	√	√	√
Drill and sample GZ (8 weeks)	√	√	√
September 2014 to August 2015			
Removal of remaining autonomous instruments	√	√	√
Data analysis, interpretation, synergy, and publication of results	√	√	√

[a]Abbreviations are as follows: SLW, Subglacial Lake Whillans; GZ, grounding zone. Check mark indicates that noted activity applies.

column and subglacial materials. Above this scale, the SIR will provide water column and bed sampling and sub-bed geophysical data in the ocean cavity. Surface geophysical observations will form the next level of resolution, tying seismic observations into the high-frequency sub-bed profiling. Satellite-based observations will form the top tier of data, bringing the in situ and surface geophysics into a large-scale regional context.

6.3. Drilling Equipment and Transport

WISSARD will use a new NSF-funded hot water drill under construction by Caltech. This drill design is based on previous hot water drilling technology ranging from that used to complete 119 boreholes through West Antarctic Ice Streams with thickness up to nearly 1300 m [*Kamb*, 2001; *Engelhardt*, 2005] to the IceCube system.

Table 2. Measurements to be Made During WISSARD

Measurement and Techniques	Purpose
LISSARD/RAGES/GBASE: Grounded Ice	

Glaciology

ice flow: imagery, GPS velocity, strain rate grids	site survey
ice thickness: radar, seismic	characterize modern and paleoice flow
surface and englacial flow features: satellite, radar	spatial context of borehole measurements
thermal profile: borehole observations, thermistors, radar attenuation	basal thermal regime, melt/freeze rates

Bed conditions

till properties: cores, probes	continuity of bed properties along flow lines
sediment transport and flux: cores, ice, water samples, borehole observations	influence of bed properties on ice velocity and sediment flux
hydrology: water samples	lake/grounding line (GL) role in transient ice stream behavior
geothermal flux and thermodynamics: cores, probes, seismic, radar	basal thermal regime, GL melt/freeze rates
water, sediment, ice and gas chemistry: samples, probes	

Ice/bed interaction (physical and chemical parameters)

ice-bed contact: borehole observations	history of basal freezing/melting
till: borehole tools and cores	present freeze/melt conditions
	quantify erosion and sediment flux

Biology (ice, water, sediment)

characterize in situ geochemistry	describe metabolic processes in subglacial environments
	determine the influence of subglacial metabolism on the rate and flux of weathered materials
characterize microbial species diversity and measure metabolic activity	evaluate metabolic and genetic diversity in subglacial environments

Paleontology, petrography, sediment chemistry

	subglacial provenance and ice sheet history
sediment: cores, composition, chemistry	subglacial weathering rates
pore waters: cores, composition, chemistry	paleosubglacial sediment flux
	sediment and nutrient flux to Southern Ocean

LISSARD/GBASE: Subglacial Lake	

Lake/ice sheet interaction (physical and chemical)

lake temperature, chemistry, and currents: moorings, water samples	constrain energy balance and melt/freeze rates
turbulent and conductive heat flux: mooring, modeling	constrain meltwater flux through lake
frazil ice concentration and particle sizes: mooring, Subglacial Lake Exploration Device	subglacial lake circulation models
natural hydrology tracers (meltwater flux): mooring, water samples	assess factors of lake position and motion
water, ice, and gas chemistry: mooring, water samples	assess ocean surface water input under Ross Ice Shelf (RIS)
	sediment and nutrient flux to Southern Ocean

Subglacial hydrology (lake and conduit flows)

water temperature and currents: moorings, water samples	lake geometry and water levels
turbulent and conductive heat flux: mooring, modeling	conduit drainage network
natural hydrology tracers (meltwater flux): mooring, water samples	sediment properties, transport and flux

Table 2. (continued)

Measurement and Techniques	Purpose
water, ice, and gas chemistry: mooring, water samples	influence of bed properties on ice velocity
particulate transport and dispersal: borehole observations, mooring	influence on subglacial water and sediment flux

Subglacial lake sediment

lake sediment properties: torvane, heat flux probe, cores	West Antarctic Ice Sheet history
paleontological record: sediment cores, basal ice cores	ice stream stability
gas isotopes and geochemistry: sediment and ice cores, water samples	paleoflow variability in basal water and ice
physical sediment properties: torvane, heat flux probe, cores	subglacial lake sedimentary processes
paleontological record: sediment cores, basal ice cores	sedimentary modeling (modern subglacial, lake)
gas isotopes and geochemistry: sediment and ice cores, water samples	interaction between lake systems and ice motion
sediment facies, assemblages, and geometries: cores, seismic	sediment provenance and mixing
geothermal heat flux: probe and modeling	subglacial geology

RAGES/GBASE: Floating Ice

Ocean/ice shelf interaction (physical and chemical)

ocean temperature, salinity, and currents: mooring, Geochemical Instrumentation Package for Sub-Ice Exploration (GIPSIE), sub-ice robotically operated vehicle (SIR)	constrain energy balance and melt/freeze rates
turbulent and conductive heat flux: mooring, GIPSIE, SIR, modeling	constrain melt water flux across GZs
frazil ice concentration and particle sizes: SIR, mooring, GIPSIE	sub-ice and global ocean circulation models
natural hydrology tracers (meltwater flux): mooring, GIPSIE, SIR water, sediment, ice and gas chemistry: mooring, GIPSIE, SIR, cores	assess factors of GL position and motion: tidal, hydrology and sedimentology
	assess ocean surface water input under RIS
	sediment and nutrient flux to Southern Ocean

Subglacial and marine sediment (release and dispersal)

particulate transport and dispersal measurements: borehole observations, mooring, SIR, AVO and seismic	GZ sedimentary processes
water and sediment chemistry: cores, mooring, GIPSIE, SIR	sedimentary modeling of modern GL systems
sea floor morphology: SIR swath and seismic	interaction between GL systems and GL motion
sedimentary facies, assemblages and geometries: cores, borehole observations, SIR, seismic	sediment provenance and mixing
geothermal heat flux: probe, SIR and modeling	subglacial geology and ice sheet history
	subglacial weathering rates
	subglacial suspended load transport
	sediment and nutrient flux to Southern Ocean

Biology (ice, water, sediment)

potential recovery of protists and macroorganisms geochemical processes, cycles, characterization genomics, metabolic rates	document genetic diversity and describe ecology

Paleontology, petrography, and sediment chemistry

filter diatoms suspended in water	identical to Grounded Ice reasons

All WISSARD equipment including the hot water drill and tower, operational structures, vehicles, and supplies will be transported to the drill sites by traverse from McMurdo station (see Plate 2a for location with respect to WIS) across the RIS at the start of the 2012–2013 season. A new fleet of vehicles and sleds will be used for this purpose in addition to a new custom-designed structure for operations around the borehole and for use of the clean-access equipment (see below).

6.4. Clean Access Drilling

WISSARD adopts protocols to ensure minimal chemical and microbial contamination to the pristine subglacial environment. "Clean access" is relevant to both the drilling fluids used to establish boreholes and the instrumentation and sampling equipment used in exploring and characterizing the sub-ice environment. A recent report [NRC, 2007] used data from Vostok glacial and accretion ice [Priscu et al., 1999; Christner et al., 2006] to recommend that microbial numbers be reduced to ~100 cell mL^{-1} within the drilling fluid in that system. The level of cellular contamination for the WISSARD project will be based on what we measure in the basal ice in the Whillans Ice Stream basal ice. During the WISSARD planning phase, WISSARD PIs (led by GBASE) are working with ICDS to establish technologies and protocols that will minimize chemical and microbial contamination of the subglacial environment and samples during LISSARD field activities. In order to reach the NRC target, the hot water drill under development will utilize a large volume microfiltration and UV treatment system to continuously remove microbial cells and organic carbon from the hot water within the borehole water. The final filter size will be 0.2 mm, which should effectively eliminate all bacterial cells and other particulate matter. The drilling water will also be maintained at a temperature >90°C, which has been shown to significantly reduce the number of viable cells within the drilling fluid [Gaidos et al., 2004]. This drill will produce a borehole of at least 20 cm diameter that will remain open for ~8 days. DNA-containing bacterial numbers will be monitored by enumeration and molecular analysis to determine the microflora present in the source water and from drilling contamination. Microfiltration and UV sterilization procedures are effective strategies for reducing the viable microbial cell concentration from large volumes and surfaces. Cables and drill hoses will be cleaned by a two-step process prior to deployment down the hole. Cables and hoses will first pass through a collar of high-pressure, hot water jets that will remove particulates, cells, and salts. The hoses and cables will then pass through a series of UV lamps for final microbial sterilization. Instruments and sampling devices sent down the hole will be cleaned with a disinfecting solution of either peroxide or hypochlorite prior to borehole deployment. Quality control of the drilling fluid and instrumentation for purposes of environmental protection will be an ongoing process that will be done in close coordination with the NSF-Office of Polar Programs environmental officer.

We intend to demonstrate a clean sampling strategy that conforms to the NRC's recommendations by conducting two intensive tests of this technology before any attempt at subglacial access drilling. The first test will occur in a lake in the United States; the second will occur on the RIS near McMurdo, which will more closely resemble conditions to be expected at the study site. This test phase will allow examination of the effectiveness at reducing microbial cells >0.2 μm in drilling fluid using pasteurization, large-scale filtration, and germicidal ultraviolet (UV-C) irradiation and the ability to remove contaminants from the exterior of down hole instruments and sampling devices using power washing and disinfectants. The testing will assess the ability of the filtering and cleaning technology to reduce the total biological cell concentration as a function of time and volume processed and for the assessment of bioburden increase associated with drill and instrument deployment in the borehole. Furthermore, the ability to maintain reduced microbiological loads in borehole water over timeframes relevant to WISSARD subglacial access requirements (7–8 days) can be evaluated, and WISSARD sampling technology can be demonstrated to be compliant with the environmental requirements.

6.5. Specialized Scientific Instrumentation

Scientific goals of WISSARD will be met in tandem with significant advances in clean-access drilling and sub-ice exploration technologies. Innovative conceptualization and engineering have led to the project being able to use complex multisensor instrumentation.

6.5.1. Sub-Ice Robotically Operated Vehicle. SIR (Plate 3) is a robotic submarine with a customized "slim line" design having a diameter of ~55 cm and a length of 8.4 m when being deployed through the ice borehole. It is rated to 1500 m depth, and its power and data are transferred from and to the surface through a neutrally buoyant, strengthened, 3 km long umbilical tether of fiber optic and power cables. Navigation is by Doppler velocity logs and a gyro compass so the SIR is used either in automated mode with this AUV technology to do spatial surveys, or is manually driven by a surface operator using real-time video imagery to investigate specific features and operate in enclosed spaces. The vehicle is highly instrumented for obtaining both remotely sensed data as well as collecting and recovering samples. The wide range of sensors and sample collectors provides data for studies in oceanography, sedimentology, stratigraphy, glaciology, biology, microbiology, geochemistry, and geophysics and includes visual imaging (four cameras), vertical scanning sonar, Doppler current meter, multibeam sonar, CHIRP sub-bottom profiler, CTD, DO meter, transmissometer, laser particle-size analyzer, triple laser beams for sizing objects, manipulator arm, thermistor probe, shear vane probe, ice corer, sediment corer, and water sampler.

6.5.2. Ocean and lake water instrumentation. A variety of modular oceanographic instruments on inductive wireline lines will be used in various combinations, depending on the mission and objective of each deployment. The basic instrumentation includes CTDs, DO meter, transmissometer, and Doppler current meter immediately below the ice to measure turbulent heat flux across the ice-water interface, termed the basal energy balance mode. Below this most commonly will be a McLane Ice Tethered Profiler that includes a conductivity-temperature-depth with added Doppler current meter that continuously profiles water column structure and currents. The McLane profiler is a Woods Hole Oceanographic Institution (Woods Hole, Massachusetts) design that has been used through sea ice deployments in the Arctic Ocean. All data are telemetered to a surface data logger for real-time monitoring. If desired, this Ice Tethered Profiler can be exchanged with

nodular oceanographic sensors set at specific depths in the water as in standard oceanographic moorings. Other forms of oceanographic instrumentation are long-term moorings that will be deployed during the last phase of science operations at a site. They will be left making their measurements for 1 year or more by relaying their data to surface recording units. A prime tool to be used in the RAGES subproject is GIPSIE (Geochemical Instrumentation Package for Sub-Ice Exploration) that is deployed on a standard oceanographic wireline with real-time telemetry system. GIPSIE consists of an array of standard oceanographic sensors and samplers that include CTD, Doppler current meter, transmissometer, laser particle-size analyzer, DO meter, automated 48-port water sampler, water column nutrient analyzer (Si, NO_3, PO_4, NH_4, CO_2, CH_4, chlorophyll), sediment pore water chemistry analyzer (T, pH, redox, O_2, H_2S, H_2, N_2O), and a down-looking color camera. Individual instruments are repacked and mounted in a profiling housing for deployment through narrow diameter ice boreholes. The design is modular, allowing mission-specific configuration and the future addition of supplemental sensors. GIPSIE provides real-time in situ measurements of critical physical, geochemical, and nutrient properties; acquires in situ water samples from water masses of particular interest; and measures the in situ redox state in subglacial and glacimarine sediment and the release of nutrients from sediment through microbes utilizing chemical energy of these processes.

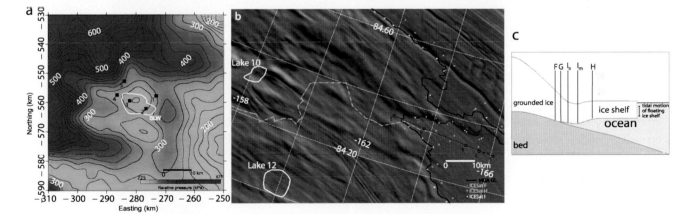

Plate 4. (a) Proposed LISSARD drill site locations on subglacial Lake Whillans (black squares); regional context is shown in Plate 2a. Background is estimated hydrostatic hydropotential. Outline of Subglacial Lake Whillans is from *Fricker and Scambos* [2009]. Black arrows show the inferred directions of water inflow and outflow to/from Subglacial Lake Whillans. (b) Proposed drill site locations at the RAGES grounding line site downstream of Subglacial Lake Whillans (red squares). Final site locations will be made once ground-based geophysical surveys are complete with the concept that sites 4 and 5 will be upstream of G and 1, 2, and 3 will be approximately equally spaced downstream of G. Inset is schematic diagram showing the key features of a typical ice shelf grounding zone, based on *Smith* [1991], *Vaughan* [1994], and *Fricker et al.* [2009]. Point F is the landward limit of ice flexure from tidal movement, point G the true GL where the grounded ice first loses contact with the bed, point I_b the break-in-slope, point I_m the local minimum in topography, and point H is the hydrostatic point where the ice first reaches approximate hydrostatic equilibrium. From *Brunt et al.* [2010], reprinted from the *Annals of Glaciology* with permission of the International Glaciological Society.

6.5.3. Sediment sampling. A variety of sediment samplers will be used for recovering different types of sediment mainly based on their stiffness. Soft sediments will be recovered using short, wide gravity core barrels for recovering the sediment-water interface. A 3 m long piston corer, proven by Caltech on Kamb Ice Stream to recover till, will be used again for WISSARD, as will a newly designed wide-barrel percussion corer with an active hammer system driven by a water pump will be used for recovery of up to 5 m long cores of stiff sediment.

6.5.4. Ice corer. WISSARD will use a modified 3 m long hot water ice corer that was used by Caltech on Kamb Ice Stream.

6.5.5. Ice strain and thermistor sensors. Borehole strings with inclinometers and thermistors will be frozen in to boreholes at WISSARD sites. Data are telemetered to the surface and will be used in combination with continuously recording GPS receivers to determine relevant terms in the three-dimensional stress tensor by using measured strain rates and ice flow law with the ice viscosity parameter estimated from vertical ice temperature determinations.

These units will be left out as long-term observatories for a year after deployment.

6.5.6. Geothermal probe and torvane. A 5 m long probe for penetrating and measuring heat flux in bottom sediment to be used in WISSARD is based on an IODP design with four thermistors at ±0.002 mK resolution. This probe will be used to determine the natural geothermal flux at each site. A torvane probe used in the past by the Caltech at Kamb Ice Stream will also be used to determine geotechnical sediment strength.

6.5.7. Subglacial Lake Exploration Device Camera. The Subglacial Lake Exploration Device (SLED) camera is a miniaturized, propelled, ice-borehole optical recording device designed to observe the borehole ice properties, water-ice interface, distribution of basal debris, suspended particles in water columns, and sediment surface properties. The development of SLED is supported by NASA. SLED will perform the following functions at both Subglacial Lake Whillans and the grounding line sites: (1) record the borehole ice properties, investigate the marine ice interface, examine distribution of entrained debris in basal ice, and observe the

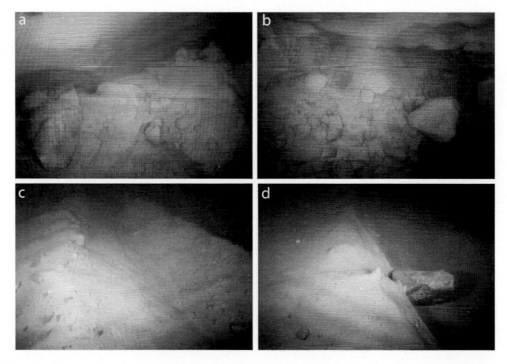

Plate 5. Previous images taken from boreholes: (a and b) grounding line till at Mackay Glacier taken by an ROV from under the floating-glacier tongue that comes away from the grounding line and perspectively comes out of the page. Largest clasts in the till are >10 cm across. (c and d) Flutes of till that were traced back to the grounding line and draped with glacimarine mud and dropped boulders. Flutes are meters high and meters in wavelength. From *Powell et al.* [1996], reprinted from the *Annals of Glaciology* with permission of the International Glaciological Society.

geometry of ice-water interface; (2) inspect the water column for suspended particles as well as possible aquatic organisms and search for visual evidence of water stratification and/or horizontal/vertical motion; and (3) investigate the sea/lake floor for evidence of erosional and sedimentary processes (glacial flutings, subaqueous sediment failures, debris flows, deltas, drainage channels, etc.) and record signs of possible bioturbation and/or evidence of benthic organisms.

6.6. WISSARD Drilling Schedule and Measurements

6.6.1. LISSARD. Drilling at Subglacial Lake Whillans will take place over an 8-week period during the 2012–2013 season. Boreholes will be of >20 cm diameter. There are five proposed drill site locations near Subglacial Lake Whillans (Plate 4a): (1) the first borehole downstream of the lake basin in the suspected outlet zone; (2) two to three subsequent boreholes on the lake itself, including the central parts of lake basins and suspected areas of sediment accumulation downstream of water inlet/s; (3) an additional borehole into the suspected inlet zones upstream of Subglacial Lake Whillans, and (4) the final control borehole of each season off to the side of each lake to explore basal and subglacial conditions in areas that are not under the direct influence of the active subglacial lake. One major reason for this sequence of drilling is that we want to move the drilling activity from downstream to upstream parts of each study region, to minimize the possibility that microbial/chemical contaminants from one site will affect subsequent sites.

The anticipated borehole measurements to be made during LISSARD are summarized in Table 2. Acquisition of samples for microbial analyses will be of the highest priority as soon as the bottom is accessed due to concerns with potential sample contamination by other operations. We will obtain separate cores for microbiology, sedimentology, micropaleontology, sediment geochemistry, paleomagnetism, and geotechnical analyses. The Jet Propulsion Laboratory (Pasadena, California)-developed SLED camera will also be deployed down each of the LISSARD boreholes to provide supporting imagery.

6.6.2. RAGES. Drilling at the grounding zone downstream from Subglacial Lake Whillans will take place over an 8-week period during the 2013–2014 season. There are five proposed sites across the grounding zone near the predicted Subglacial Lake Whillans outlet (Plate 4b): sites 1 and 2, offshore (i.e., on the ice shelf); site 3, within range of the grounding line (point G in Plate 4b inset); and sites 4 and 5, landward of G (grounded ice).

On the seaward side of the grounding line where the ice is floating, the intention is to drill through the ice shelf and sample the sub-ice-shelf cavity. Here boreholes will be at least 80 cm wide to allow deployment of SIR, the specialized robotic submarine, into the cavity (Plate 4) and GIPSIE, the oceanographic and geochemical sensor package, and the ice-tethered profiler component of oceanographic mooring systems. The SLED camera will be deployed down the narrow boreholes of RAGES borehole to provide supporting imagery (Plate 5).

Once the boreholes have been drilled to the ice base, there will be a measurement period during which time several measurements will be made by placing clean instruments down the borehole, followed by a period when samples will be collected from the borehole (Table 2). At the end of science operations at each site, longer-term monitoring instruments will be deployed.

7. EDUCATION AND OUTREACH

Climate change, melting of ice sheets, and related sea level change are contemporary issues of social relevance. WISSARD's large-scale interdisciplinary approach to studying the Whillans Ice Stream provides unparalled opportunity to highlight the interdisciplinary nature of scientific discovery and the use of novel technology in Antarctic science. A coordinated WISSARD education and public outreach effort will connect cutting-edge scientific research to public audiences and create the next generation of scientists through education and outreach initiatives that support and enhance training opportunities for undergraduate and graduate students and K-12 teachers and learners.

A balanced education and public outreach portfolio for WISSARD include three focus areas: (1) training of graduate and undergraduate students; (2) K-12 teaching and learning through standards-based curriculum and teacher professional development activities; and (3) local to national community outreach that engage the public through print and electronic media (Web, video, and social media); displays and exhibits (museums and science centers); and interaction with scientists (public lectures and activities). A fourth focus area includes program evaluation. Evaluation metrics have been designed to assess progress, adjust programming to reach target audiences effectively, and to maximize resources most efficiently. These focus areas are being addressed by (1) the development of K-12 teacher professional development opportunities and inquiry-based activities at participating universities; (2) involvement of undergraduate and graduate students in activity development and delivery; (3) our K-12 EPO programming is stratified to serve teachers locally and nationally, as well as underserved learning communities; and (4) the initiation of Web-based professional development for teachers in partnership the National Science Teachers Association, which will expand the national reach of the project.

Print and electronic media efforts will be catalyzed through a variety of deliverables including a dynamic Webpage (http://wissard.org), print and digital media opportunities with national outlets such as National Geographic and Discovery, and incorporation into the Red Planet; Blue Planet exhibit at the Museum of Science and Industry in Chicago, Illinois.

8. SUMMARY

Following more than a decade of promoting and planning subglacial research, we believe these systems have gone from a curiosity to a focus for scientific research [Priscu et al., 2005]. Largely as the result of these efforts, we enter a new decade with three funded projects for drilling into subglacial environments (Vostok Subglacial Lake (Russia), Ellsworth Subglacial Lake (United Kingdom), and WISSARD (United States)). These three projects will be the first to explore these novel subglacial environments in a comprehensive and environmentally friendly manner.

The WISSARD project is unique among the three projects in that it targets a lake and the grounding zone downstream using a multidisciplinary scientific approach designed to understand the subglacial-ice-ocean system as an integrated whole. In so doing, WISSARD will advance knowledge in a variety of scientific questions at the cutting edge of glaciology, geology, and microbiology. Results from these three projects will not only test new theories about these environments, but also provide important new data about the role of subglacial environments on a global scale. While a major commitment of resources is necessary to implement such ambitious programs of exploration and research, the potential payoff in scientific and educational returns is immense.

REFERENCES

Alley, R. B., S. Anandakrishnan, T. K. Dupont, B. R. Parizek, and D. Pollard (2007), Effect of sedimentation on ice-sheet grounding-line stability, Science, 315(5820), 1838–1841, doi:10.1126/science.1138396.

Anandakrishnan, S., G. A. Catania, R. B. Alley, and H. J. Horgan (2007), Discovery of till deposition at the grounding line of Whillans Ice Stream, Science, 315(5820), 1835–1838, doi:10.1126/science.1138393.

Bamber, J. L., R. E. M. Riva, B. L. A. Vermeersen, and A. M. LeBrocq (2009), Reassessment of the potential sea level rise from a collapse of the West Antarctic Ice Sheet, Science, 324(5929), 901–903, doi:10.1126/science.1169335.

Bell, R. E. (2008), The role of subglacial water in ice-sheet mass balance, Nat. Geosci., 1, 297–304.

Bindschadler, R. A., M. A. King, R. B. Alley, S. Anandakrishnan, and L. Padman (2003), Tidally controlled stick-slip discharge of a West Antarctic Ice, Science, 301(5636), 1087–1089, doi:10.1126/science.1087231.

Bowman, J. P., and R. D. McCuaig (2003), Biodiversity, community structural shifts, and biogeography of prokaryotes within Antarctic continental shelf sediment, Appl. Environ. Microbiol., 69, 2463–2483.

Brunt, K. M., H. A. Fricker, L. Padman, T. A. Scambos, and S. O'Neel (2010), Mapping the grounding zone of the Ross Ice Shelf, Antarctica, using ICESat laser altimetry, Ann. Glaciol., 51(55), 71–79.

Christner, B. C., G. Royston-Bishop, C. M. Foreman, B. R. Arnold, M. Tranter, K. A. Welch, W. B. Lyons, A. I. Tsapin, and J. C. Priscu (2006), Limnological conditions in Subglacial Lake Vostok, Antarctica, Limnol. Oceanogr., 51, 2485–2501.

Clough, J. W., and B. L. Hansen (1979), The Ross Ice Shelf Project, Science, 203(4379), 433–434.

Craven, M., I. Allison, H. A. Fricker, and R. C. Warner (2009), Properties of a marine ice layer under the Amery Ice Shelf, J. Glaciol., 55(192), 717–728.

Dawber, M., and R. D. Powell (1998), Epifaunal distributions at marine-ending glaciers: Influences of ice dynamics and sedimentation, in The Antarctic Region: Geological Evolution and Processes (Proceedings VII International Symposium on Antarctic Earth Sciences, Siena, Italy, 1995), edited by C. A. Ricci, pp. 875–884, Terra Antarctica, Siena, Italy.

Dowdeswell, J. A., D. Ottesen, J. Evans, C. Ó Cofaigh, and J. B. Anderson (2008), Submarine glacial landforms and rates of ice-stream collapse, Geology, 36(10), 819–822, doi:10.1130/G24808A.1.

Dunbar, G. B., T. Naish, P. J. Barrett, C. F. Fielding, and R. D. Powell (2008), Constraining the amplitude of Late Oligocene bathymetric changes in western Ross Sea during orbitally-induced oscillations in the East Antarctic Ice Sheet: (1) Implications for glacimarine sequence stratigraphic models, Palaeogeogr. Palaeoclimatol. Palaeoecol., 260(1–2), 50–65.

Engelhardt, H. (2005), Thermal regime and dynamics of the West Antarctic Ice Sheet, Ann. Glaciol., 39, 85–92.

Engelhardt, H., N. Humphrey, B. Kamb, and M. Fahnestock (1990), Physical conditions at the base of a fast moving Antarctic Ice Stream, Science, 248(4951), 57–59, doi:10.1126/science.248.4951.57.

Fricker, H. A., and T. Scambos (2009), Connected subglacial lake activity on lower Mercer and Whillans ice streams, West Antarctica, 2003–2008, J. Glaciol., 55(190), 303–315.

Fricker, H. A., T. A. Scambos, R. A. Bindschadler, and L. Padman (2007), An active Subglacial Water System in West Antarctica mapped from space, Science, 315(5818), 1544–1548, doi:10.1126/science.1136897.

Fricker, H. A., R. Coleman, L. Padman, T. A. Scambos, J. Bohlander, and K. M. Brunt (2009), Mapping the grounding zone of the Amery Ice Shelf, East Antarctica using InSAR, MODIS and ICESat, Antarct. Sci., 21(5), 515–532, doi:10.1017/S095410 200999023X.

Gaidos, E. J., B. D. Lanoil, T. Thorsteinsson, A. Graham, M. L. Skidmore, S. Han, T. Rust, and B. Popp (2004), A viable microbial community in a subglacial volcanic crater lake, Iceland, *Astrobiology*, *4*, 327–344.

Gray, L., I. Joughin, S. Tulaczyk, V. Spikes, R. Bindschadler, and K. Jezek (2005), Evidence for subglacial water transport in the West Antarctic Ice Sheet through three-dimensional satellite radar interferometry, *Geophys. Res. Lett.*, *32*, L03501, doi:10.1029/2004GL021387.

Hodson, A., A. M. Anesio, M. Tranter, A. Fountain, M. Osborn, J. C. Priscu, J. Laybourn-Parry, and B. Sattler (2008), Glacial ecosystems, *Ecol. Monogr.*, *78*(1), 41–67.

Holland, D. M., and A. Jenkins (2001), Adaptation of an isopycnic coordinate ocean model for the study of circulation beneath ice shelves, *Mon. Weather Rev.*, *129*, 1905–1927.

Holland, D. M., S. S. Jacobs, and A. Jenkins (2003), Modeling Ross Sea ice shelf-ocean interaction, *Antarct. Sci.*, *15*, 13–23.

Holland, P. R., A. Jenkins, and D. M. Holland (2008), The response of ice shelf basal melting to variations in ocean temperature, *J. Clim.*, *21*, 2558–2572, doi:10.1175/2007JCLI1909.1.

Hulbe, C. L., and M. A. Fahnestock (2007), Century-scale discharge stagnation and reactivation of the Ross ice streams, West Antarctica, *J. Geophys. Res.*, *112*, F03S27, doi:10.1029/2006JF000603.

Jacobel, R. W., B. C. Welch, D. Osterhouse, R. Pettersson, and J. A. MacGregor (2009), Spatial variation of radar-derived basal conditions on Kamb ice stream, *Ann. Glaciol.*, *50*(51), 10–16.

Jacobs, S. S. (2006), Observations of change in the Southern Ocean, *Philos. Trans. R. Soc. A*, *364*, 1657–1681.

Jenkins, A., and D. M. Holland (2002), A model study of ocean circulation beneath Filchner Ronne Ice Shelf, Antarctica: Implications for bottom water formation, *Geophys. Res. Lett.*, *29*(8), 1193, doi:10.1029/2001GL014589.

Joughin, I., S. Tulaczyk, R. Bindschadler, and S. F. Price (2002), Changes in west Antarctic ice stream velocities: Observation and analysis, *J. Geophys. Res.*, *107*(B11), 2289, doi:10.1029/2001JB001029.

Joughin, I., et al. (2005), Continued deceleration of Whillans Ice Stream, West Antarctica, *Geophys. Res. Lett.*, *32*, L22501, doi:10.1029/2005GL024319.

Joughin, I., S. B. Das, M. A. King, B. E. Smith, I. M. Howat, and T. Moon (2008), Seasonal speedup along the western flank of the Greenland Ice Sheet, *Science*, *320*(5877), 781–783, doi:10.1126/science.1153288.

Kamb, B. (2001), Basal zone of the West Antarctic Ice Streams and its role in lubrication of their rapid motion, in *The West Antarctic Ice Sheet: Behavior and Environment, Antarct. Res. Ser.*, vol. 77, edited by R. B. Alley and R. A. Bindschadler, pp. 157–199, AGU, Washington, D. C.

Lanoil, B., M. Skidmore, J. C. Priscu, S. Han, W. Foo, S. W. Vogel, S. Tulaczyk, and H. Engelhardt (2009), Bacteria beneath the West Antarctic Ice Sheet, *Environ. Microbiol.*, *11*(3), 609–615.

LeBrocq, A. M., A. J. Payne, M. J. Siegert, and R. B. Alley (2009), A subglacial water-flow model for West Antarctica, *J. Glaciol.*, *55*(193), 879–888, doi:10.3189/002214309790152564.

Lemke, P., et al. (2007), Observations: Changes in snow, ice and frozen ground, in *Climate Change 2007: The Physical Science Basis. Contribution of Working Group I to the Fourth Assessment Report of the Intergovernmental Panel on Climate Change*, edited by S. Solomon et al., pp. 337–383, Cambridge Univ. Press, New York.

Makinson, K. (2002), Modeling tidal current profiles and vertical mixing beneath Filchner–Ronne Ice Shelf, Antarctica, *J. Phys. Oceanogr.*, *32*, 202–215.

McKay, R., et al. (2009), The stratigraphic signature of the Late Cenozoic Antarctic ice sheets in the Ross Embayment, *Geol. Soc. Am. Bull.*, *121*(11/12), 1537–1561, doi:10.1130/B26540.1.

Mikucki, J. A., and J. C. Priscu (2007), Bacterial diversity associated with Blood Falls: A subglacial outflow from the Taylor Glacier, Antarctica, *Appl. Environ. Microbiol.*, *73*(12), 4029–4039.

Mikucki, J. A., C. M. Foreman, B. Sattler, W. B. Lyons, and J. C. Priscu (2004), Geomicrobiology of Blood Falls: An iron-rich saline discharge at the terminus of the Taylor Glacier, Antarctica, *Aquat. Geochem.*, *10*, 199–220.

Mikucki, J. A., A. Pearson, D. T. Johnston, A. V. Turchyn, J. Farquhar, D. P. Schrag, A. D. Anbar, J. C. Priscu, and P. A. Lee (2009), A contemporary, microbially-maintained, ferrous subglacial "ocean", *Science*, *324*(5925), 397–400.

Mitchell, A. C., and G. H. Brown (2008), Modelling geochemical and biogeochemical reactions in subglacial environments, *Arct. Antarct. Alp. Res.*, *40*, 531–547.

Mitchell, A. C., G. H. Brown, and R. Fuge (2006), Minor and trace elements as indicators of chemical weathering and flow routing in subglacial environments, *Hydrol. Processes*, *20*, 877–897.

Naish, T., et al. (2009), Obliquity-paced Pliocene West Antarctic Ice Sheet oscillations, *Nature*, *458*, 322–328, doi:10.1038/nature07867.

National Research Council (2003), *Frontiers in Polar Biology in the Genomic Era*, 166 pp., Natl. Acad. Press, Washington, D. C.

National Research Council (NRC) (2007), *Exploration of Antarctic Subglacial Aquatic Environments: Environmental and Scientific Stewardship*, 166 pp., Natl. Acad. Press, Washington, D. C.

Neal, A. L., K. M. Rosso, G. G. Geesey, Y. A. Gorby, and B. J. Little (2003), Surface structure effects on direct reduction of iron oxides by *Shewanella oneidensis*, *Geochim. Cosmochim. Acta*, *67*(23), 4489–4503.

Nemergut, D. R., T. Barkay, and J. Coombs (2007), Mobile gene elements in environmental microbial communities, in *Manual of Environmental Microbiology*, 3rd ed., edited by C. J. Hurst et al., pp. 758–768, ASM Press, Washington, D. C.

Oppenheimer, M. (1998), Global warming and the stability of the West Antarctic Ice Sheet, *Nature*, *393*(6683), 325–332.

Peters, L. E., S. Anandakrishnan, R. B. Alley, and A. M. Smith (2007), Extensive storage of basal meltwater in the onset region of a major West Antarctic ice stream, *Geology*, *35*(3), 251–254, doi:10.1130/G23222A.1.

Pollard, D., and R. M. DeConto (2009), Modelling West Antarctic Ice Sheet growth and collapse through the past 5 million years, *Nature*, *458*, 329–332, doi:10.1038/nature07809.

Powell, R. D., and R. B. Alley (1997), Grounding line systems: Processes, glaciological inferences and the stratigraphic record, in *Geology and Seismic Stratigraphy of the Antarctic Margin, Part 2, Antarct. Res. Ser.*, vol. 71, edited by A. K. Cooper and P. F. Barker, pp. 169–187, AGU, Washington, D. C.

Powell, R. D., and J. M. Cooper (2002), A sequence stratigraphic model for temperate, glaciated continental shelves, in *Glacier-Influenced Sedimentation on High Latitude Continental Margins: Ancient and Modern*, edited by J. A. Dowdeswell and C. Ó Cofaigh, *Geol. Soc. Spec. Publ.*, *203*, 215–244.

Powell, R. D., M. Dawber, N. McInnes, and A. R. Payne (1996), Observations of the grounding-line area at a floating glacier terminus, *Ann. Glaciol.*, *22*, 217–223.

Priscu, J. C., and B. Christner (2004), Earth's icy biosphere, in *Microbial Diversity and Prospecting*, edited by A. T. Bull, pp. 130–145, ASM Press, Washington, D. C.

Priscu, J. C., et al. (1999), Geomicrobiology of sub-glacial ice above Vostok Station, *Science*, *286*, 2141–2144.

Priscu, J. C., M. C. Kennicutt, II, R. E. Bell, S. A. Bulat, J. C. Ellis-Evans, V. V. Lukin, J.-R. Petit, R. D. Powell, M. J. Siegert, and I. Tabacco (2005), Exploring subglacial Antarctic lake environments, *Eos Trans. AGU*, *86*(20), 193–197.

Priscu, J. C., S. Tulaczyk, M. Studinger, M. C. Kennicutt, II, B. C. Christner, and C. M. Foreman (2008), Antarctic subglacial water: Origin, evolution and ecology, in *Polar Lakes and Rivers*, edited by W. Vincent and J. Laybourn-Parry, pp. 119–135, Oxford Univ. Press, Oxford, U. K.

Rignot, E., and S. S. Jacobs (2002), Rapid bottom melting widespread near Antarctic ice shelf grounding lines, *Science*, *296*(5575), 2020–2023.

Scherer, R. P., C. M. Sjunneskog, N. Iverson, and T. Hooyer (2004), Assessing subglacial processes from diatom fragmentation patterns, *Geology*, *32*(7), 557–560.

Scherer, R. P., S. Bohaty, R. Dunbar, O. Esper, J.-A. Flores, R. Gersonde, D. Harwood, A. Roberts, and M. Taviani (2008), Antarctic records of precession-paced insolation-driven warming during early Pleistocene Marine Isotope Stage 31, *Geophys. Res. Lett.*, *35*, L03505, doi:10.1029/2007 GL032254.

Sharp, M., J. Parkes, B. Cragg, I. J. Fairchild, H. Lamb, and M. Tranter (1999), Widespread bacterial populations at glacier beds and their relationship to rock weathering and carbon cycling, *Geology*, *27*, 107–110.

Shepherd, A., D. Wingham, and E. Rignot (2004), Warm ocean is eroding West Antarctic Ice Sheet, *Geophys. Res. Lett.*, *31*, L23402, doi:10.1029/2004GL021106.

Siegert, M. J., R. Hindmarsh, H. Corr, A. Smith, J. Woodward, E. C. King, A. J. Payne, and I. Joughin (2004), Subglacial Lake Ellsworth: A candidate for in situ exploration in West Antarctica, *Geophys. Res. Lett.*, *31*, L23403, doi:10.1029/2004 GL021477.

Siegert, M. J., S. P. Carter, I. E. Tabacco, S. Popov, and D. D. Blankenship (2005), A revised inventory of Antarctic subglacial lakes, *Antarct. Sci.*, *17*, 453–460.

Sjunneskog, C. S., and R. P. Scherer (2005), Mixed diatom assemblages in Ross Sea (Antarctica) glacigenic facies, *Palaeogeogr. Palaeoclimatol. Palaeoecol.*, *218*(3–4), 287–300.

Smith, A. M. (1991), The use of tiltmeters to study the dynamics of Antarctic ice-shelf grounding lines, *J. Glaciol.*, *37*(125), 51–58.

Smith, B., H. A. Fricker, I. Joughin, and S. Tulaczyk (2009), An inventory of active subglacial lakes in Antarctica detected by ICESat (2003–2008), *J. Glaciol.*, *55*(102), 573–595.

Stearns, L. A., B. E. Smith, and G. S. Hamilton (2008), Subglacial floods cause increased flow speeds on a major East Antarctic outlet glacier, *Nat. Geosci.*, *1*, 827–831, doi:10.1038/ngeo356.

Steffen, K., R. Thomas, S. Marshall, G. Cogley, D. Holland, E. Rignot, and P. Clark (2009), Rapid changes in glaciers and ice sheets and their impacts on sea level, *Abrupt Climate Change, USGS/ CCSP SAP 3.4*, 52 pp., U.S. Geol. Surv., Reston, Va.

Steig, E. J., D. P. Schneider, S. D. Rutherford, M. E. Mann, J. C. Comiso, and D. T. Shindell (2009), Warming of the Antarctic ice-sheet surface since the 1957 International Geophysical Year, *Nature*, *457*, 459–462, doi:10.1038/nature07669.

Studinger, M., et al. (2003), Ice cover, landscape setting, and geological framework of Lake Vostok, East Antarctica, *Earth Planet. Sci. Lett.*, *205*(3–4), 195–210.

Tranter, M., M. Skidmore, and J. Wadham (2005), Hydrological controls on microbial communities in subglacial environments, *Hydrol. Processes*, *19*, 995–998.

Tulaczyk, S., R. Pettersson, N. Quintana-Krupinski, H. Fricker, I. Joughin, and B. Smith (2008), Do dynamic subglacial lakes impact temporal behavior of fast-flowing ice streams? GPS and radar investigations on two West Antarctic Ice Streams, *Geophys. Res. Abstr.*, *10*, EGU2008-A-11565.

Vail, P. R., R. M. Mitchum, R. G. Todd, J. M. Widmier, S. Thompson, J. B. Sangree, J. N. Bubb, and W. G. Hatlelid (1977), Seismic stratigraphy and global changes of sea level, in *Seismic Stratigraphy—Applications to Hydrocarbon Exploration, Part 3*, edited by C. E. Payton, *Mem. Am. Assoc. Pet. Geol.*, *26*, 63–81.

Vail, P. R., F. Audemard, S. A. Bowman, P. N. Eisner, and C. Perez-Cruz (1991), The stratigraphic signatures of tectonics, eustasy and sedimentology—An overview, in *Cycles and Events in Stratigraphy*, edited by G. Einsele, W. Ricken, and A. Seilacher, pp. 615–659, Springer, Berlin, Germany.

van de Wal, R. S. W., W. Boot, M. R. van den Broeke, C. J. P. P. Smeets, C. H. Reijmer, J. J. A. Donker, and J. Oerlemans (2008), Large and rapid melt-induced velocity changes in the ablation zone of the Greenland Ice Sheet, *Science*, *321*, 111–113.

Vaughan, D. G. (1994), Investigating tidal flexure on an ice shelf using kinematic GPS, *Ann. Glaciol.*, *20*, 372–376.

Vaughan, D., and R. Arthern (2007), Why is it so hard to predict the future of the ice sheets?, *Science*, *315*, 1503–1504, doi:10.1126/science.1141111.

Wadham, J. L., M. Tranter, M. Skidmore, A. J. Hodson, J. Priscu, W. B. Lyons, M. Sharp, P. Wynn, and M. Jackson (2010), Biogeochemical weathering under ice: Size matters, *Global Biogeochem. Cycles*, *24*, GB3025, doi:10.1029/2009GB003688.

Walker, R., and D. M. Holland (2007), A two-dimensional coupled model for ice shelf-ocean interaction, *Ocean Modell.*, *17*, 123–139.

Walker, R. T., T. K. Dupont, B. R. Parizek, and R. B. Alley (2008), Effects of basal-melting distribution on the retreat of ice-shelf grounding lines, *Geophys. Res. Lett.*, *35*, L17503, doi:10.1029/2008GL034947.

Waller, R. I. (2001), The influence of basal processes on the dynamic behaviour of cold-based glaciers, *Quat. Int.*, *86*, 117–128.

Wingham, D., M. Siegert, A. Shepherd, and A. Muir (2006), Rapid discharge connects Antarctic subglacial lakes, *Nature*, *440*, 1033–1036.

Woodward, J., A. M. Smith, N. Ross, M. Thoma, H. F. J. Corr, E. C. King, M. A. King, K. Grosfeld, M. Tranter, and M. J. Siegert (2010), Location for direct access to Subglacial Lake Ellsworth: An assessment of geophysical data and modeling, *Geophys. Res. Lett.*, *37*, L11501, doi:10.1029/2010GL042884.

S. Anandakrishnan and H. Horgan, Department of Geosciences, Pennsylvania State University, University Park, PA 16802, USA.

B. Christner, Department of Biological Sciences, Louisiana State University, Baton Rouge, LA 70803, USA.

A. T. Fisher and S. Tulaczyk, Department of Earth and Planetary Sciences, University of California, Santa Cruz, Santa Cruz, CA 95064, USA.

H. A. Fricker and J. Severinghaus, Scripps Institution of Oceanography, 9500 Gilman Drive, La Jolla, CA 92093-0244, USA. (hafricker@ucsd.edu)

D. Holland, Courant Institute of Mathematical Sciences, New York University, New York, NY 10012, USA.

R. Jacobel, Department of Physics, St. Olaf College, Northfield, MN 55057, USA.

J. Mikucki, Earth Science Department and Environmental Studies Program, Dartmouth College, 6105 Fairchild, Hanover, NH 03755 USA.

A. Mitchell, Center for Biofilm Engineering, Montana State University, Bozeman, MT 59717, USA.

R. Powell and R. Scherer, Department of Geological and Environmental Sciences, Northern Illinois University, DeKalb, IL 61125, USA.

J. Priscu, Department of Land Resources and Environmental Sciences, Montana State University, Bozeman, MT 59717, USA.

Ellsworth Subglacial Lake, West Antarctica:
A Review of Its History and Recent Field Campaigns

N. Ross,[1] M. J. Siegert,[1] A. Rivera,[2] M. J. Bentley,[3] D. Blake,[4] L. Capper,[4] R. Clarke,[4] C. S. Cockell,[5] H. F. J. Corr,[4] W. Harris,[6] C. Hill,[4] R. C. A. Hindmarsh,[4] D. A. Hodgson,[4] E. C. King,[4] H. Lamb,[7] B. Maher,[8] K. Makinson,[4] M. Mowlem,[9] J. Parnell,[10] D. A. Pearce,[4] J. Priscu,[11] A. M. Smith,[4] A. Tait,[4] M. Tranter,[6] J. L. Wadham,[6] W. B. Whalley,[12] and J. Woodward[13]

Ellsworth Subglacial Lake, first observed in airborne radio echo sounding data acquired in 1978, is located within a long, deep subglacial trough within the Ellsworth Subglacial Highlands of West Antarctica. Geophysical surveys have characterized the lake, its subglacial catchment, and the thickness, structure, and flow of the overlying ice sheet. Covering 28.9 km^2, Ellsworth Subglacial Lake is located below 2.9 to 3.3 km of ice at depths of -1361 to -1030 m. Seismic reflection data have shown the lake to be up to 156 m deep and underlain by unconsolidated sediments. Ice sheet flow over the lake is characterized by low velocities (<6 m yr^{-1}), flow convergence, and longitudinal extension. The lake appears to be in steady state, although the hydrological balance may vary over glacial-interglacial cycles. Direct access, measurement, and sampling of Ellsworth Subglacial Lake are planned for the 2012/2013 Antarctic field season. The aims of this access experiment are to determine (1) the presence, character, and maintenance of microbial life in Antarctic subglacial lakes and (2) the Quaternary history of the West Antarctic ice sheet. Geophysical data have been used to define a preferred lake access site. The factors that make this location suitable for exploration are (1) a relatively thin overlying ice column (~3.1 km), (2) a significant measured water depth (~143 m), (3) >2 m of sediment below the lake floor,

[1]School of GeoSciences, University of Edinburgh, Edinburgh, UK.

[2]Centro de Estudios Científicos, Valdivia, Chile.

[3]Department of Geography, Durham University, Durham, UK.

[4]British Antarctic Survey, Natural Environment Research Council, Cambridge, UK.

[5]CEPSAR, Open University, Milton Keynes, UK.

[6]School of Geographical Sciences, University of Bristol, Bristol, UK.

[7]Institute of Geography and Earth Sciences, Aberystwyth University, Aberystwyth, UK.

[8]Lancaster Environment Centre, Lancaster University, Lancaster, UK.

[9]National Oceanography Centre, University of Southampton, Southampton, UK.

[10]Department of Geology and Petroleum Geology, University of Aberdeen, Aberdeen, UK.

[11]Department of Land Resources and Environmental Sciences, Montana State University, Bozeman, Montana, USA.

[12]School of Geography, Archaeology and Palaeoecology, Queen's University Belfast, Belfast, UK.

[13]School of Built and Natural Environment, Northumbria University, Newcastle, UK.

Antarctic Subglacial Aquatic Environments
Geophysical Monograph Series 192
Copyright 2011 by the American Geophysical Union.
10.1029/2010GM000936

(4) water circulation modeling suggesting a melting ice-water interface, and (5) coring that can target the deepest point of the lake floor away from marginal, localized sediment sources.

1. INTRODUCTION

Deep continental Antarctic subglacial lakes are one of the few remaining unexplored environments on Earth. Found beneath both the East and West Antarctic ice sheets, it has been hypothesized that these lakes are extreme, yet viable habitats, which may host unique microbial assemblages that have been potentially isolated for millions of years [*Siegert et al.*, 2001]. Furthermore, the sediments that have accumulated at the bottom of subglacial lakes may contain important records of Antarctic ice sheet history.

Ellsworth Subglacial Lake (hereafter referred to as "Lake Ellsworth") is a small subglacial lake (14.7 km long by 3.05 km wide) in West Antarctica. The lake, which lies beneath 2.9 to 3.3 km of ice, is located within the catchment of Pine Island Glacier in a deep subglacial trench ~30 km northwest of the central ice divide of the West Antarctic Ice Sheet (WAIS), and some ~70 km west of the Ellsworth Mountains (Figure 1). Lake Ellsworth has been identified as a prime candidate for direct measurement and sampling [*Siegert et al.*, 2004; *Vaughan et al.*, 2007; *Lake Ellsworth Consortium*, 2007]. Recent geophysical surveys have demonstrated that the lake is up to 156 ± 1.5 m deep [*Woodward et al.*, 2010] and that the lake floor is draped with over 2 m of unconsolidated sediments [*Smith et al.*, 2008]. A new program, the "Lake Ellsworth Consortium," plans to sample Lake Ellsworth in the 2012/2013 Antarctic field season. This U.K.-led program involves scientists from 11 U.K. universities and research institutions, along with collaborators from Chile and the United States. The Lake Ellsworth experiment aims to determine (1) the presence, origin, evolution, and maintenance of life in an Antarctic subglacial lake through direct measurement, sampling, and analysis of this extreme environment and (2) the palaeoenvironment and glacial history of the WAIS including, potentially, the date of its last decay, by recovering a sedimentary record from the lake floor. The case for the exploration of Lake Ellsworth has been outlined previously [*Siegert et al.*, 2004; *Lake Ellsworth Consortium*, 2007]. This article reviews the history of Lake Ellsworth research and summarizes recent geophysical reconnaissance studies that underpin the lake access experiment.

2. LAKE ELLSWORTH: DISCOVERY AND REDISCOVERY

Lake Ellsworth was first observed in a single 60 MHz airborne radio echo sounding (RES) survey line acquired during the U.K. Scott Polar Research Institute, U.S. National Science Foundation, and the Technical University of Denmark (SPRI-NSF-TUD) 1977/1978 airborne campaign [*Jankowski and Drewry*, 1981; *Siegert et al.*, 2004]. Although the exact positions of the survey platform were uncertain owing to the navigational techniques available at the time, it is clear, on the basis of comparison with recent data sets, that these RES data were acquired directly over the center of Lake Ellsworth, in an orientation lying along the lake's long axis, subparallel to the direction of ice flow. However, it was not the first time an important scientific campaign had traveled over the lake; the route of the 1957/1958 Sentinel Range (Marie Byrd Land) traverse [*Bentley and Ostenso*, 1961] also crossed the lake in an orientation roughly perpendicular to the long axis and the later RES transect. Owing to the spacing of the seismic and gravity measurements and the small target body, however, no observations of the lake were made during the traverse. The position of Lake Ellsworth was first mapped by *McIntyre* [1983]. Despite this, the lake remained unrecognized in the broader literature and was not included in the first official inventory of subglacial lakes [*Siegert et al.*, 1996]. In 2003, the lake was "rediscovered" in McIntyre's thesis, and the original RES data were reexamined, revealing a typical lake-like reflector ~10 km long, in hydrostatic equilibrium with the overlying ice sheet. Lake Ellsworth was then identified as a suitable target for exploration [*Siegert et al.*, 2004] and was named and included in the updated subglacial lake inventory [*Siegert et al.*, 2005].

3. RECENT GEOPHYSICAL SURVEYS

As part of an extensive airborne geophysical survey of the Pine Island Glacier catchment in 2004/2005, the British Antarctic Survey (BAS) acquired ice thickness and surface elevation data over Lake Ellsworth. This was followed in January 2006 by a ground-based traverse from Patriot Hills to Lake Ellsworth, via the Institute Ice Stream, undertaken by researchers from the Centro de Estudios Científicos (CECS). Radio echo sounding and GPS measurements were made along the traverse route and over and around the lake. Both the CECS and BAS surveys acquired ice thickness data using radars with a frequency of 150 MHz. The combined results of these surveys were reported by *Vaughan et al.* [2007].

These surveys indicated that Lake Ellsworth was located at the bottom of a long, deep subglacial trough, but that the lake was much narrower than previously estimated, with a maximum width of only ~2.7 km. This led to a reduction in

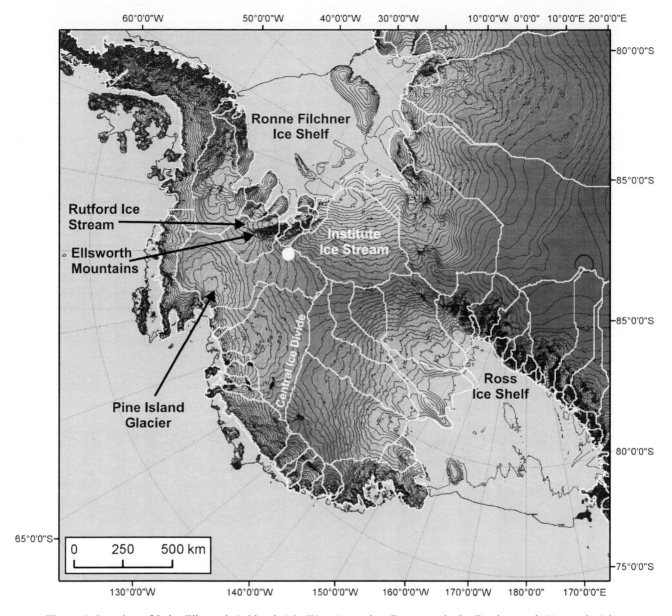

Figure 1. Location of Lake Ellsworth (white dot) in West Antarctica. Base map is the Bamber et al. (Antarctic 1 km Digital Elevation Model [DEM] from Combined ERS-1 Radar and ICESat Laser Satellite Altimetry, data set, National Snow and Ice Data Center, Boulder, Colorado, 2009, available at http://nsidc.org/data/nsidc-0422.html) satellite altimetry–derived DEM with contours at 100 m intervals. White lines mark major ice divides.

the estimated total area of the lake to ~18 km^2. This was rather less than the 100 km^2 suggested by *Siegert et al.* [2004] because the lake is constrained topographically, making it elongated rather than circular. Using the RMS variability of the hydrological head, derived from the uncertainty in ice thickness measurements, and assuming an ice density of 920 kg m^{-3}, *Vaughan et al.* [2007] isolated the possible density of fluid in the lake. They concluded that the fluid

body of Lake Ellsworth has a density between 950 and 1013 kg m^{-3}, indicative of fresh water, with little likelihood of the lake containing substantial concentrations of materials that would result in a fluid body of greater density (e.g., acids, salts or heavy clathrates).

Data from the BAS airborne survey of Pine Island Glacier catchment revealed that Lake Ellsworth is one of a series of subglacial lakes located in deep, SE-NW trending, subglacial valleys within the Ellsworth Subglacial Highlands [*Vaughan et al.*, 2007]. These valleys are probably erosional features developed during periods of less extensive glaciation at some stage prior to the formation of the current ice sheet [*Vaughan et al.*, 2007]. Analysis of the regional hydrological regime indicated that these subglacial troughs act as conduits for subglacial water flow from the ice divide (between Pine Island Glacier and the Institute Ice Stream) to the Byrd Subglacial Basin. *Vaughan et al.* [2007] suggested that Lake Ellsworth was part of a well-connected drainage system, with well-defined upstream and downstream hydrological pathways through its deep subglacial catchment, making it effectively an "open system" in terms of its hydrology. From the RES data, they inferred there to be no hydrological barrier that could block water flow out of the bottom end of the lake and that the outflow probably drains into a second lake only 15 km downstream. Based on this, *Vaughan et al.* [2007] suggested that the likelihood of Lake Ellsworth hosting long-isolated microbiological life was reduced, but still recommended that Lake Ellsworth remained a "strong candidate for *in situ* exploration."

4. GEOPHYSICAL RESULTS

A full geophysical characterization of Lake Ellsworth was undertaken during the Austral summers of 2007/2008 and 2008/2009. The aims of this fieldwork were to (1) determine lake water depth and bathymetry, (2) map the outline of the lake and the topography of its catchment, (3) produce a detailed map of ice flow over the lake, (4) characterize the nature of the ice-water and water-bed interfaces, (5) establish sediment thickness beneath the lake floor, (6) map the geometry of englacial layering within the overlying ice sheet, and (7) measure any detectable tidal signatures.

4.1. Radio Echo Sounding Surveys

A detailed grid of RES survey lines was acquired over Lake Ellsworth and its surrounding area using the ground-based ~1.7 MHz pulsed DEep-Look-Radar-Echo-Sounder (DELORES) radar system (Figures 2 and 3a). Just over 1000 km of RES data (including BAS and CECS data) have now been acquired over, and in the vicinity of, Lake Ellsworth. The bed was identified in more than 95% of the survey line data, and an ice thickness map and digital elevation model (DEM) of the bed were constructed (Figures 3c and 3d). Roving GPS data acquired along the RES profile lines during the surveys were used to generate a DEM of the ice surface (Figure 3b).

4.1.1. Ice sheet surface. The center of Lake Ellsworth is located approximately 30 km from the ice divide between Pine Island Glacier and the Institute Ice Stream; a major

divide of the WAIS (Figures 1 and 3b). The ice sheet surface around Lake Ellsworth is strongly influenced by the presence of the lake and the geomorphology of the lake's subglacial catchment. Owing to low basal shear stresses, the gradient of the ice sheet surface over the lake is low (<0.003) when compared to the surrounding ice surface (~0.006). Upstream of the lake, the ice sheet surface has a semiamphitheater-like morphology, the result of an abrupt transition in basal conditions as the ice sheet flows into the deep subglacial trough from the surrounding subglacial highlands. Subglacial morphology is also likely to play an important role in maintaining the ice surface saddle that characterizes the ice divide southeast of Lake Ellsworth (Figure 3b).

4.1.2. Bed topography and lake surface. The map of basal topography (Figure 3d) integrates the DELORES RES data with the BAS and CECS RES data and the five seismic reflection profiles. This map confirms that Lake Ellsworth is located within a deep, broad subglacial overdeepening in the Ellsworth Subglacial Highlands [*Siegert et al.*, 2004; *Vaughan et al.*, 2007], which runs for at least 45 km north-westward from the ice divide. This trough is constrained on either side by high, rugged subglacial topography, and the maximum peak-to-trough amplitude is of the order of 2300 m. In the upper reaches of the catchment, the trough is relatively narrow (2.5–3.5 km across), with the valley floor generally at elevations between −800 and −950 m. Although impounded by high topography on both sides (at elevations generally >400 m), to the southeast of the lake there is a particularly pronounced area of subglacial mountains with peak elevations between 1200 and 1400 m. In the vicinity of Lake Ellsworth, the trough widens and deepens, becoming 5.5 to 6.5 km across, with the valley floor attaining a minimum elevation of approximately −1360 m.

The new RES data show that the extent of Lake Ellsworth (Figure 4) is greater than that mapped by *Vaughan et al.* [2007], particularly in its northeast sector. The total area is now estimated to be ~28.9 km². The lake surface, between the minimum and maximum elevations of −1361 and −1030 m, has a pronounced along-flow lake surface gradient (~0.03), markedly steeper than other reported subglacial lake gradients (e.g., Vostok Subglacial Lake). Steep gradients are more likely to result in differential melting and freezing along the lake axis and therefore enhanced water circulation [*Siegert et al.*, 2001, 2004; *Woodward et al.*, 2010]. The lake surface is also twisted across its long axis; in the upstream half of the lake, the surface slopes toward the northeast, whereas in the downstream half, the slope is toward the southwest (Figure 5b). This lake surface morphology is probably caused by flexural support of the overlying ice by the steep bedrock walls that flank the lake.

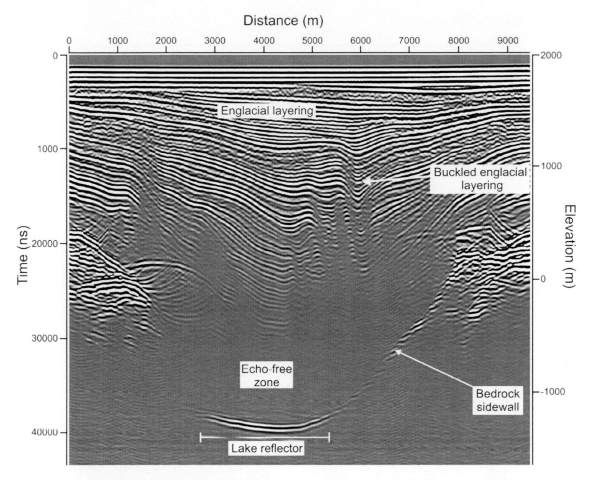

Figure 2. DEep-Look-Radar-Echo-Sounder (DELORES) radio echo sounding (RES) data across Lake Ellsworth (Line D7.5). A prominent lake-like reflector is observed between 2600 and 5400 m along the profile between depths of approximately −1200 to −1310 m. Buckled englacial layers, generated by ice flow over and around subglacial mountains upstream of the lake (see Figure 3), are annotated. Ice flow is roughly into the page. Elevations are relative to WGS-84 ellipsoid. See Figure 3 for location.

Downstream of Lake Ellsworth, the bed topography is marked by a pronounced bedrock ridge, which trends obliquely across the lake outlet zone and across the valley (Figure 3d). This ridge, which rises ~150 to 200 m above the elevation of the lake surface, appears to determine the downstream boundary of Lake Ellsworth and is likely to play a key role in controlling the nature and timing of drainage from the lake. The subglacial hydrological implications of this landform are currently being explored [*Ross et al.*, 2009], but it is a distinct possibility that a hydrological sill exists. This would be contrary to the conclusions of *Vaughan et al.* [2007], who suggested that the lake was part of an open hydrological system.

4.1.3. Englacial layering and modeling. The DELORES RES data are characterized by strong, well-defined englacial

reflections in all profiles (e.g., see Figure 2). Throughout the majority of the Lake Ellsworth catchment, however, imaged englacial layers are predominantly restricted to the upper 2000 m of the ice column. Below this depth, most of the radargrams are markedly free of internal reflections. This echo-free zone [*Drewry and Meldrum*, 1978] was also a characteristic of the SPRI-NSF-TUD, BAS, and CECS RES data sets in this part of West Antarctica [*Siegert et al.*, 2004; *Vaughan et al.*, 2007]. It is not clear whether the echo-free zone around Lake Ellsworth is a property of the ice or simply a consequence of the absorption of electromagnetic energy in the warmer parts of the ice at depth.

Englacial layers within the RES profiles (Figure 2) have been picked and transformed into three-dimensional (3-D) surfaces. These data have been integrated with the DEM of

(a) RES data acquisition
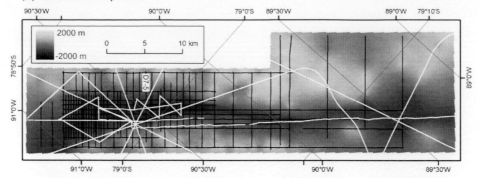

(b) Ice sheet surface
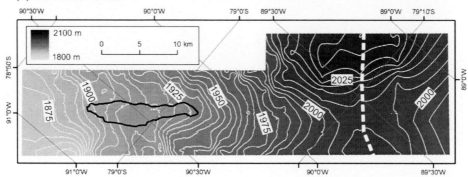

(c) Ice thickness
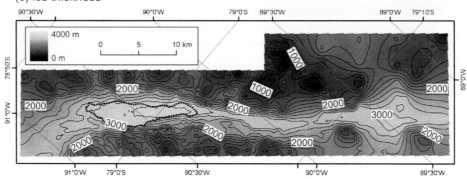

(d) Bed elevation
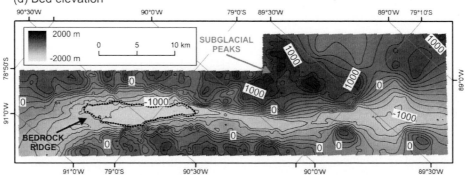

the subglacial bed to facilitate 3-D numerical modeling of ice flow and basal melting over Lake Ellsworth [*Ross et al.*, 2009]. Initial results show that, over the lake, some anomalies in the layering near the steeper bedrock wall can be understood in terms of perturbations to the velocity field caused by ice sheet flow into the deep subglacial trough from the surrounding subglacial highlands, as well as being caused by melt anomalies.

4.2. Seismic Reflection Surveys and Water Circulation Modeling

At least 387 Antarctic subglacial lakes have now been identified using data from ground-based, airborne, and satellite platforms [*Wright and Siegert*, this volume], but water depth measurements from such lakes are few in number. Gravity data have been used to estimate the depth, and in some cases the bathymetry, of a series of lakes [*Studinger et al.*, 2004; *Bell et al.*, 2006; *Filina et al.*, 2008]. However, prior to the Lake Ellsworth field campaign in 2007/2008, seismic measurements had only been made at Vostok Subglacial Lake and South Pole Lake [*Masolov et al.*, 2006; *Peters et al.*, 2008].

Five seismic reflection lines spaced ~1.4 km apart were acquired across the long axis of Lake Ellsworth (Figures 4, 5) [*Woodward et al.*, 2010]. The ice base and lake bed reflections were identified and picked in all five of the processed seismic profiles. By applying seismic velocities to the reflection picks and combining the product with surface GPS data, reflector elevations were established. These data were gridded, along with picked RES data where appropriate, to produce 3-D surfaces of the ice-water interface, lake bed, and the water column thickness (Figures 5c–5e) [*Woodward et al.*, 2010].

The seismic data reveal that Lake Ellsworth has a broad, generally U-shaped, lake bed morphology (Figure 5e). The water column progressively increases down-lake (SE to NW) from a maximum thickness of 52 ± 1.5 m on line A to 156 ± 1.5 m on line E (Figure 5d) [*Woodward et al.*, 2010]. Based on the gridded data sets, the estimated volume of Lake Ellsworth is 1.37 km^3, although this figure is subject to an error margin of ± 0.2 km^3 because of uncertainties in the gridding of the lake's bathymetry in the upstream and downstream parts of the lake beyond the end two seismic lines [*Woodward et al.*, 2010].

The 3-D numerical model Rombax [*Thoma et al.*, 2007], using the seismic and RES-derived gridded data sets as input data, has been used to describe water circulation in Lake Ellsworth and to evaluate the potential mass balance characteristics of the overlying basal ice [*Woodward et al.*, 2010]. This model assumes a closed hydrological system (i.e., no water flows into or out of the lake); any melt input into the lake is balanced by accretion ice formation elsewhere. The modeling suggests that the ice-water interface in the upstream half of the lake is characterized by basal melting (mean melt rate of 3.8 ± 0.7 cm yr^{-1}). However, in the downstream half of the lake, the model predicts basal freezing and the development of a thin layer of accretion ice (mean thickness of 12.5 ± 3.5 m) [*Woodward et al.*, 2010]. At this stage, no evidence for accretion ice has been observed in the RES data from Lake Ellsworth.

Because Lake Ellsworth is overlain by a range of ice thicknesses (3280–2930 m) (Figure 3c), it is characterized by an unusual thermodynamic regime [*Woodward et al.*, 2010]. The critical pressure (p_c) [*Wüest and Carmack*, 2000] occurs at the intersection between the freezing point of water (T_f) and the temperature of maximum density (T_{md}) [*Woodward et al.*, 2010]. When the overburden pressure (ice and water) is greater than p_c, warmed water will rise through buoyancy, but when the overburden pressure is less than p_c, warmed water will tend to sink. Because the critical pressure boundary intersects Lake Ellsworth [*Woodward et al.*, 2010] (Figure 5b), the water column straddles two thermodynamic states with the potential for the development of zones with distinct water circulation patterns (convective or stratified). Modeling, however, suggests that this does not occur. Driven by the steeply sloping ice-water interface, water circulation occurs throughout the lake irrespective of the presence of the critical pressure boundary [*Woodward et al.*, 2010].

Preliminary analysis, using the seismic reflection data to establish the acoustic impedance of the water-sediment interface, suggests that the lake bed is composed of high-porosity, low-density sediments [*Smith et al.*, 2008]. These sediments have acoustic properties very similar to material found on the deep ocean floor, indicative of deposition in a low-energy environment. Analysis suggests that this sedimentary sequence is a minimum of 2 m thick.

Figure 3. (opposite) (a) RES data acquired around Lake Ellsworth. Black lines represent DELORES data; white lines represent British Antarctic Survey and Centro de Estudios Cientificos 150 MHz data [*Vaughan et al.*, 2007] also used in the gridding of ice thickness and subglacial topography. RES line D7-5 (Figure 2) is indicated. Background layer is bed elevation (Figure 3d). (b) Ice sheet surface, with contours at 5 m intervals. White dashed line shows approximate position of ice divide between Pine Island Glacier (to left) and the Institute Ice Stream (to right). Black polygon is the outline of Lake Ellsworth. (c) Ice thickness, with contours at intervals of 200 m. (d) Subglacial topography, with contours at intervals of 200 m. Elevations in (a), (b), and (d) are all relative to WGS-84 ellipsoid. Ice flow across the lake is roughly right to left throughout Figure 3.

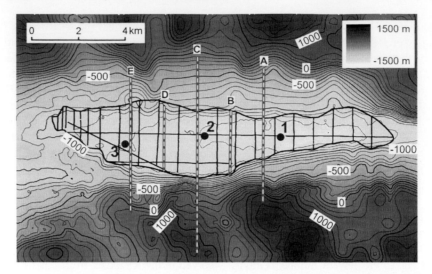

Figure 4. Outline of Lake Ellsworth mapped from "lake-like" reflectors identified in RES (thin black lines) and seismic data. Backdrop grid is subglacial topography with contour lines at 100 m intervals. White dashed lines represent acquired seismic lines (labeled A–E). Black numbered circles mark the locations of GPS base stations over the lake (1, uplake; 2, midlake; 3, lowlake). Elevations are relative to WGS-84 ellipsoid. Ice flow is roughly right to left.

4.3. GPS Measurements

During the 2007/2008 field season, four continuously recording Global Positioning System (GPS) base stations were deployed above, and in the vicinity of, Lake Ellsworth [*Woodward et al.*, 2010], with the primary role of monitoring any tidal signal in the lake, determining the ice sheet flow regime (velocity and direction), and for use as base stations when processing kinematic GPS data (Figure 6a). Two of the base stations ("offlake" and "midlake") were reoccupied during the 2008/2009 field season. Owing to proximity to the ice divide, ice flow velocities around Lake Ellsworth are low (<10 m yr^{-1}). During the 2007/2008 season, measured ice flow at the offlake base station, situated over relatively thin (~1 km) ice beyond the limits of the lake, was 1.9 ± 0.1 m yr^{-1}. Measurements over the lake, where the ice was thicker and subject to zero basal shear stresses, showed higher velocities, between 4.5 ± 0.2 m yr^{-1} ("uplake" station) and 5.5 ± 0.1 m yr^{-1} ("lowlake" station) [*Woodward et al.*, 2010].

In addition to these base station data, ice flow data were also acquired from measurements of the positions of a series of temporary stakes installed on the snow surface. This "stake network" consisted of 58 aluminum poles installed over the lake during the 2007/2008 field season and 8 wooden stakes installed downstream of Lake Ellsworth in January 2006. GPS measurements of all stakes were made during both field seasons. From the measured changes in the positions of the markers in the stake net-

work between the 2007/2008 and 2008/2009 field seasons, the direction and rate of ice flow at the ice surface were calculated. This has been used to produce a map of the rate and direction of ice flow over the lake (Figure 6a). The GPS data show that ice flow in the vicinity of Lake Ellsworth is characterized by the following:

1. There is convergent ice flow over the lake as the ice flows into the depression in the underlying topography. Convergence is greatest at the top end of Lake Ellsworth, with a noticeable decrease apparent down-lake. Data from GPS stake lines D6 and D5 (Figure 6a) show a shift from convergent flow (D10-D6) to a normal flow regime (D5) near the downstream end of the lake.

2. Ice flow velocity increases down the length of the lake (extensional flow). The measured rate of ice flow over Lake Ellsworth increases linearly from 4.09 m yr^{-1} near the head of the lake to 5.98 m yr^{-1} at its downstream end (Figure 6a). The maximum flow velocity measured was 6.60 m yr^{-1}, some 4.5 km downstream of the lake.

3. Ice flow is faster over the central axis of the lake (relative to other measurements in this survey). Rates of flow decrease toward both lateral lake margins (GPS stake lines D9 to D6) (Figure 6a).

4. There is westward rotation of ice flow in the lower parts of, and downstream of, the lake. Comparison between the GPS measurements in the area downstream of Lake Ellsworth with the GPS data over the lake clearly shows a pronounced local westward rotation of flow (Figure 6a).

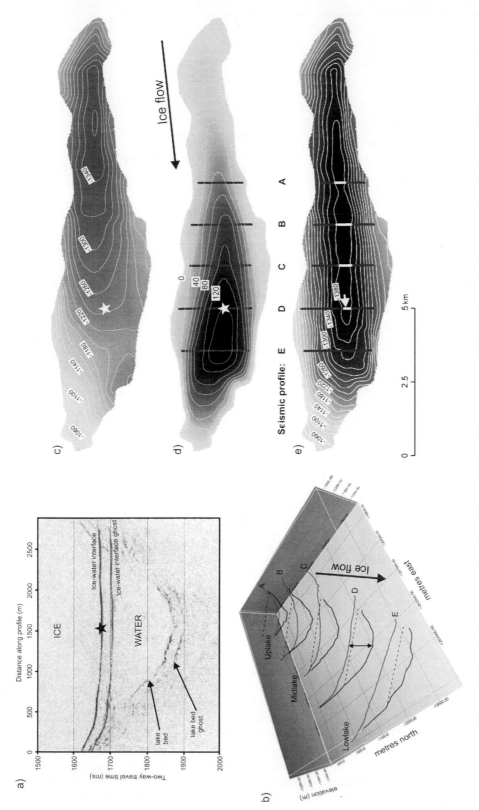

Figure 5. Seismic reflection data from Lake Ellsworth [from *Woodward et al.*, 2010]. (a) Example of seismic reflection data (profile D). Main reflections and corresponding ghosts (data acquisition artifacts) are identified. View is uplake with ice flow out of the page. The black star marks the proposed point of lake access. (b) A three-dimensional visualization of the lake surface (light gray lines) and lake bed (dark gray lines) identified from the five seismic reflection profiles. Dashed lines represent the critical pressure boundary (ice thickness of ~3170 m) for each seismic line. The proposed point of access is marked by the vertical arrowed line on profile D. View is uplake (into ice flow). Coordinates are meters polar-stereographic with a true scale at −71S. (c) Ice-water interface (m relative to WGS-84 ellipsoid). White stars and arrows indicate the proposed access location. Gray lines across the lake in (d) and (e) indicate the measured positions of the lake bed (and water column thickness) from the seismic data; the parts highlighted white in (e) represent the areas of the bed below −1380 m. For (c) to (e), contours are at 20 m intervals.

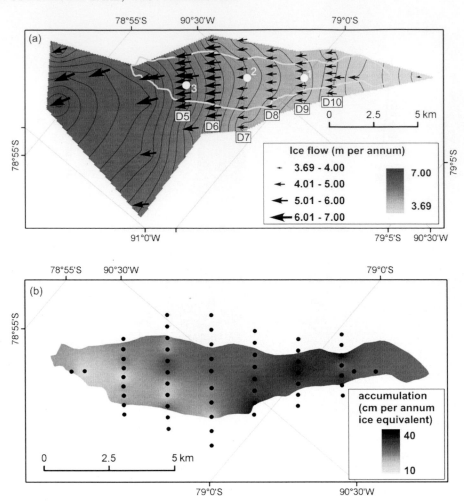

Figure 6. (a) Direction and rate of present-day ice flow from GPS measurements of a network of surface markers. Size of arrows denotes rate of flow (larger arrows equal faster flow). Base image is the gridded ice flow velocity with contours at 10 cm yr^{-1} intervals. White circles mark the locations of GPS base stations over the lake (1, uplake; 2, midlake; 3, lowlake). A fourth base station (off lake) was positioned 11 km northeast of the lake. (b) Surface accumulation (cm yr^{-1} ice equivalent) over Lake Ellsworth between 2008 and 2009. Black dots mark observation points.

4.4. Ice Cores and Lake Geochemistry

Three shallow (<20 m) ice cores were recovered from the ice sheet surface above Lake Ellsworth in the 2007/2008 field season. One core was analyzed in the field for measurements of snow and firn density, whereas the other two were returned to U.K. for laboratory analysis (one for oxygen isotope-derived accumulation rates and the other for biogeochemistry). A temperature of −31.9°C ±0.2°C was measured at a depth of 20 m at the base of one of these core holes [*Barrett et al.*, 2009].

Provisional geochemical data for the average composition of firn and ice in the top 20 m of one of the recently acquired surface ice cores from above Lake Ellsworth ($n = 11$) are given in Table 1. The *Lake Ellsworth Consortium* [2007] assumed that the chemistry of meteoric ice melt into Lake Ellsworth is equivalent to that of the average chemistry recorded in the Byrd Ice Core and used the ice core chemistry to calculate the expected chemistry of the lake waters (Table 1). The near-surface ice core values from Lake Ellsworth are higher in most species than the average chemistry of the Byrd core, which may be a consequence of aeolian inputs of crustal debris from the nearby Ellsworth Mountains and factors such as proximity to sources of sea salt aerosol and the relative amounts of sublimation of snow prior to deposition. The provisional geochemical data from the Lake Ellsworth ice core have been combined with the revised dimensions of the lake [*Woodward et al.*, 2010], melt and accretion rates of 4 cm yr^{-1}, and a residence time of 2370 years to calculate revised values of inferred water chemistry in the lake (Table 1). These figures suggest that given closed conditions, the lake waters may be more solute-rich than first estimated, with overall solute concentrations being comparable to basal meltwaters sampled to date from beneath smaller warm- and polythermal-based glaciers in the Northern Hemisphere [*Skidmore et al.*, 2010]. It is important to note that these calculations assume a closed system with no input from basal meltwaters derived from the subglacial hydrological catchment. Inflow of such water, which *Vaughan et al.* [2007] suggested may be the dominant input source for Lake Ellsworth, could alter the lake water composition significantly. Under such a scenario, the chemical composition of the lake would be strongly influenced by (1) any input of meltwater derived from sediment porewaters with high solute concentrations [*Skidmore et al.*, 2010]; (2) the degree of stratification of the lake (particularly in relation to how much denser solute-rich water may accumulate beneath the surface layer); and (3) the extent of mixing of inflow with preexisting lake water.

4.5. Surface Accumulation

Surface accumulation (in m yr^{-1}) in the vicinity of Lake Ellsworth is relatively high, as demonstrated by the inability of the 2007/2008 field party to locate 7 of the 15 wooden stakes installed less than 2 years previously. Based on simple vertical measurements of the position of the snow surface at each of the 58 aluminum poles in 2007/2008 and 2008/2009, a rough approximation of the spatial distribution of accumulation around Lake Ellsworth has been established (Figure 6b). Despite localized anomalies caused by the heavily sastrugied snow surface, it is clear that the overall trend is for decreasing accumulation over the lake in the downstream direction (Figure 6b). Accumulation is generally within a range of 0.25 to 0.38 m yr^{-1} ice equivalent upstream of the lake's midpoint and within a range of 0.12 to 0.25 m yr^{-1} ice equivalent downstream of this point (Figure 6b).

4.6. Implications for Lake Access

The results of the geophysical investigations are critical for determining the location best suited to access and direct measurement of Lake Ellsworth. On the basis of the geophysical surveys and water circulation modeling, a preferred lake access location (78°58.1′S 90°34.5′W) (Figure 5) has now been established [*Woodward et al.*, 2010]. This site has been chosen because at this location, (1) the ice column (~3155 m) is thin relative to the majority of the lake; (2) the measured water depth is significant (~143 m), providing an opportunity to acquire a comprehensive profile of the water column; (3) more than 2 m of sediment is present on the lake floor; (4) modeling suggests a melting ice-water interface; (5) coring activities can target a deep point of the lake floor (−1386 m), near to the depositional center of the lake (coring further downstream would occur on the upslope to the lake margin); and (6) the sedimentation rate is likely lowest, given it is distal from the likely input sources of sediment upstream (the sediments are therefore likely to get older with depth more rapidly than in other places, increasing the chance of a longer record of ice sheet history, and are more likely to contain an undisturbed, continuous sedimentary sequence).

One factor that has important implications for the lake access experiment is whether the lake is an open or closed system. At this stage, the analyzed data do not allow us to make unequivocal statements concerning this issue. So far, however, no recognizable signal of ice sheet surface elevation change (indicative of subglacial water movement) has been reported over Lake Ellsworth in studies using ICESat data [*Smith et al.*, 2009; *Pritchard et al.*, 2009]. Although small-scale ice surface elevation changes can be identified within the GPS data, these are believed to lie within the uncertainties associated with snowpack compaction, accumulation, and base station movement. The lack of evidence for significant (i.e., >1 m) changes in ice surface elevation would tend to suggest that the lake is currently in hydrological steady state, which is possible in either a closed or an

Table 1. Estimates of the Chemical Composition of Water in Lake Ellsworth Making the Simple Assumption That the Lake Is a Closed System and That All Solute From Melting Meteoric Ice Accumulates in the Lake[a]

	H$^+$	pH	Ca^{2+}	Mg^{2+}	Na$^+$	K$^+$	NH$_4^+$	Cl$^-$	SO$_4^{2-}$	NO$_3^-$	HCO$_3^-$
Average Byrd ice core	1.8	5.7	~1.0	0.4	1.5	0.05	0.13	2.0	1.0	0.7	~1.2
Provisional surface firn and ice concentrations	NA	NA	8.9	0.65	2.8	0.57	NA	5.3	1.4	0.83	~8.2
Inferred Lake Ellsworth (from Byrd Core)	300	>3.52	170	68	250	8.5	<22	340	170	<120	~200
Inferred Lake Ellsworth (from provisional surface data)	NA	NA	1500	110	470	96	NA	900	240	<140	~1400

[a]Estimates are rounded to two significant figures and assume that the lake has been in existence for 400,000 years, with a residence time of 2370 years. Units are μeq/L. NA, data are not available.

open system configuration (input from melt of the overlying ice and basal water flow is matched by accretion ice formation and/or outflow). Preliminary analysis of the geomorphology of the area immediately downstream of Lake Ellsworth [*Ross et al.*, 2009] indicates that a prominent bedrock ridge impounds the lake. Such a feature is very likely to inhibit water outflow, at least to some degree. However, even if it is unable to exit the lake by overflow, subglacial water could exploit groundwater pathways (e.g., fractures, bedding planes, substrate permeability). Moreover, the gross hydrology of the lake may well change during periods when the ice sheet was thicker in this area (e.g., during the Last Glacial Maximum); thicker ice increases the likelihood of an open system so that an open system configuration may well be the predominant, i.e., glacial state, configuration for Lake Ellsworth.

5. PLANS FOR EXPLORATION

Direct access, measurement, and sampling of lake water and sediment from Lake Ellsworth are planned for the 2012–2013 field season. The most effective means of obtaining rapid, clean access to Lake Ellsworth through more than 3 km of overlying ice is by hot water drilling. Measurement and sampling of the water column and the upper lake bed sediments will be undertaken using a probe deployed and instructed from the ice sheet surface. Once in the lake, the probe will have the capability to directly measure pressure, temperature, conductivity, oxygen concentration, redox potential, acoustic velocity, and pH of the lake's water body. Underwater imagery will also be acquired, while the lake floor will be mapped using a profiling sonar mounted on the forward face of the probe; equipment for profiling the underside of the ice is also under consideration. Sampling will be initiated once the probe reaches the water-lake bed interface. Here a narrow-diameter push corer on the tip of the probe will sample a key target for microbiological life, the first few centimeters of sediment below the lake floor. The probe will then acquire lake water samples and filtrate at various points through the water column as it is winched back up to the access hole in the overlying ice. After retrieval of the probe at the ice surface, the access hole will be used to deploy the sediment corer. Current designs suggest that the corer will be lowered to the sediment surface and hammered into the sediment using a remote percussion hammer operated from the surface. Anticipated penetration depths are in the order of ~1 to 3 m, but this will be dependent on the physical properties of the sediment (e.g., grain size and water content).

Following recovery, water and sediment samples will initially be transferred to Rothera Research Station on Adelaide Island, Antarctic Peninsula for preliminary analysis and sample splitting. Water and sediment samples will then be transferred to U.K. laboratories for analysis of their (1) microbiology (lake water and sediment), (2) organic geochemistry (lake water and sediment), (3) hydrogeochemistry (lake water and sediment), and (4) sedimentology and palaeoenvironmental reconstruction (sediment core only). These laboratory analyses will be integrated with the direct measurements of the lake's water column (made by the probe) to determine the biological, chemical, and physical processes in the lake. The sediment core analyses will reveal how environmental conditions in the lake have changed through time, potentially providing important new information regarding the glacial history of West Antarctica. The exploration of Ellsworth Subglacial Lake may therefore produce profound scientific discoveries regarding both the life in extreme environments and WAIS history.

Acknowledgments. This paper is a contribution by the Lake Ellsworth Consortium. This work was funded by NERC-AFI (NE/D00875/1, NE/D009200/1, NE/D008638/1) and NERC Consortium (NERC NE/ G00465X/1) grants. We thank BAS for logistics support, NERC-GEF for equipment (loans 838, 870), Dan Fitzgerald and Dave Routledge for excellent support in the field, and Mark Maltby for building the DELORES radar system. Two anonymous reviewers are thanked for reviews that significantly focused and improved this contribution. David Vaughan is thanked for providing data from the BAS airborne survey of Lake Ellsworth. The CECS is financed by the Millennium Science Initiative and the Basal Financing Program for Excellence Centres of CONICYT. The CECS acknowledges the contribution made by Antarctic Logistics and Expeditions LLC.

REFERENCES

Barrett, B. E., K. W. Nicholls, T. Murray, A. M. Smith, and D. G. Vaughan (2009), Rapid recent warming on Rutford Ice Stream, West Antarctica, from borehole thermometry, *Geophys. Res. Lett.*, *36*, L02708, doi:10.1029/2008GL036369.

Bell, R. E., M. Studinger, M. A. Fahnestock, and C. A. Shuman (2006), Tectonically controlled subglacial lakes on the flanks of the Gamburtsev Subglacial Mountains, East Antarctica, *Geophys. Res. Lett.*, *33*, L02504, doi:10.1029/2005GL025207.

Bentley, C. R., and N. A. Ostenso (1961), Glacial and subglacial topography of West Antarctica, *J. Glaciol.*, *3*, 882–911.

Drewry, D. J., and D. T. Meldrum (1978), Antarctic airborne radio echo sounding, 1977–78, *Polar Rec.*, *19*, 267–273.

Filina, I. Y., D. D. Blankenship, M. Thoma, V. V. Lukin, V. N. Masolov, and M. K. Sen (2008), New 3D bathymetry and sediment distribution in Lake Vostok: Implication for pre-glacial origin and numerical modelling of the internal processes within the lake, *Earth Planet. Sci. Lett.*, *276*, 106–114.

Jankowski, E. J., and D. J. Drewry (1981), The structure of West Antarctica from geophysical studies, *Nature*, *291*, 17–21.

Lake Ellsworth Consortium (2007), Exploration of Ellworth Subglacial Lake: A concept paper on the development, organisation and execution of an experiment to explore, measure and sample the environment of a West Antarctic subglacial lake, *Rev. Environ. Sci. Biotechnol.*, *6*, 1569–1705.

McIntyre, N. F. (1983), The topography and flow of the Antarctic ice sheet, Ph.D. thesis, 198 pp., Univ. of Cambridge, Cambridge, U. K.

Masolov, V. N., S. V. Popov, V. V. Lukin, A. N. Sheremetyev, and A. M. Popkov (2006), Russian geophysical studies of Lake Vostok, Central East Antarctica, in *Antarctica: Contributions to Global Earth Sciences*, edited by D. K. Fütterer et al., pp. 135–140, Springer, New York.

Peters, L. E., S. Anandakrishnan, C. W. Holland, H. J. Horgan, D. D. Blankenship, and D. E. Voigt (2008), Seismic detection of a subglacial lake near the South Pole, Antarctica, *Geophys. Res. Lett.*, *35*, L23501, doi:10.1029/2008GL035704.

Pritchard, H. D., R. J. Arthern, D. G. Vaughan, and L. A. Edwards (2009), Extensive dynamic thinning on the margins of the Greenland and Antarctic ice sheets, *Nature*, *461*, 971–975.

Ross, N., A. M. Smith, J. Woodward, M. J. Siegert, R. C. A. Hindmarsh, H. F. J. Corr, E. C. King, D. G. Vaughan, F. Gillet-Chaulet, and M. Jay-Allemand (2009), Ice flow dynamics and outlet zone morphology of Subglacial Lake Ellsworth, *Eos Trans. AGU*, *90*(52), Fall Meet. Suppl., Abstract U53C-02.

Siegert, M. J., J. A. Dowdeswell, M. R. Gorman, and N. F. McIntyre (1996), An inventory of Antarctic sub-glacial lakes, *Antarct. Sci.*, *8*, 281–286.

Siegert, M. J., J. C. Ellis-Evans, M. Tranter, C. Mayer, J.-R. Petit, A. Salamatin, and J. C. Priscu (2001), Physical, chemical and biological processes in Lake Vostok and other Antarctic subglacial lakes, *Nature*, *414*, 603–609.

Siegert, M. J., R. C. A. Hindmarsh, H. F. J. Corr, A. M. Smith, J. Woodward, E. C. King, A. J. Payne, and I. Joughin (2004), Subglacial Lake Ellsworth: A candidate for in situ exploration in West Antarctica, *Geophys. Res. Lett.*, *31*, L23403, doi:10.1029/2004GL021477.

Siegert, M. J., S. Carter, I. Tabacco, S. Popov, and D. D. Blankenship (2005), A revised inventory of Antarctica subglacial lakes, *Antarct. Sci.*, *17*, 453–460.

Skidmore, M., M. Tranter, S. Tulaczyk, and S. Lanoil (2010), Hydrochemistry of ice stream beds—Evaporitic or microbial effects?, *Hydrol. Processes*, *24*, 517–523.

Smith, A. M., J. Woodward, N. Ross, M. J. Siegert, H. F. J. Corr, R. C. A. Hindmarsh, E. C. King, D. G. Vaughan, and M. A. King (2008), Physical conditions in Subglacial Lake Ellsworth, *Eos Trans. AGU*, *89*(53), Fall Meet. Suppl., Abstract C11A-0467.

Smith, B. E., H. A. Fricker, I. R. Joughin, and S. Tulaczyk (2009), An inventory of active subglacial lakes in Antarctica detected by ICESat (2003–2008), *J. Glaciol.*, *55*, 573–595.

Studinger, M., R. E. Bell, and A. A. Tikku (2004), Estimating the depth and shape of subglacial Lake Vostok's water cavity from aerogravity data, *Geophys. Res. Lett.*, *31*, L12401, doi:10.1029/2004GL019801.

Thoma, M., K. Grosfeld, and C. Mayer (2007), Modelling mixing and circulation in subglacial Lake Vostok, *Antarctica*, *57*, 531–540.

Vaughan, D. G., A. Rivera, J. Woodward, H. F. J. Corr, J. Wendt, and R. Zamora (2007), Topographic and hydrological controls on Subglacial Lake Ellsworth, West Antarctica, *Geophys. Res. Lett.*, *34*, L18501, doi:10.1029/2007GL030769.

Woodward, J., A. M. Smith, N. Ross, M. Thoma, H. F. J. Corr, E. C. King, M. A. King, K. Grosfeld, M. Tranter, and M. J. Siegert (2010), Location for direct access to subglacial Lake Ellsworth: An assessment of geophysical data and modeling, *Geophys. Res. Lett.*, *37*, L11501, doi:10.1029/2010GL042884.

Wright, A., and M. J. Siegert (2011), The identification and physiographical setting of Antarctic subglacial lakes: An update based on recent discoveries, in *Antarctic Subglacial Aquatic Environments*, *Geophys. Monogr. Ser.*, doi: 10.1029/2010GM000933, this volume.

Wüest, A., and E. Carmack (2000), A priori estimates of mixing and circulation in the hard-to-reach water body of Lake Vostok, *Ocean Modell.*, *2*, 29–43.

M. J. Bentley, Department of Geography, Durham University, Durham DH1 3LE, UK.

D. Blake, L. Capper, R. Clarke, H. F. J. Corr, C. Hill, R. C. A. Hindmarsh, D. A. Hodgson, E. C. King, K. Makinson, D. A. Pearce, A. M. Smith, and A. Tait, British Antarctic Survey, Natural Environment Research Council, Cambridge CB3 0ET, UK.

C. S. Cockell, CEPSAR, Open University, Milton Keynes MK7 6AA, UK.

W. Harris, M. Tranter, and J. L. Wadham, School of Geographical Sciences, University of Bristol, Bristol BS8 1SS, UK.

H. Lamb, Institute of Geography and Earth Sciences, Aberystwyth University, Aberystwyth SY23 3DB, UK.

B. Maher, Lancaster Environment Centre, Lancaster University, Lancaster LA1 4YQ, UK.

M. Mowlem, National Oceanography Centre, University of Southampton, Southampton SO14 3ZH, UK.

J. Parnell, Department of Geology and Petroleum Geology, University of Aberdeen, Aberdeen AB24 3UE, UK.

J. Priscu, Department of Land Resources and Environmental Sciences, Montana State University, Bozeman, MT 59717-3120, USA.

A. Rivera, Centro de Estudios Científicos, Arturo Prat 514, Valdivia, Chile.

N. Ross and M. J. Siegert, School of GeoSciences, University of Edinburgh, Edinburgh EH8 9XP, UK. (neil.ross@ed.ac.uk)

W. B. Whalley, School of Geography, Archaeology and Palaeoecology, Queen's University Belfast, Belfast BT7 1NN, UK.

J. Woodward, School of Built and Natural Environment, Northumbria University, Newcastle NE1 8ST, UK.

AGU Category Index

Index

Note: Page numbers with italicized *f* and *t* refer to figures and tables.